Magnetism in Heavy Fermion Systems

Series in Modern Condensed Matter Physics – Vol. 11

Editor

Harry B. Radousky

Materials Research Institute,
Lawrence Livermore National Laboratory

Magnetism in Heavy Fermion Systems

World Scientific

Singapore • New Jersey • London • Hong Kong

Published by

World Scientific Publishing Co. Pte. Ltd.

P O Box 128, Farrer Road, Singapore 912805

USA office: Suite 1B, 1060 Main Street, River Edge, NJ 07661

UK office: 57 Shelton Street, Covent Garden, London WC2H 9HE

British Library Cataloguing-in-Publication Data
A catalogue record for this book is available from the British Library.

ISBN 981-02-4348-0

Printed in Singapore by World Scientific Printers

Dedication –
This book is dedicated in memory of my mother, Esther Radousky, who early on taught me the love of books and of learning.

PREFACE

In a book on a broad topic such as heavy fermions, it is impossible to include all aspects of the subject. The choices I have made reflect the original request from the editors at World Scientific to create a book which looked at magnetic behavior in heavy fermion systems. In particular I made a decision to include a major chapter on superconductivity in heavy fermion materials, both because of the intimate link between superconductivity and magnetic interactions, and my own intrinsic interest in the subject. I have also included a short chapter which touches on the link between heavy fermion behavior and the Pr containing cuprates, such as $PrBa_2Cu_3O_7$ and related structures. The two core chapters on magnetism cover two important and complimentary ways of studying magnetic behavior, neutron scattering and muon spin rotation. While the book is weighted towards experimental articles, I chose to include one theory article which details a particular set of theoretical models for f-electron systems. While references are made to many other approaches and results, this is by no means an attempt to do justice to the extensive literature of theoretical work in this area. The introductory article serves to overview the role of magnetism in heavy fermion materials, and provides a summary of the contents of each chapter.

There are many topics which could have been included, but were left out for reasons of space, availability of the right authors to contribute a chapter, and the existence of recent review articles on that topic. Three particular examples of topics not covered are high-pressure effects, heavy fermion behavior in very high magnetic fields and non-Fermi liquid behavior. The Appendix lists a hopefully useful set of references to review articles on heavy fermions, as well as articles which have a broad perspective relevent to magnetic behavior of heavy fermion systems.

I would like to take this opportunity to thank everyone who has contributed comments and suggestions. These include: Michael McElfresh, Greg Boebinger, Wayne Cooke, Lynda Soderholm, Jeff Lynn, Jon Lawrence, Dan Cox, Nick Kioussis, Jim Schilling, Joe Thompson and Carel Boekema. I want to thank Marsha McInnis, who provided the technical editing which turned the collection of manuscripts into a coherent book. I especially want to thank my wife Judith for her patience and understanding of the time and energy spent on making this idea into a reality.

This work was performed under the auspices of the U.S. Department of Energy by University of California Lawrence Livermore National Laboratory under contract No. W-7405-Eng-48, through the LLNL Materials Research Institute.

Harry B. Radousky
Livermore, CA
March, 2000

CONTENTS

1

INTRODUCTION

Douglas E. MacLaughlin

U-C Riverside Department of Physics
Riverside, California 92521, U.S.A.

1

I. HEAVY FERMION PHYSICS

The term "heavy fermion systems" is used to designate metallic compounds and alloys which contain $4f$ or $5f$ ions (usually Ce, Pr, Yb, or U) and which exhibit enormously enhanced effective conduction electron masses at low temperatures. These large effective masses characterize low-lying excitations from the many-body ground state, and are qualitatively different from the nearly free local-moment behavior observed at high temperatures (indicated by, e.g., a Curie-Weiss temperature dependence of the susceptibility). The crossover temperature between the local-moment and heavy fermion regimes, called the characteristic or "Kondo" temperature T_K, is inversely proportional to the effective mass and can be as low as a few kelvin; heavy fermion behavior is a low-temperature phenomenon.

The existence of heavy itinerant quasiparticles has been confirmed by de Haas-van Alphen measurements on a number of compounds, where the Fermi surface is found to be well described by band theory, but the mass is considerably enhanced over the band value. The origin of this enhancement, together with the rich variety of other phenomena exhibited by heavy fermion systems, has puzzled and fascinated condensed-matter physicists for almost three decades.[1-8]

It is expected on general grounds that in the absence of a phase transition to a cooperative state the low-lying excitations of the system should be well described by the Fermi liquid theory of Landau.[9] Conforming to this expectation, some heavy fermion compounds and alloys exhibit no phase transition and appear to obey the predictions of Fermi liquid theory down to very low temperatures: the magnetic susceptibility $\chi(T)$ tends to a constant Pauli-like value, the specific heat $C(T)$ depends linearly on temperature [the Sommerfeld coefficient $\gamma(T) = C(T)/T$ tends to a constant], and the electrical resistivity varies as T^2 from its zero-temperature value.

Phase transitions are, however, more common than not in heavy fermion materials. In broad terms the low-temperature states are two in kind, superconducting and magnetically ordered; these are not always mutually exclusive. In heavy fermion systems both superconducting and magnetic states have novel and unique properties. The ordered magnetism often exhibits extremely small ordered magnetic moments ($\sim 10^{-3}$–$10^{-2}\,\mu_B$/ion), and usually coexists with heavy band electrons. Heavy fermion superconductivity is definitely a property of the massive electrons (as opposed to a parallel channel of more conventional band states), as evidenced by the large jump in specific heat at the superconducting transition temperature T_c. There is considerable evidence that heavy fermion superconductivity is unconventional, in the sense that the superconducting order parameter is complex and describes internal degrees of freedom of the Cooper pairs.

Recently it has been discovered that low-temperature behavior in some heavy fermion alloys can disagree qualitatively with the predictions of Fermi liquid theory yet not involve a phase transition, at least at nonzero temperatures. This so-called "non-Fermi liquid" behavior is a considerable challenge to theory.

Although the derivation of Fermi liquid behavior from the many-body effects that give rise to heavy fermion phenomena is a difficult problem, strict obedience to Fermi liquid predictions, when observed, is currently considered to be fairly well understood. The same can be said for superconductivity, at least in general outline; the evidence for unconventional Cooper pairing is persuasive, although the exact mechanism for the pairing and symmetry of the Cooper pairs remains controversial. Our understanding of heavy fermion magnetism, on the other hand, remains incomplete, and the theoretical situation regarding non-Fermi liquid behavior is evolving rapidly.

A. Heavy Fermion Superconductivity

In the well-studied heavy fermion superconductor UPt_3 a consistent picture is obtained[10] by positing an anisotropic pairing mechanism of undetermined origin, which gives rise to a complex superconducting order parameter (gap function). In the "standard model" of this compound, interaction between the superconducting electrons and the weak-moment antiferromagnetism which sets in at a Néel temperature $T_N \approx 5$ K leads to two zero-field superconducting transitions and three separate phases in the H–T plane below $T_{c(\mathrm{max})} \approx 0.5$ K. Thorium-doped UBe_{13} also exhibits a complex phase diagram, in which one superconducting phase possesses static magnetism (even though undoped UBe_{13} shows no sign of magnetic order); there is also evidence for such magnetism in the low-temperature zero-field phase of UPt_3.

These unusual properties of heavy fermion superconductors are evidence that the superconducting pairing is unconventional, in that symmetries other than gauge symmetry are broken in the superconducting state. These additional broken symmetries lead to structure and internal degrees of freedom associated with the Cooper pairs. It is quite difficult to determine the superconducting-state symmetry directly (although tunneling experiments on high-temperature superconductors seem to have done this). Indirect evidence for unconventional pairing is, however, found from a number of features.

Historically the first to be noted was the fact that, in contrast to conventional pairing, the additional degrees of freedom available in symmetry-broken pairing states lead to structure in the superconducting energy gap $\Delta(\mathbf{k})$. Specifically, $\Delta(\mathbf{k})$ can exhibit nodes at points or along lines on the Fermi surface, and in the vicinity of these nodes low-lying excitations exist with energies extending down to $E = 0$. This means that thermal

4

excitations are no longer activated over an energy gap, so that the temperature dependence of any quantity associated with such excitations is drastically modified. Early evidence for unconventional superconductivity was found in the power-law temperature dependence of transport coefficients in the superconducting state, and a large number of other properties have been found to exhibit non-activated behavior. These studies are described in detail in Chapter 2.

Under appropriate circumstances unconventional superconductivity can even involve breaking of time-reversal symmetry. Then some form of superconducting-state magnetism, involving either orbital or spin degrees of freedom and perhaps requiring the presence of impurities, is expected on theoretical grounds. Muon spin rotation (μSR) is an ideal probe of such magnetism, which may be responsible for the static magnetism of (U,Th)Be$_{13}$ and (possibly) UPt$_3$. The status of the question of time-reversal-symmetry breaking in heavy fermion superconductors is discussed in Chapter 3.

B. Heavy Fermion Magnetism

Magnetic order in heavy fermion materials, on the other hand, has proved difficult to account for in other than in qualitative terms. The so-called "Kondo necklace" model introduced by Doniach[11] considers the competing effects of Kondo compensation (singlet formation) of local moments in a metal and Ruderman-Kittel-Kasuya-Yosida (RKKY) coupling between these moments. The former is characterized by a Kondo temperature $T_K \propto \exp(-1/\rho J)$, where ρ is the host density of states at the Fermi energy and J is the local-moment/conduction-electron exchange interaction energy. On the other hand, the RKKY coupling energy in temperature units is of the order of $T_{\mathrm{RKKY}} \approx \rho J^2/k_B$; this is also of the order of the magnetic ordering temperature T_M.[12]

The RKKY interaction therefore dominates for small J such that $T_{\mathrm{RKKY}} \gg T_K$, but for values near a critical value J_c the RKKY and Kondo characteristic energies become comparable. For $J \gg J_c$ ($T_K \gg T_{\mathrm{RKKY}}$) the Kondo effect compensates the local moments and no magnetic order is expected. The simplest picture for weak-moment magnetism in heavy fermion metals is, therefore, that for such systems J is only slightly smaller than J_c, and the Kondo effect compensates all but a small fraction of each local moment. This scenario predicts that the entropy increase associated with warming such a system from $T = 0$ through T_M should be only a small fraction of the total spin entropy $k_B \ln(2S+1)$, since the majority of the spin entropy will appear only at higher temperatures $T^* \approx T_K$ where the Kondo compensation is broken up and local-moment behavior restored. Such a small entropy increase is indeed observed in many (but not all) small-moment heavy fermion magnets. In this case one might conceive of the magnetic order as due to band magnetism of heavy quasiparticles formed above T_M. If on the other hand $T_M > T_K$, then the local moments will not have been appreciably compensated for $T > T_M$, and one expects RKKY ordering of these moments with relatively minor effects due to Kondo compensation.

So far the discussion has been on a very crude "mean-field" level. No account has been taken of spin fluctuations, due to either Kondo or RKKY couplings. Spin fluctuations affect the behavior of heavy fermion solids in many ways. They are directly responsible for the dynamical behavior as observed, for example, in neutron scattering and nuclear spin-lattice relaxation; they modify and in some cases suppress completely the magnetic ordering temperature; and in the most widely held theories of heavy fermion superconductivity they are responsible for the attractive pairing interaction.

C. Non-Fermi liquid Behavior

Another possibly related breakdown of the canonical Fermi liquid picture of heavy fermion behavior is encountered in a number of heavy fermion systems, where there is no evidence for a phase transition (at least at nonzero temperature) but the low-temperature thermodynamic and transport properties deviate substantially from the predictions of the Fermi liquid theory. The two most striking examples of this "non-Fermi liquid" behavior are a weak divergence of the specific heat coefficient C/T and a linear variation of the electrical resistivity with temperature from its zero-temperature value (rather than the T^2 variation expected in the Fermi liquid picture). In most of these alloys there is no sign of static magnetism (see Chapter 3), so the anomalous behavior cannot be ascribed to some kind of spin freezing or spin-glass state. In the absence of a phase transition the Fermi liquid picture for the behavior of (nonmagnetic and non-superconducting) metallic systems is expected to be very robust, so that its breakdown in some materials has given rise to a good deal of theoretical effort.[13]

II. SUMMARY OF CHAPTERS

The chapters of this book approach the problems associated with magnetism in heavy fermion systems in a variety of ways. Some, such as the reviews of neutron scattering by R. Robinson (Chapter 4) and muon spin rotation by R. H. Heffner (Chapter 3), view the issues through the prism of a particular experimental technique. Others concentrate on general behavior, such as the survey of heavy fermion superconductivity by R. N. Shelton and Rose Zhang (Chapter 2). Some chapters review lessons learned from a specific class of systems, such as the discussion of Pr-doped cuprates by H. B. Radousky (Chapter 6). In Chapter 5, B. R. Cooper summarizes the situation concerning the band-

6

theoretical approach to heavy fermion phenomena. The subject matter of each chapter is briefly introduced below.

A. Heavy Fermion Superconductors (Chapter 2)

A wide range of experimental techniques have been used to study the phenomenon of heavy fermion superconductivity. Specific heat measurements showed the enhanced discontinuity at T_c mentioned above, and the temperature dependence of the specific heat at low temperatures $T \ll T_c$ exhibits a power law, consistent with a superconducting energy gap which possesses point or line nodes on the Fermi surface. The entropy released in the transition, the dependence of the specific heat on impurity doping, and the effects of applied magnetic field and hydrostatic pressure have been studied. Aspects of the magnetic susceptibility and electrical resistivity, both normal-state properties, have been examined for their relevance to superconductivity.

The behavior of the superconducting upper critical field $H_{c2}(T)$ is a valuable probe of the superconducting state. In heavy fermion superconductors both the zero-temperature value $H_{c2}(0)$ and the slope $dH_{c2}(T)/dT$ at $T = T_c$ are extremely large; these properties are directly associated with the enhanced quasiparticle mass. The temperature dependence of H_{c2} is sometimes very anomalous, exhibiting a maximum or kink at a nonzero tempera-ture. The crystalline anisotropy of H_{c2} has also been studied in great detail, and in at least one case (UPt_3) bears on the nature (singlet vs. triplet) of the superconducting pairing.

Thermal transport and dissipative behavior in the superconducting state can yield important clues as to the nature of the quasiparticle excitations, and has been extensively investigated in heavy fermion superconductors. The temperature dependence, behavior near zero temperature, and anisotropy of the thermal conduc-tivity are all related to the nature of the superconducting pairing, as is the corre-sponding behavior of the ultrasonic attenuation coefficient. Indeed, it was this property which first showed the power-law behavior characteristic of a supercon-ducting gap function with nodes.

Other experimental methods have also been useful in furthering our understanding of heavy fermion superconductivity. Magnetic resonance techniques, principally nuclear magnetic resonance (NMR) and the related muon spin rotation (μSR) technique, have been extensively employed; μSR and its application to heavy fermion magnetism are reviewed in Chapter 3. The Meissner effect and the London penetration depth that characterizes it have been measured directly in the low-field Meissner state, but also in the type-II superconducting mixed-state vortex lattice using NMR and μSR. In the mixed state the NMR or μSR lineshape gives rather directly the distribution function for the local magnetic field in the vortex lattice. This in turn is related to the London penetration depth λ_L: the shorter λ_L relative to the spacing between superconducting vortices, the greater the spread of local fields. The corresponding broadening of the resonance line can be used to measure λ_L.

Tunneling experiments have proved difficult to implement in heavy fermion superconductors, at least in part because the surfaces of these materials are notoriously hard to prepare in a clean condition and to characterize. Some important results have nevertheless been obtained, and are summarized in this Chapter.

The perturbative effect of impurity doping on superconductivity is often very informative, and doping studies have been carried out on all the known heavy fermion superconductors. Uncontrolled defects are, however, the bane of the experimenter's existence; sample dependence of many properties has plagued the study of heavy fermion superconductivity from the beginning. It seems that the low energy scales involved (no heavy fermion superconductor has a transition temperature greater than a few kelvin) render heavy fermion superconductivity particularly sensitive to the presence of defects. Numerous strategies have been devised to overcome these problems and to yield experimental results which are believed to be characteristic of ideal systems.

B. Muon Spin Relaxation Studies of Small-Moment Heavy Fermion Systems (Chapter 3)

Muon spin rotation (μSR) is a magnetic resonance technique that has been used extensively as a probe of local magnetism in a wide variety of solids. In the μSR technique spin-polarized muons, obtained from a so-called "meson factory" accelerator (examples are TRIUMF, Vancouver, and PSI, Zurich), are implanted in the sample. The direction of emission of the subsequent decay positron, which is correlated with the muon spin orientation at the time of decay, is monitored using standard techniques of experimental nuclear physics.[14] In the majority of experiments positive muons are used, which can be considered short-lived impurities after they come to rest at interstitial sites in the sample. The experimental details of μSR and NMR are very different, but the results are analyzed in similar ways and yield similar information.

The application of μSR to heavy fermion materials has centered on three important phenomena: small-moment heavy fermion magnetism, heavy fermion superconductivity, and non-Fermi liquid behavior. These are not independent; as mentioned above the "standard model" of UPt_3 involves interaction between a complex superconducting order parameter and small-moment antiferromagnetism, and one proposed mechanism for non-Fermi liquid behavior invokes critical fluctuations associated with a quantum critical point at zero temperature. μSR studies have addressed questions in each of these areas; indeed, the antiferromagnetic transition in UPt$_3$ was discovered by μSR.[15] A specific advantage of μSR is its extreme sensitivity to small magnetic moments; in typical heavy fermion compounds f-ion moments as small as $0.001\mu_B$ can be detected. As discussed previously μSR, like NMR, can be used to measure the superconducting London penetration depth. The applicability of μSR to the non-Fermi liquid problem resides in the technique's sensitivity to the effect of disorder. Specifically, the μSR (and NMR) linewidth is a direct measure of the inhomogeneous distribution of local spin polarization, which expected to be anomalously large at low temperatures if the so-called "Kondo disorder" mechanism for non-Fermi liquid behavior is applicable.

C. Neutron Scattering From Heavy Fermions (Chapter 4)

The neutron scattering technique uses the interchange of energy and momentum between incident neutrons and electronic spins to study both static and dynamic aspects of magnetic systems via elastic and inelastic scattering respectively. As mentioned above the massiveness of elementary excitations in heavy fermion systems is related to the low characteristic energies of their excitation spectra; these excitations can be absorbed or emitted by neutrons and thus studied.

Static magnetism, if spatially ordered, leads to well-defined Bragg diffraction peaks in elastic neutron scattering which signal the presence of a phase transition. The scattering angles also yield the magnetic structure of the ordered state and the temperature dependence of its order parameter. Of particular interest is the question of whether antiferromagnetic order is characterized by one (single-$\mathbf{q}$) or several (multiple-$\mathbf{q}$) propagation vectors. The strength of the Bragg scattering is related to the size of the ordered moment which, as mentioned above, can be quite small in magnetic heavy fermion systems.

In inelastic neutron scattering the excitations probed can be collective in nature, such as spin waves in ordered magnets; single-ion excitations, such as crystalline electric field transitions; and incoherent spin fluctuations in paramagnets without static magnetic order. All of these are important in heavy fermion materials, and all have been studied by magnetic neutron scattering experiments. As is well known, neutrons can also scatter from atomic nuclei in a crystal (via the nuclear strong force); inelastic nuclear scattering probes the phonon structure of the system.

Properties of the vortex lattice in the mixed state of Type-II heavy fermion superconductors have recently been studied in some detail by neutron Bragg scattering. The weak scattering amplitude and large size scale of the vortex structure makes these very difficult experiments which, if successful, yield valuable information on the nature of the various superconducting phases of heavy fermion superconductors.

D. f-Electronic and Magnetic Behavior Between Atomlike and Fully Itinerant: Random Localized Magnetism and Heavy Fermion (Chapter 5)

A number of fundamental questions remain concerning the role of the Kondo effect in a "Kondo lattice" of f ions.[16] One of the most important of these is whether the essential features of the Kondo effect can be captured in a band theory with a sufficiently realistic treatment of electron-electron correlations; in particular, as a refinement of the local-density functional approximation. This Chapter describes a picture, applicable to both d-electron oxides and light (Ce or light actinide) f-electron systems, in which local chemical environment effects control restructuring of the d or f shell. Depending on local hybridization, the d or f ions occupy a coherent highly correlated narrow band or localized states. Successful quantitative predictions can be made in the case of random alloys, where Anderson localization in the disordered lattice drives the process, and the model shows generally the connection between localized and itinerant-electron magnetism.

E. Magnetism in the Pr Containing Cuprates (Chapter 6)

At first sight this topic might seem somewhat out of place in a book on the physics of heavy fermion materials, which are not usually taken to include transition-metal oxides and superconducting cuprates. The latter are, however, (unconventional) metallic systems, and praseodymium is unique among the rare-earth elements in that it can exhibit characteristics of strong hybridization, Kondo-like behavior, and can exhibit large effective masses at low T when doped into either high-temperature superconducting cuprates or related cuprate materials. A wide range of experimental investigations has been carried out on these systems.

One of the most perplexing questions in this field is the remarkably rapid depression by Pr doping of the superconducting transition temperature T_c of high-T_c cuprates. Indeed, in Pr-doped $YBa_2Cu_3O_7$ exhibits a metal-insulator transition with increasing Pr doping after T_c has been depressed to zero. This behavior is remarkable only in comparison with the ineffectiveness of other rare-earth dopants in depressing T_c which, it has been argued, is due to the weak exchange interaction between localized f-ion moments and CuO_2 itinerant electron states. A number of explanations have been reported, all of which invoke strong Pr-ion/CuO_2-band-state hybridization but differ in the assumed consequences of this hybridization.

In contrast to the majority of heavy fermion alloys, a number of sample preparation methods are available for the layered cuprates. These include ceramic powder techniques, single crystal growth, and fabrication of single and multilayer thin films. The effect of preparation method on properties and behavior of Pr-doped cuprates is considerable. A wide range of experimental techniques, including a number of spectroscopies, has been employed to study these materials. In Chapter 6, the magnetic properties of the Pr containing cuprates which are insulating and exhibit anomalous magnetism only in the Pr containing versions of a structure are summarized.

REFERENCES

1.   G. R. Stewart, Rev. Mod. Phys. **56**, 755 (1984).

2.   N. B. Brandt and V. V. Moschalkov, Adv. Phys. **33**, 373 (1984).

3.   C. M. Varma, Comments Solid State Phys. **11**, 221 (1985).

4.   F. Steglich, Springer Series in Solid-State Sciences **62**, 23 (1985).

5.   P. A. Lee, T. M. Rice, J. W. Serene, L. J. Sham, and J. W. Wilkins, Comments Condens. Matter Phys. **12**, 99 (1986).

6.   H. R. Ott, in *Progress in Low Temperature Physics*, edited by D. F. Brewer (Elsevier, Amsterdam, 1987), vol. XI, p. 215.

7.   P. Fulde, J. Keller, and G. Zwicknagl, Solid State Phys. **41**, 1 (1988).

8.   N. Grewe and F. Steglich, in *Handbook on the Physics and Chemistry of Rare Earths*, edited by K. A. Gschneidner and L. Eyring (North-Holland, Amsterdam, 1991), vol. 14, p. 3

9.   L. D. Landau, Sov. Phys.—JETP **3**, 920 (1956); *ibid.* **5**, 101 (1957); *ibid.* **8**, 70 (1959).

10.   See, e.g., R. H. Heffner and M. R. Norman, Comments Cond. Matt. Phys. **17**, 361 (1996).

11.   S. Doniach, Physica (Amsterdam) **B91**, 231 (1977).

12.   In dilute alloys T_{RKKY} will be reduced by a factor of the order of the local-moment impurity concentration

13.   For reviews of topics in this area see the Special Issue of J. Phys.: Condens. Matter (vol. 8, no. 48, 1996), which contains papers presented at the Institute for Theoretical Physics Conference on Non-Fermi Liquid Behavior in Metals, Santa Barbara, June 1996.

14.   For a review of the μSR technique, see A. Schenck, *Muon Spin Rotation Spectroscopy* (Adam Hilger, Bristol, 1985).

15.   D. W. Cooke *et al.*, Hyperfine Interact. **31**, 425 (1986).

16.   The term "Kondo" is vigorously opposed by some in the field, on the grounds that it should be reserved for the single-ion Kondo effect.

2

HEAVY FERMION SUPERCONDUCTORS

Lu Zhang and R.N. Shelton

UC-Davis Department of Physics,
Davis, CA 95616, USA.

I. INTRODUCTION

The study of heavy fermion superconductivity began in 1979 with the discovery by Steglich *et al.* 1979 of superconductivity in the ternary compound $CeCu_2Si_2$. Today, the field of heavy fermion superconductivity has been expanded to include six compounds. The original material, $CeCu_2Si_2$, has been joined by five compounds featuring uranium as the common element. Each of these compounds displays a significant electronic mass enhancement. Table I lists these six heavy fermion superconductors, referencing the year of discovery, the crystal structure and the value of the superconducting transition temperature, T_c, at ambient pressure.

Table I. Six heavy Fermion Superconductors

Compound	Reference	Structure	T_c (K)
$CeCu_2Si_2$	Steglich *et al.* 1979[1]	Tetragonal	0.6
UBe_{13}	Ott *et al.* 1983[2]	Cubic	0.9
UPt_3	Stewart *et al.* 1984b[3]	Hexagonal	0.5
URu_2Si_2	Schlabitz *et al.* 1984[4]	Tetragonal	1.2
UNi_2Al_3	Geibel *et al.* 1991b[5]	Hexagonal	1
UPd_2Al_3	Geibel *et al.* 1991a[6]	Hexagonal	2

General features for the heavy fermion superconductors include: a superconducting transition temperature in the range of 0.5 to 2 K, a large linear coefficient of the normal state specific heat of 70 - 1000 mJ/moleK (indicative of a large effective mass), a large upper critical field, and an unusual temperature dependence of many physical properties in the superconducting state.

One of the most exciting aspects of heavy fermion superconductivity is the possibility that some or all of these six compounds may be unconventional superconductors in terms of the pairing mechanism responsible for the superconducting state. This possibility serves as the primary focus of many research programs today, encompassing both experimental and theoretical studies.

The conventional superconducting state is described by an order parameter that is a complex scalar function having the symmetry of the point group of the underlying crystalline lattice. In contrast, in an unconventional superconductor, one or more of these symmetries may be broken.[7] Superconductivity in the heavy fermion systems differs markedly from that described by the conventional theory by

Bardeen, Cooper and Schrieffer (BCS) or extensions thereof. Some manifestations of these differences include non-exponential temperature dependence of various physical quantities, large critical fields, strong impurity effects and multiphase transitions in the superconducting state. The difference between heavy fermion and conventional s-wave isotropic superconductors will be presented in this chapter in terms of experimental evidence to indicate the possibility of unconventional superconductivity in heavy fermion systems.

Extensive experimental studies have been carried out in the field of heavy fermion superconductivity. This chapter presents a collection of some experimental findings of superconductivity occurring in these systems. Collectively these experimental results display the rich variety of unexpected properties of heavy fermion superconductors, which have enhanced our understanding, not only of superconductivity, but also of the general behavior of highly correlated electron systems.

The low temperature behavior in the superconducting state determined from various experiments strongly suggests that the gap function vanishes for some region on the Fermi surface. The temperature dependencies of the specific heat, thermal conductivity and ultrasonic attenuation are presented in sections II, III and IV, respectively.

Heavy fermion superconductors are type-II superconductors with κ as large as 50 or more. Section V presents the upper critical fields of heavy fermion superconductors, determined by transition curves from dc or ac susceptibility, electrical resistivity and specific heat determined in applied magnetic fields.

The heavy fermion superconducting state is very sensitive to impurities and defects. Chemical substitution by either nonmagnetic or magnetic impurities is presented in section VI. This section covers not only the depression of the superconducting transition temperatures and superconducting properties, but particularly interesting, the resulting complicated phase diagrams with superconducting multiphases or magnetic phase transitions in the superconducting state.

In addition to the impurities and defects, the application of high pressure has a strong influence on the superconducting properties of the heavy fermion superconductors. Section VII also presents some recent studies of the Ce-based heavy fermion compounds where the superconductivity was induced by high pressure near the suppression of the antiferromagnetic ordering state.

II. SPECIFIC HEAT EXPERIMENT

A. Introduction

In BCS theory, the quasiparticles at the Fermi surface of a conventional super-conductor are unstable to the formation of Cooper pairs at the superconducting transition temperature, T_c. Correlations between Cooper pairs generally lead to the formation of a gap in the quasiparticle dispersion relation. This results in a thermally activated temperature dependence (i.e., exponential) for many physical properties at sufficiently low temperature below T_c.[8] The weak-coupling BCS theory also predicts the normalized specific heat discontinuity, $\Delta C/\gamma T_c = 1.43$, for a conventional superconductor.[9]

The peculiar features observed in the specific heat in the heavy fermion super-conductors might be indicative of a non-conventional pairing mechanism in the superconducting state. In particular, the observation of a non-exponential temperature dependence of specific heat, at temperatures below the superconducting transition temperature, could be taken as evidence for some intrinsic gapless regions existing on the Fermi surface (see Thermal Conductivity: introduction). Taking into account the anomalous specific heat discontinuity (a noticeable deviation from $\Delta C/\gamma T_c = 1.43$), it might suggest a deviation from BCS conventional superconductivity such as gapless superconductivity ($\Delta C/\gamma T_c = 0$), or strong electron-electron coupling ($\Delta C/\gamma T_c > 1.43$).

B. Specific Heat Discontinuity

The specific heat discontinuity is defined as $\Delta C/\gamma T_c$ where $\Delta C = C_s(T_c) - C_n(T_c)$ is the specific heat jump at the superconducting transition temperature, T_c and $C_n(T_c) = \gamma T_c$. C_s and C_n represent specific heat values in the superconducting state and normal state, respectively. $\Delta C/\gamma T_c$ of six heavy fermion superconductors are listed in the Table II.

Table II. Specific heat discontinuity, $\Delta C / \gamma T_c$, in six heavy fermion superconductors.

Compound	Sample	$\Delta C / \gamma T_c$	Reference
$CeCu_2Si_2$	unannealed polycrystals	1.4	Steglich et al. 1979[1]
	annealed polycrystal	1.1	Lieke et al. 1982[10]
	annealed polycrystals	0.0 - 1.3	Bredl et al. 1983[11]
	crystal	1.27	Assmus et al. 1984[12]
UBe_{13}	polycrystal	2.5	Ott et al. 1984[13]
	polycrystal	2.40	Mayer et al. 1986[14]
	polycrystal	2.3	Ravex et al. 1987[15]
UPt_3	annealed crystal	1.0	Stewart et al. 1984[3]
	polycrystal	0.45	Franse et al. 1985[16]
	annealed polycrystal	0.86	Sulpice et al. 1986[17]
	unannealed crystals	0.30	de Visser et al. 1987[18]
	annealed polycrystals	0.35	de Visser et al. 1987[18]
	polycrystals	0.65, 0.78	Fisher et al. 1989[19]
	crystal	0.6	Schuberth et al. 1990[20]
URu_2Si_2	polycrystal	1.3	Palstra et al. 1985[21]
	annealed polycrystal	0.63	Schlabitz et al. 1986[22]
	polycrystal	0.8	Fisher et al. 1990[23]
	crystal	0.8	Hasselbach et al. 1991[24]
UNi_2Al_3	polycrystal	0.4	Geibel et al. 1991[5]
UPd_2Al_3	polycrystal	1.2	Geibel et al. 1991[6]
	annealed polycrystals	1.0, 1.2	Caspary et al. 1993[25]

1. $CeCu_2Si_2$

For the compound $CeCu_2Si_2$, the $\Delta C / \gamma T_c$ ratio of about 1.4 was first reported by Steglich et al. (1979)[1] in two unannealed polycrystalline samples (see inset of Fig.1). This experimental observation indicated the possibility that this phenomenon might be describable in terms of well-known BCS theory applied to the heavy Fermi-liquid.[9]

The results, however, were complicated by the delicate dependence on sample preparation, sample quality and data analysis (see table II). For polycrystalline samples, the proper heat treatment generally improved sample quality by the

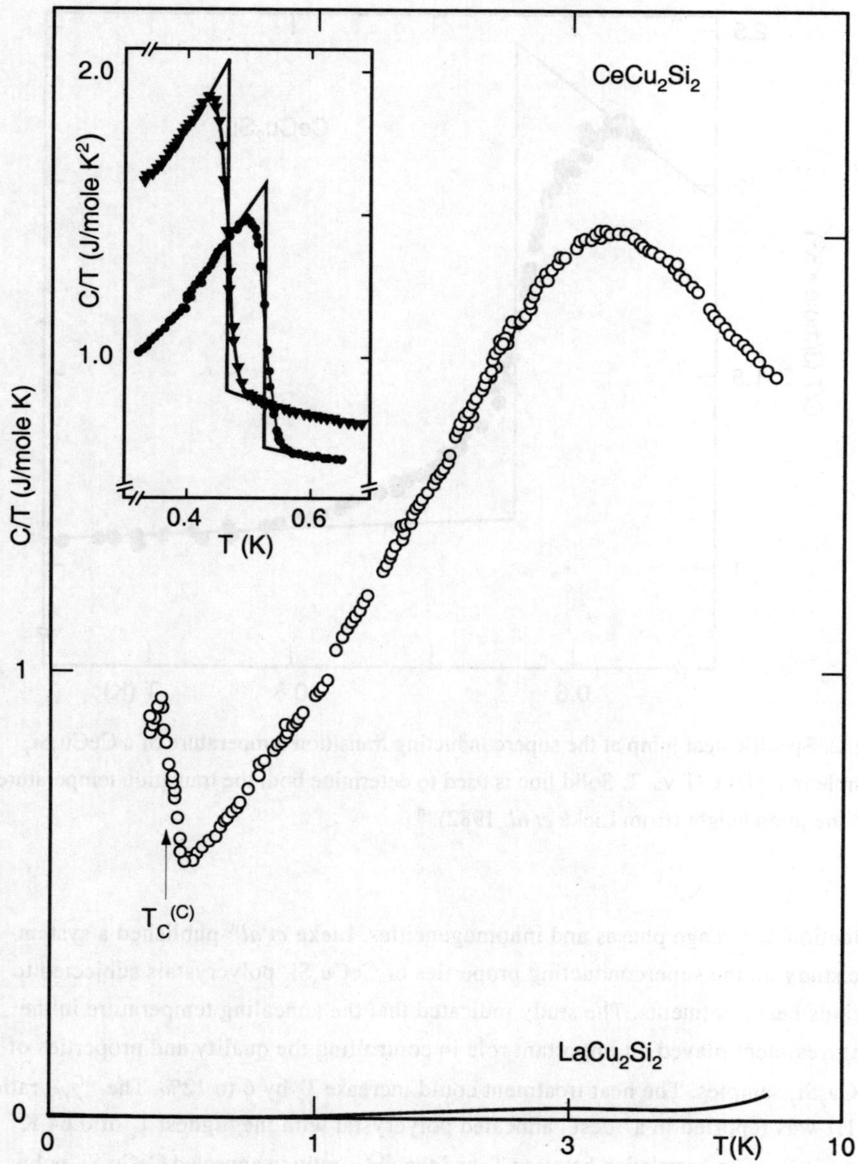

Fig. 1. Molar specific heat of CeCu$_2$Si$_2$ at B=0 as function of temperature on logarithmic scale. Arrow marks the transition temperature $T_c^{(C)} = 0.51 \pm 0.04$ K. Inset shows the specific heat jumps of two other CeCu$_2$Si$_2$ samples in a C/T vs. T plot (from Steglich et al. 1979).[1]

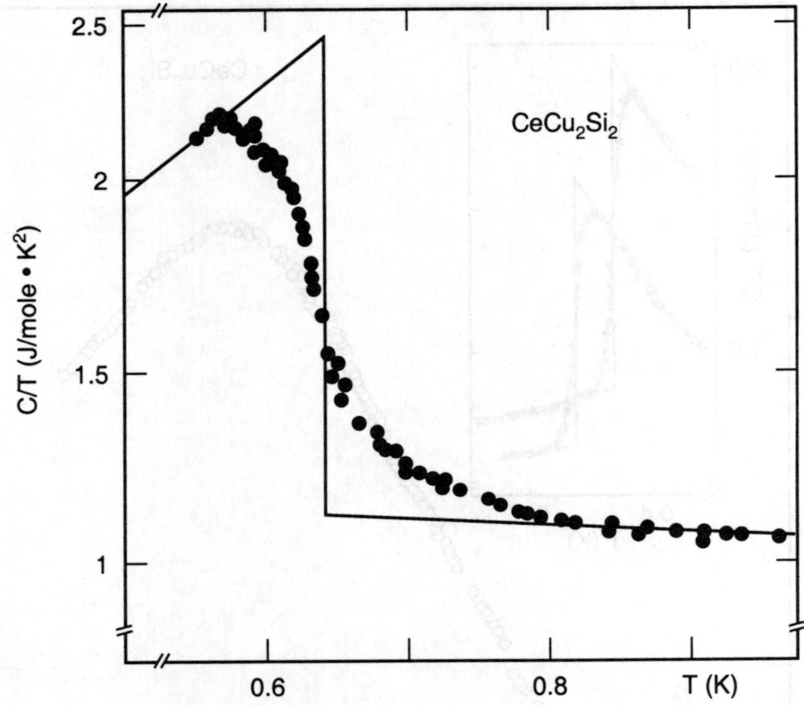

Fig. 2. Specific heat jump at the superconducting transition temperature of a $CeCu_2Si_2$ sample in a plot C/T vs. T. Solid line is used to determine both the transition temperature and the jump height (from Lieke *et al.* 1982).[10]

reduction of strange phases and inhomogeneities. Lieke *et al*[10] published a systematic study on the superconducting properties of $CeCu_2Si_2$ polycrystals subjected to various heat treatments. The study indicated that the annealing temperature in the heat treatment played an important role in controlling the quality and properties of $CeCu_2Si_2$ samples. The heat treatment could increase T_c by 6 to 12%. The $\Delta C/\gamma T_c$ ratio of 1.1 was reported in a "best" annealed polycrystal with the highest T_c of 0.64 K (see Fig.2). The correlation between T_c and the $\Delta C/\gamma T_c$ ratio in annealed $CeCu_2Si_2$ polycrystalline samples was investigated by Bredl *et al.*[11] The results showed $\Delta C/\gamma T_c$ varied from 0.0 to 1.3 and suggested that the higher T_c yielded a higher $\Delta C/\gamma T_c$ ratio, as shown in the inset of Fig.3. $CeCu_2Si_2$ single crystals prepared with excess Cu exhibited bulk superconductivity at 0.65 - 0.69 K and a specific heat discontinuity of 1.27 (see Fig.4b) and the same crystals prepared from stoichiometric melts indicated no superconductivity.

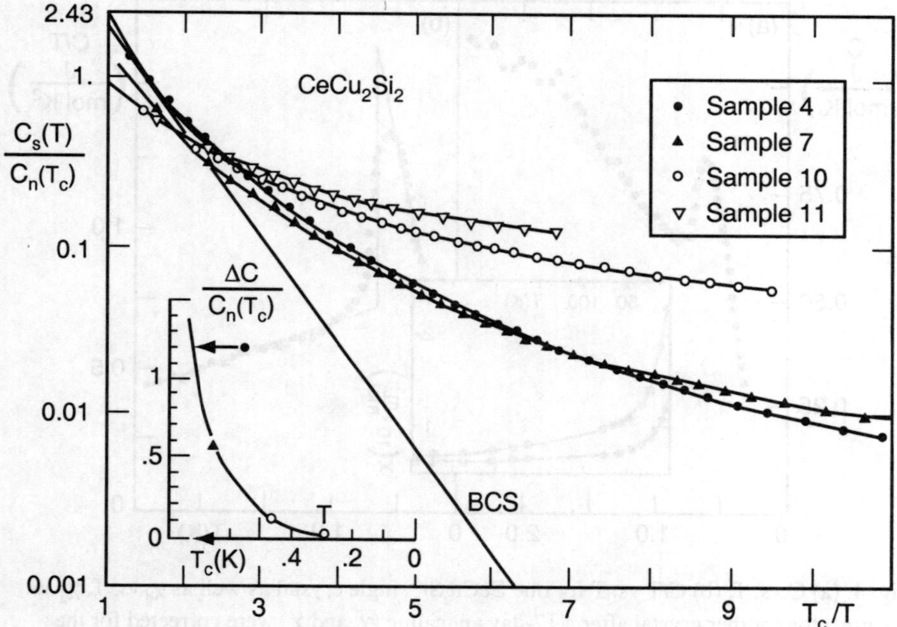

Fig. 3. Specific heat in the superconducting state of several CeCu$_2$Si$_2$ samples in a semilog plot C$_s$(T)/C$_n$(T$_c$) vs. T$_c$/T. Inset shows normalized specific heat jump at T$_c$ as a function of T$_c$ for the same samples (from Bredl *et al.* 1983).[11]

Under the same sample preparation and heat treatment, the $\Delta C/\gamma T_c$ ratio could still vary drastically.[11] Among the six heavy fermion superconductors, the compound CeCu$_2$Si$_2$ had the most extreme variation in ΔC with sample (see Impurity: sample dependence). The method used to determine $\Delta C/\gamma T_c$ from specific heat data could affect $\Delta C/\gamma T_c$. The superconducting anomaly in C was broad and γ showed temperature dependence. Some researchers determined ΔC by from C$_{max}$ - C$_n$ extrapolated to the lower temperature. This method differed from the estimation of an idealized, sharp ΔC.[26] This difference could be as large as 30 - 40%. In most studies (see references in Table II), the largest value for the $\Delta C/\gamma T_c$ ratio were used.

2. UBe$_{13}$

In contrast to the compound CeCu$_2$Si$_2$, UBe$_{13}$ showed an enhanced $\Delta C/\gamma T_c$ ratio of 2.5 in a small polycrystal, as displayed in Fig.5.[13] Similar $\Delta C/\gamma T_c$ ratios of 2.40 and 2.3 were

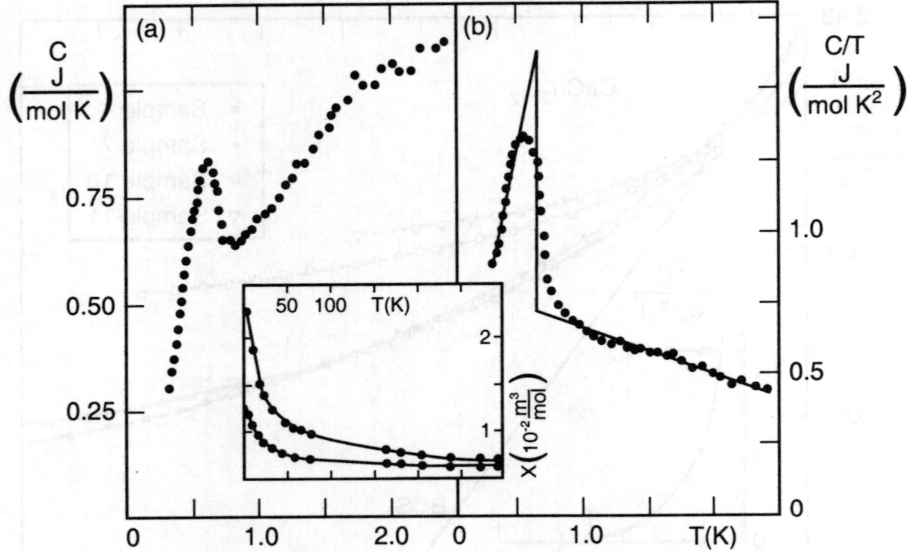

Fig. 4. (a) C vs. T, (b) C/T vs. T for one $CeCu_2Si_2$ single crystal as well as $\chi_\parallel$ vs. T, $\chi_\perp$ vs. T (inset) for another crystal after a 17-day annealing. $\chi_\parallel$ and $\chi_\perp$ were corrected for the demagnetization factors, $N_\parallel$ =0.68, $N_\perp$ =0.15 (from Assmus *et al.* 1984).[12]

obtained in polycrystalline samples by Mayer *et al.*[14] and Ravex *et al.*,[15] respectively (see Figs. 6 and 7). The large $\Delta C/\gamma T_c$ value observed in the compound UBe_{13} showed a noticeable deviation from BCS theory. This system has values comparable to those obtained for Pb which have been ascribed to strong coupling involving lattice phonons.[27]

3. UPt₃

The $\Delta C/\gamma T_c$ ratio of 1.0 was first reported in a annealed flux-grown crystal of UPt_3 by Stewart *et al.*[3] (see Fig.8). A smaller value of 0.48 was estimated from the specific heat data[16] in a polycrystalline UPt_3 (see Fig.9). Sulpice *et al.*[17] analyzed the specific heat measurements for an annealed polycrystalline sample and obtained the $\Delta C/\gamma T_c$ value of 0.86 (see Fig.10). de Visser *et al.*[18] presented experimental results of the specific heat discontinuity for different samples in the same compound UPt_3.[18] The specific heat data taken on an unannealed Czochralski-grown crystal and the annealed polycrystalline samples revealed a broad anomaly below T_c. The $\Delta C/\gamma T_c$ ratios of 0.30 and 0.35 were

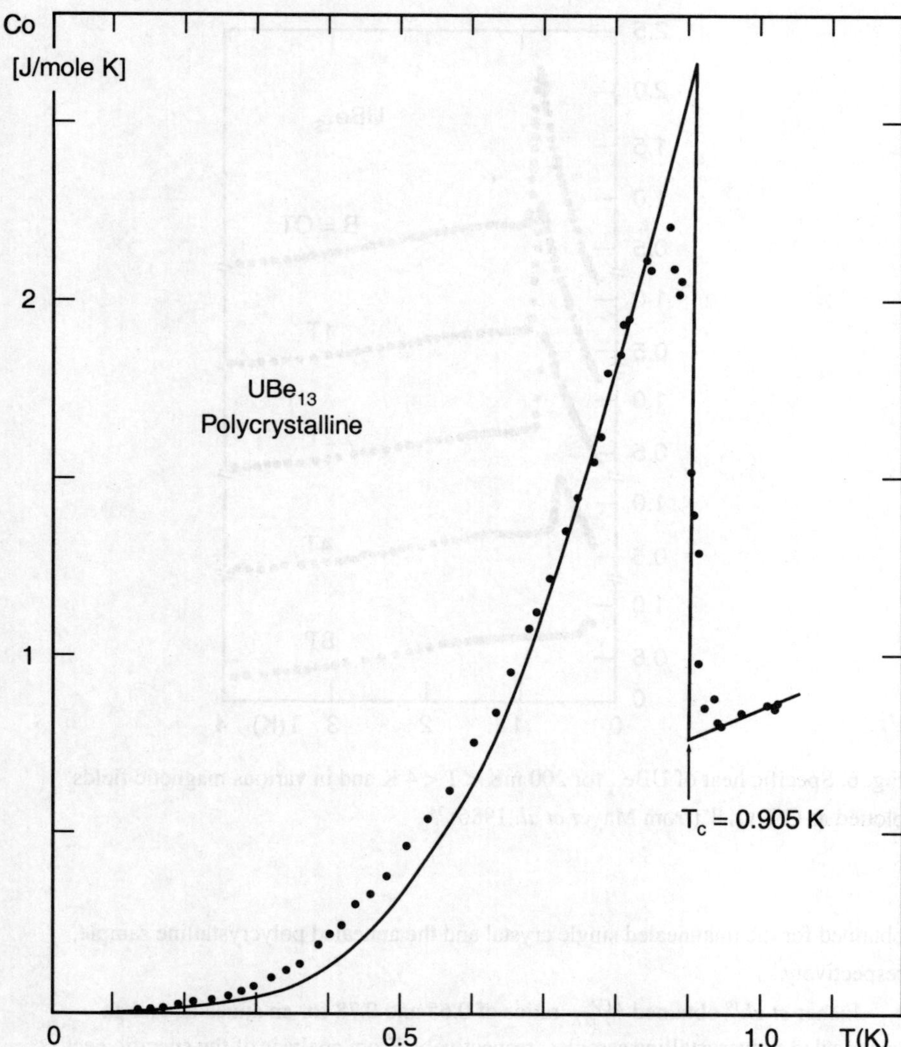

Fig. 5. Low temperature specific heat of UBe$_{13}$ between 0.07 and 1.1 K. The solid line is calculated for an Anderson-Brinkman-Morel p-wave superconductor with strong-coupling (from Ott *et al*. 1984).[13]

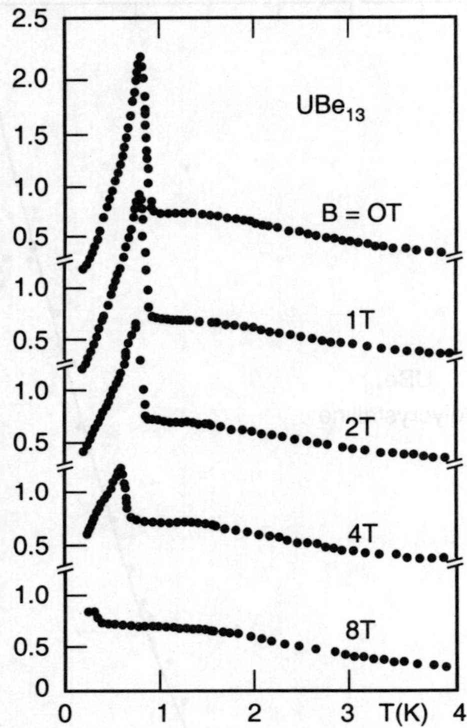

Fig. 6. Specific heat of UBe$_{13}$ for 200 mK < T < 4 K and in various magnetic fields plotted as C/T vs. T (from Mayer *et al.* 1986).[14]

obtained for the unannealed single crystal and the annealed polycrystalline sample, respectively.

Fisher *et al.*[19] obtained $\Delta C/\gamma T_c$ ratios of 0.65 and 0.78 for an annealed and an unannealed polycrystalline samples, respectively. From analysis of the specific heat discontinuity in the compound UPt$_3$, Fisher *et al.*[19] concluded that the $\Delta C/\gamma T_c$ ratio increased from 0.38 to 1.00 while T$_c$ increased from 0.37 to 0.59 K (see Fig. 11). Combined with other published data, the $\Delta C/\gamma T_c$ ratio in the compound UPt$_3$ is given in Table II, which is in the range of 0.30 to 1.00, smaller than one expected from BCS theory of 1.43. The reduced $\Delta C/\gamma T_c$ ratios in UPt$_3$ were explained by assuming that some heavy-mass quasiparticles did not participate in superconductivity down to very low temperatures. This explanation was inferred from a large linear contribution $\gamma_s T$ to the specific heat in the superconducting state. Compared with CeCu$_2$Si$_2$, the sample variation plays a

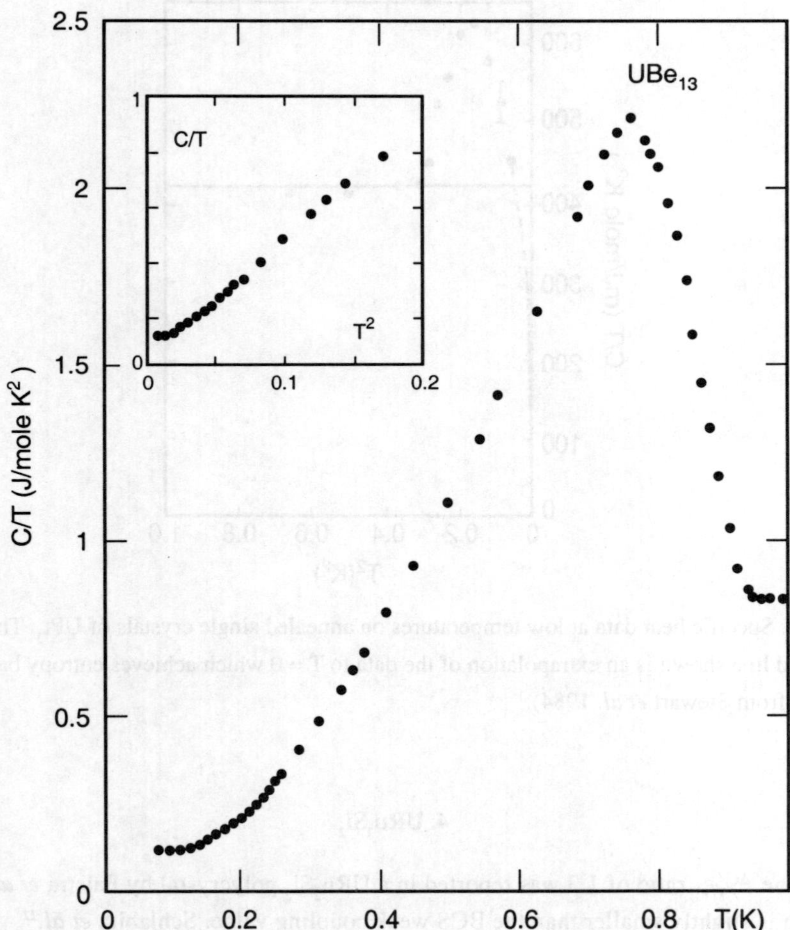

Fig. 7. C/T vs. T of UBe$_{13}$. Inset represents C/T vs. T^2 (from Ravex *et al*. 1987).[15]

slight role in the superconducting properties of the compound UPt$_3$. However, the sample quality plays an important role for superconductivity. In the high purity UPt$_3$ samples, a higher T$_c$, sharper transition, smaller residual resistivity and absence of remnance in magnetization were observed, which led to a higher $\Delta C/\gamma T_c$ as pointed out by Stewart *et al*.[3] and Fisher *et al*.[19]

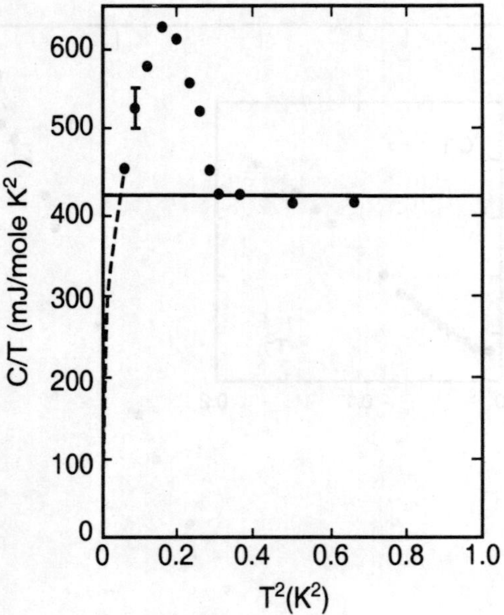

Fig. 8. Specific heat data at low temperatures on annealed single crystals of UPt$_3$. The dashed line shown is an extrapolation of the data to T = 0 which achieves entropy balance at T$_c$ (from Stewart *et al.* 1984).[3]

4. URu$_2$Si$_2$

The $\Delta C/\gamma T_c$ ratio of 1.3 was reported in a URu$_2$Si$_2$ polycrystal by Palstra *et al.*21 which is slightly smaller than the BCS weak coupling value. Schlabitz *et al.*[22] published a much smaller value of 0.63 in an annealed polycrystalline sample (see Fig.12). and explained that the specific heat data in URu$_2$Si$_2$ was similar to the findings in UPt$_3$. Fisher *et al.*[23] obtained a $\Delta C/\gamma T_c$ value of 0.8 in a polycrystalline sample. Single crystal studies of the specific heat in the compound URu$_2$Si$_2$ were performed by Hasselbach *et al.*[24] We estimated the $\Delta C/\gamma T_c$ ratio to be 0.8 form the best URu$_2$Si$_2$ crystal as shown in Fig.13.

5. UNi$_2$Al$_3$ and UPd$_2$Al$_2$

The study of specific heat in polycrystalline samples of UNi$_2$Al$_3$ and UPd$_2$Al$_2$ by Geibel *et al.*[5-6] showed the $\Delta C/\gamma T_c$ ratio of 0.4 and 1.2, respectively (see Figs.14 and 15).

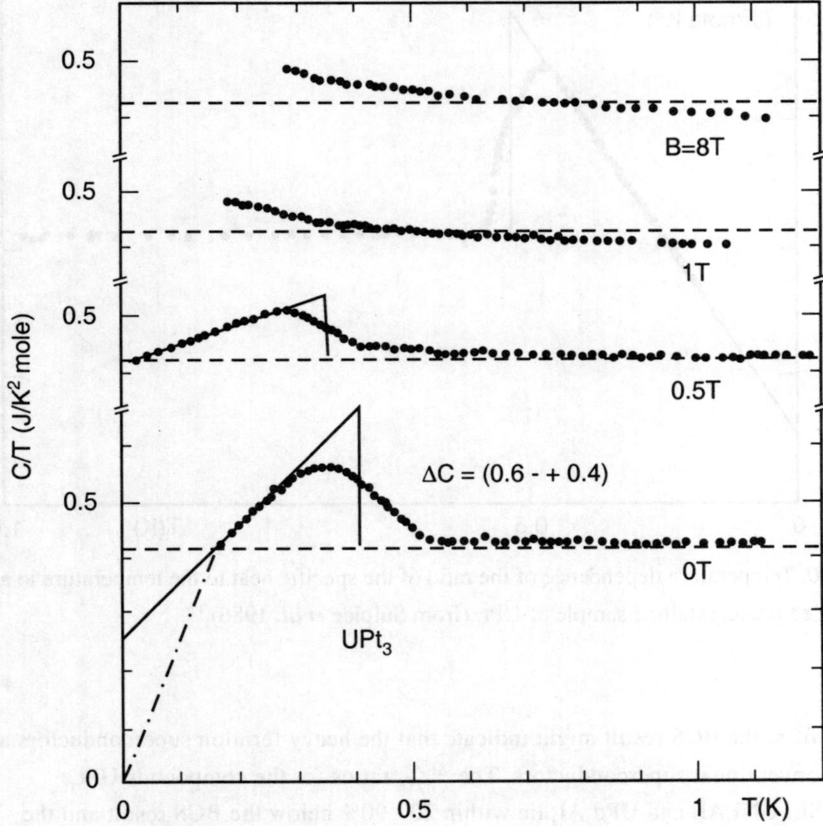

Fig. 9. The specific heat of polycrystalline UPt$_3$ at temperatures below 1.1 K in fields up to 8 T. The dashed curve represents the equation $C/T = \gamma^* + \beta^* T^2 + \delta T^2 \ln(T/1K)$, with γ^* = 422 mJ/K^2 mol, β^* = - 4.18 mJ/K^4 mol and δ = 1.54 mJ/K^4. Schematic extrapolation of zero field data by dash-dotted line satisfies entropy balance (from Franse *et al.* 1985).[16]

Specific heat measurements were performed as a function of temperature in the heavy fermion superconductor UPd$_2$Al$_3$ by Caspary *et al.*[25] The idealized $\Delta C / \gamma T_c$ values are 1.0 and 1.2 for two annealed polycrystalline samples estimated from Fig.16.

6. Conclusion

Most heavy fermion superconductors show smaller specific heat discontinuities than the BCS weak coupling value of 1.43. The exception is UBe$_{13}$. The deviation of

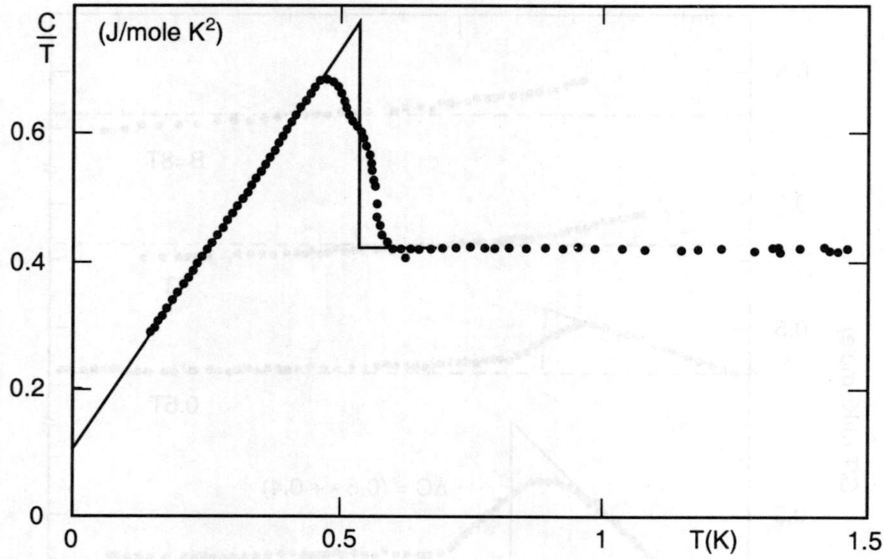

Fig. 10. Temperature dependence of the ratio of the specific heat to the temperature in an annealed polycrystalline sample of UPt$_3$ (from Sulpice *et al.* 1986).[17]

$\Delta C/\gamma T_c$ from the BCS result might indicate that the heavy fermion superconductors are non-conventional superconductors. The $\Delta C/\gamma T_c$ ratios for the compounds UPt$_3$, URu$_2$Si$_2$, UNi$_2$Al$_3$ and UPd$_2$Al$_2$ lie within 20 - 90% below the BCS result and the specific heat feature marking the transitions into the superconducting state at T$_c$ is often broad. The large variations of $\Delta C/\gamma T_c$ in the compound CeCu$_2$Si$_2$ are mainly due to sample preparation, heat treatment and sample quality. The compound UBe$_{13}$ shows the sharpest superconducting transition and the largest specific heat discontinuity among the class of heavy fermion superconductors. Larger values around 2.4 observed in both UBe$_{13}$ polycrystalline and single crystalline samples might occur as a result of strong coupling to lattice phonons.

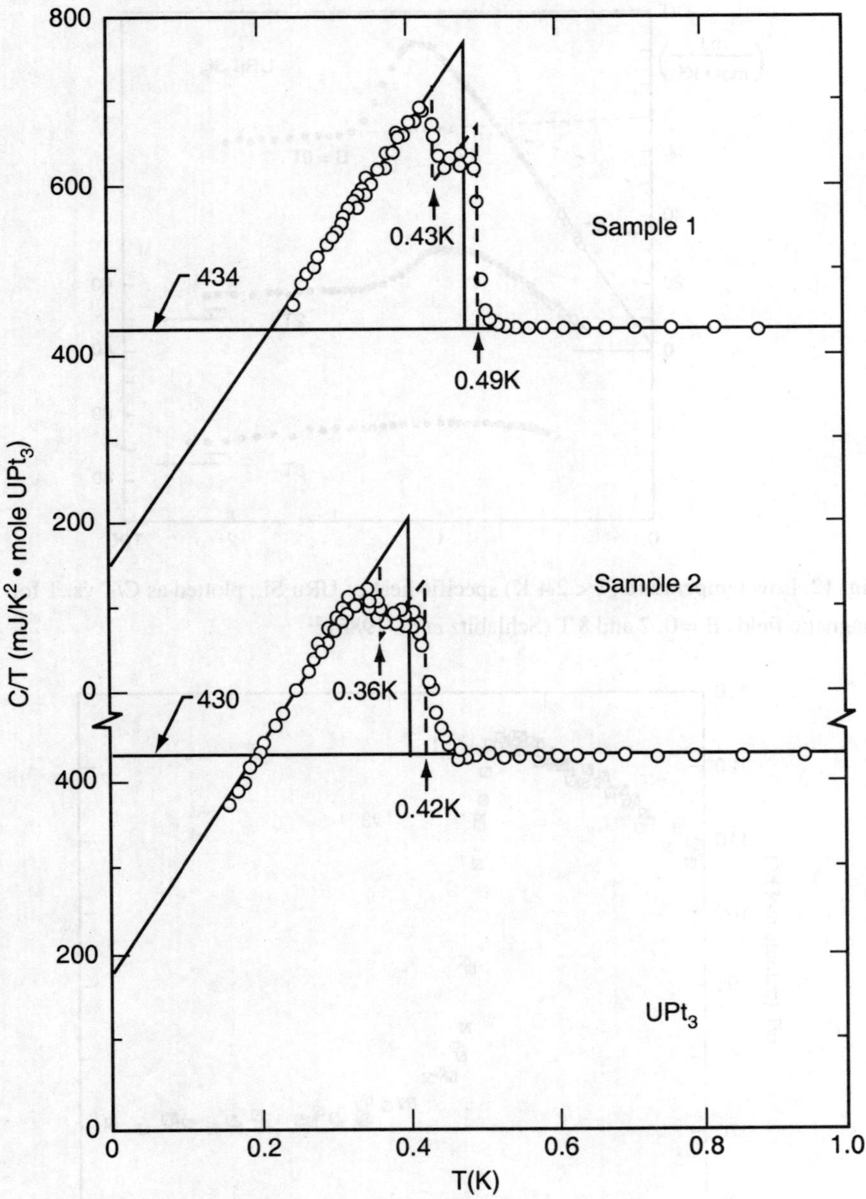

Fig. 11. The specific heat of UPt$_3$ in the vicinity of the superconducting transition for two samples. The dashed lines represent two ideally sharp transitions at T_a and T_b; the solid lines represent an ideally sharp single transition, with the same total entropy, at T_c (from Fisher *et al*. 1989).[19]

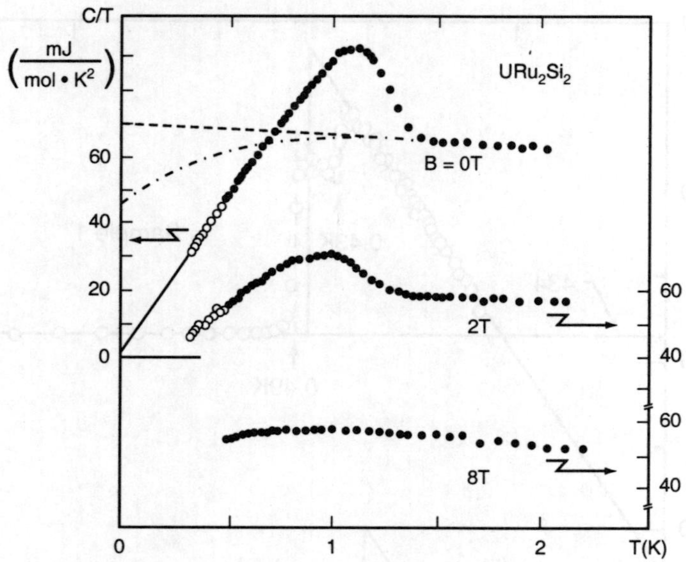

Fig. 12. Low temperature (T < 2.4 K) specific heat of URu_2Si_2, plotted as C/T vs. T for magnetic fields B = 0, 2 and 8 T (Schlabitz *et al.* 1986).[22]

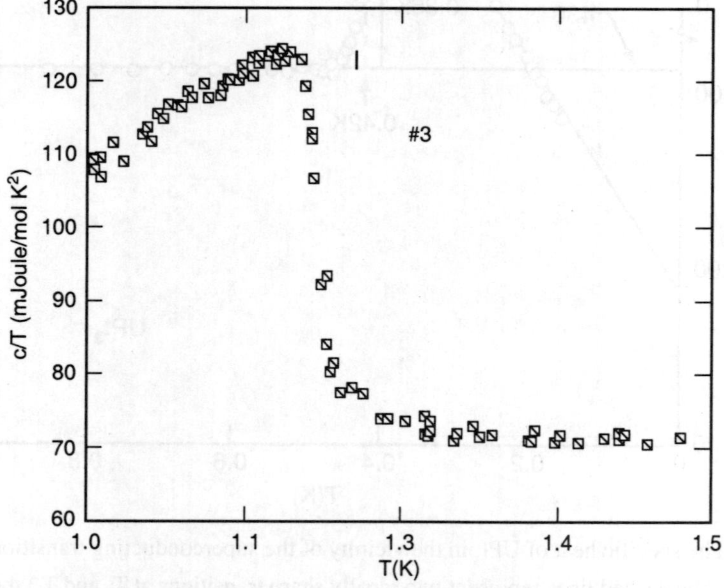

Fig.13. Search for a double-transition in URu_2Si_2 (from Hasselbach *et al.* 1991).[24]

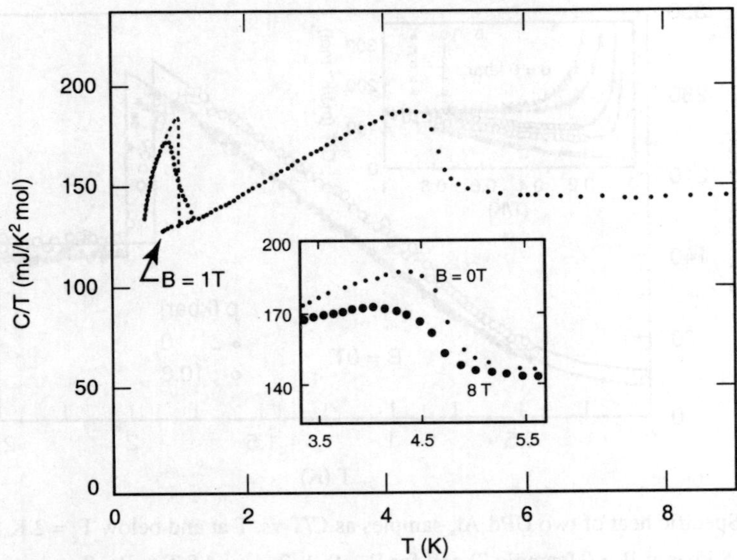

Fig. 14. Specific heat of UNi_2Al_3 in a plot C/T vs. T between 0.5 K and 9 K. B = 1 T data shown represent normal state. Dashed lines are used to replace, under conservation of total entropy, broadened anomaly at T_c by idealized jump. Inset shows slight shift of antiferromagnetic transition in a field of 8 T (from Geibel *et al.* 1991).[5]

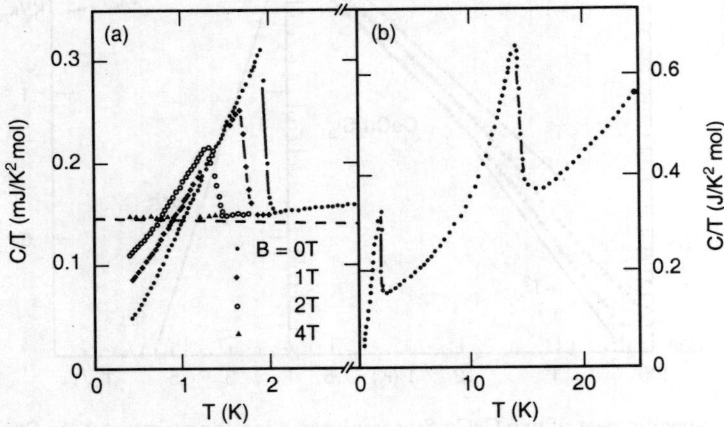

Fig.15. C/T vs. T for a polycrystalline sample of UPd_2Al_3 measured between 0.4 K and 3 K at B = 0 and in several magnetic fields (a) as well as up to 25 K at B = 0 (b). Dashed curve in (a) represents electronic contribution, $C_{el}(T)/T$. The solid lines are guides to the eye (from Geibel *et al.* 1991).[6]

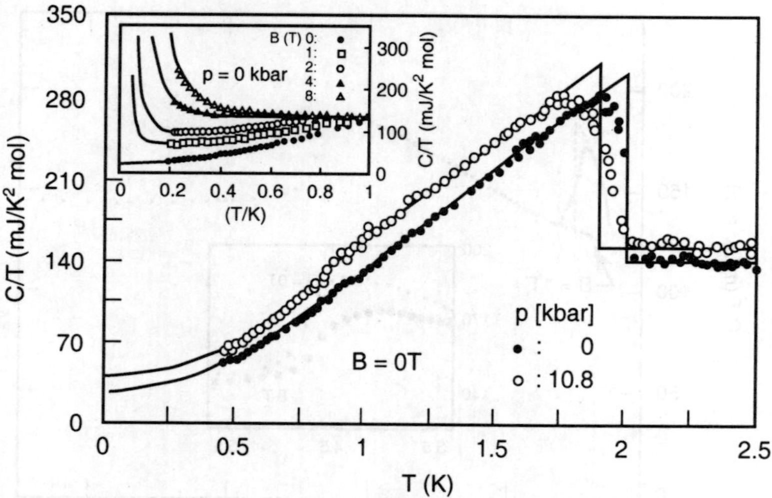

Fig. 16. Specific heat of two UPd_2Al_3 samples as C/T vs. T at and below $T_c = 2$ K for P = 0 and 10.8 kbar at B = 0 (sample 2) and for B = 0, 1, 2, 4, and 8 T at P = 0 (sample 1); see inset. Solid lines represent idealized jumps to determine T_c or a fit to the data by $C = \alpha/T^2 + \gamma_s T + \delta_s T^3$ (from Caspary et al. 1993).[25]

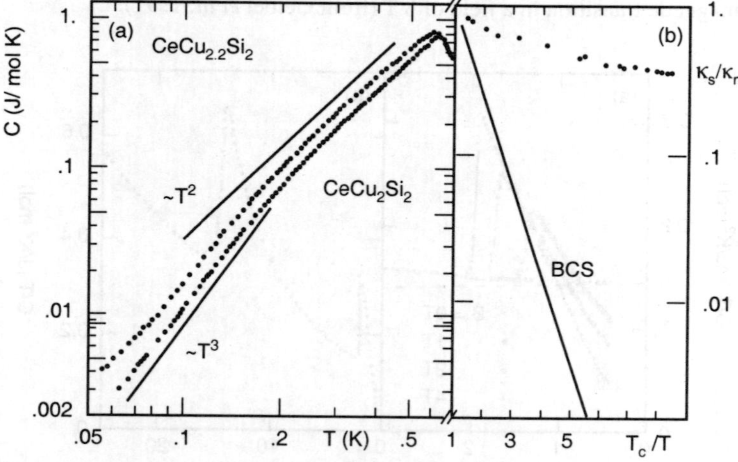

Fig. 17. (a) Specific heat of two $CeCu_2Si_2$ samples in a log-log representation. Straight lines indicate simple power laws for comparison. (b) Thermal conductivity of $CeCu_{2.02}Si_{1.98}$ in the superconducting state, $\kappa_s(T)$, normalized to the normal state values $\kappa_n(T)$, plotted logarithmically as a function of T_c/T. Solid line is BCS result for comparison (from Steglich et al. 1984).[28]

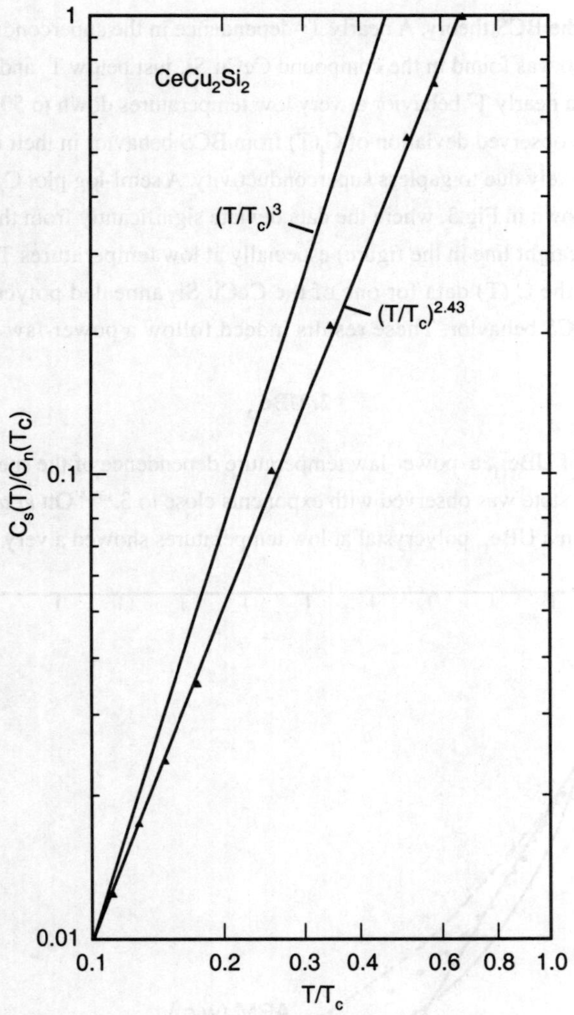

Fig. 18. Log-log plot of $C_s(T)/C_n(T)$ vs. T/T_c for CeCu$_2$Si$_2$. The two lines shown correspond to two different power laws (from Stewart 1984).[26]

C. Temperature Dependence

1. CeCu$_2$Si$_2$

The specific heat of two CeCu$_2$Si$_2$ samples are shown in Fig.17. where the data could not fitted by an exponential curve (i.e. thermally activated temperature dependence) as predicted by the BCS theory. A nearly T^2-dependence in the superconducting specific heat curve, $C_s(T)$, was found in the compound $CeCu_2Si_2$ just below T_c and a continuous change towards a nearly T^3 behavior at very low temperatures down to 50 mK.[28] Bredl *et al.*[11] ascribed the observed deviation of $C_s(T)$ from BCS behavior in their $CeCu_2Si_2$ polycrystals as likely due to gapless superconductivity. A semi-log plot $C_s(T)/C_n(T_c)$ versus T_c/T is shown in Fig.3. where the data deviate significantly from the BCS prediction (see solid straight line in the figure) especially at low temperatures $T < 0.4T_c$. Fig.18 exhibits the $C_s(T)$ data for one of the $CeCu_2Si_2$ annealed polycrystalline samples from BCS behavior. These results indeed follow a power-law of T.[2.43]

2. UBe$_{13}$

In the case of UBe_{13}, a power-law temperature dependence of the specific heat in the superconducting state was observed with exponents close to 3.[13-15] Ott *et al*[13] claimed their $C_s(T)$ data in a UBe_{13} polycrystal at low temperatures showed a very clear deviation

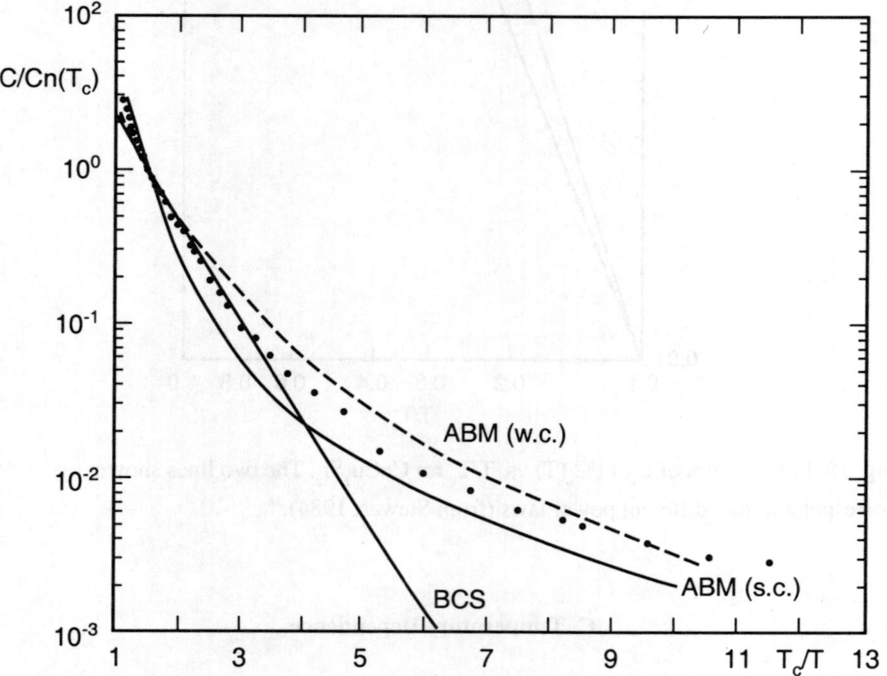

Fig. 19. $C_s/C_n(T_c)$ for superconducting UBe_{13}. Dashed line: weak-coupling Anderson, Brinkman, Morel (ABM) state; solid lines: BCS and strong-coupling ABM state (from Ott *et al.* 1984).[13]

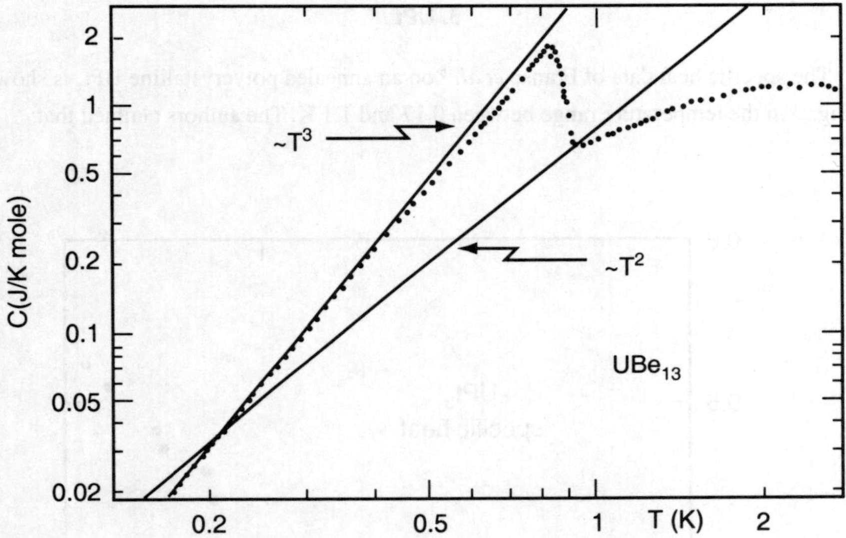

Fig. 20. Specific heat of UBe$_{13}$ for T < 3 K. The zero-field data are compared in a log-log representation with sample power laws (from Mayer *et al.* 1986).[14]

from the BCS form and followed a power-law form: $C_s(T)/C_n(T_c) \approx 2.8$ $(T/T_c)^3$ (see Fig.19). The re-analysis by Stewart[26] showed the $C_s(T)$ data of Ott *et al.*[25] appeared to follow a power-law of $T^{3.15\pm0.15}$. Mayer *et al.*[14] measured the superconducting specific heat in a UBe$_{13}$ polycrystalline sample where the best fit was found close to $C_s(T) \propto T^{2.9}$ in a large temperature range $0.23T_c < T < 0.97T_c$ (i.e. 200 - 850 mK) (see Fig.20). Ravex *et al.*25 again confirmed a $T^{2.9}$-dependence between 0.2 and 0.7 K with a non-vanishing linear T term of 0.11 J/molK^2 at T → 0 (see Fig.7). However, a clear change of the specific heat behavior (i.e. deviation from linear line in C_s/T) was found down to very low temperature of about 60 mK.

Ott *et al.*[13] explained that the presence of a power-law was taken as evidence for triplet (p-wave) superconductivity such as occurs in ^{3}He.[29-32] By assuming that the superconducting state of UBe$_{13}$ was an Anderson-Brinkman-Morel (AMB) p-wave superconductor, Ott *et al.*[13] suggested that the gap function went to zero someplace on the Fermi surface and the zeros of the gap gave rise to a T^3-dependence and the $\Delta C/\gamma T_c$ ratio of 1.18, compared with their observations of $T^{3.15}$-dependence and $\Delta C/\gamma T_c$ of 2.5. As pointed out by Volovik and Gor'kov,[35] a T^3 power-law for the superconducting electronic specific heat may also occur in a singlet, or BCS superconductor, as well as for p-wave superconductors.

3. UPt$_3$

The specific heat data of Franse *et al.*[16] on an annealed polycrystalline UPt$_3$ is shown in Fig. 9 in the temperature range between 0.17 and 1.1 K. The authors claimed that

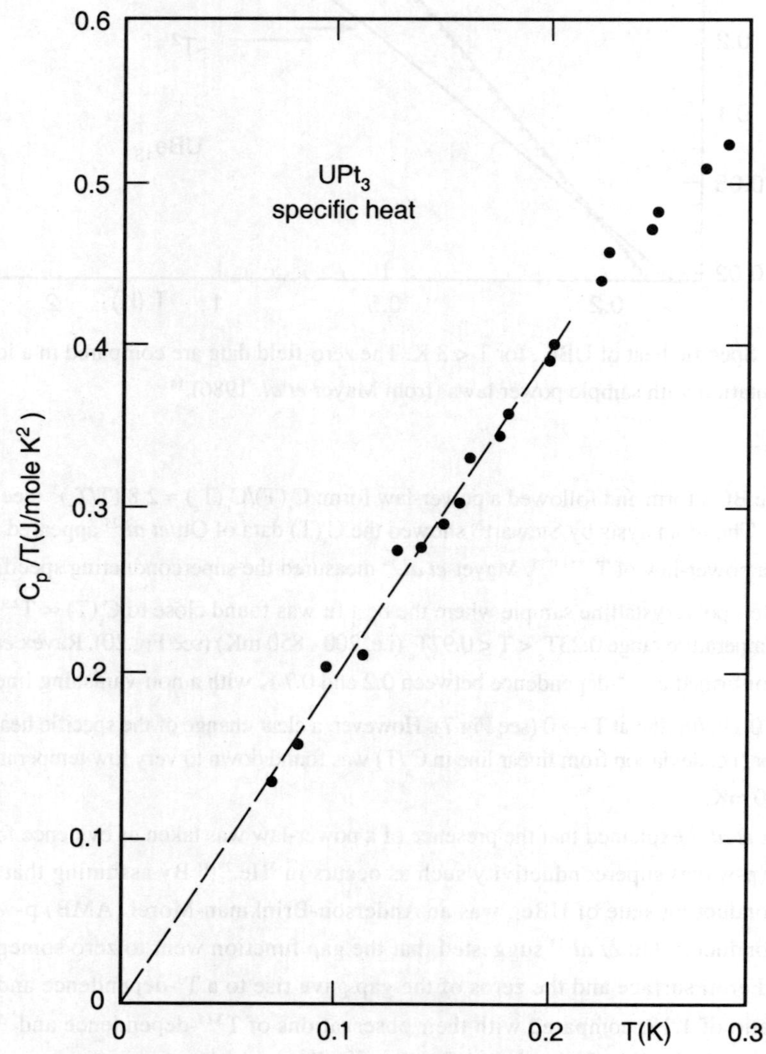

Fig. 21. Specific heat of superconducting UPt$_3$ below 0.3 K. The broken line is a guide to the eye (from Ott *et al.* 1987).[34]

$C_s(T)/T$ was almost linear in temperature at low temperatures well below T_c with a finite intersect of 260 mJ/K^{2}mole at T=0. Sulpice *et al.*[17] repeated the specific heat measurement of Franse *et al.*[16] down to 146 mK on an annealed UPt$_3$ polycrystal. A beautiful quadratic temperature dependence of $C_s(T)$ was found between 146 mK and T_c (see Fig.10): $C_s(T) = \gamma_s T + \beta_s T^2$ with $\beta_s = 1.25$ J/moleK^3 and $\gamma_s = 0.11$ J/moleK^2. Very low temperature specific heat studies were given by Ott *et al.*[34] where the low limit temperature was 70 mK. The sample used in the experiment was claimed as a very high quality polycrystal because this sample had successfully been used for de Haas-van Alphen studies.[34] As shown in Fig. 21, a broken line was plotted as a conventional extrapolation to T = 0 K and fitted well to $C_s(T) = \beta_s T^2$ with $\beta_s = 1.975$ J/moleK^3. Fisher *et al.*[19] reported specific heat measurements on two samples of UPt$_3$: one annealed and another unannealed polycrystals (see Fig. 8). Their $C_s(T)$ data were best fitted to the quadratic temperature dependence: $\gamma_s T + \beta_s T^2$. Fisher *et al.*[19] summarized the experimental result in UPt$_3$ up to date and suggested a correlation between sample quality and fitting coefficients γ_s and β_s. The linear $\gamma_s T$ term might exist due to poor quality or impurities in the samples. The $\beta_s T^2$-term of the specific heat in the superconduct-

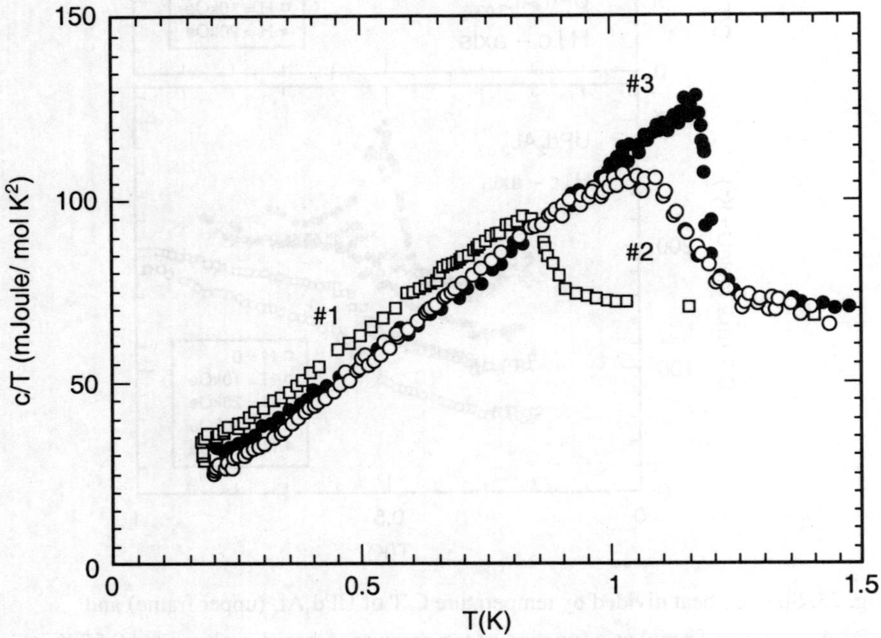

Fig. 22. Specific heat data of three monocrystalline samples of URu$_2$Si$_2$ represented as C/T plotted as a function of T (from Hasselbach *et al.* 1991).[24]

ing state could be qualitatively explained by a superconductor with the polar-like gap function vanished at lines on the Fermi surface (see Thermal Conductivity: introduction).

4. URu_2Si_2

Quadratic $C_s(T)$ dependence in a URu_2Si_2 polycrystal could be followed from slightly below T_c to 0.3 K as shown in Fig. 12.[22] However, a γ_s value of 4 mJ/K^2mol

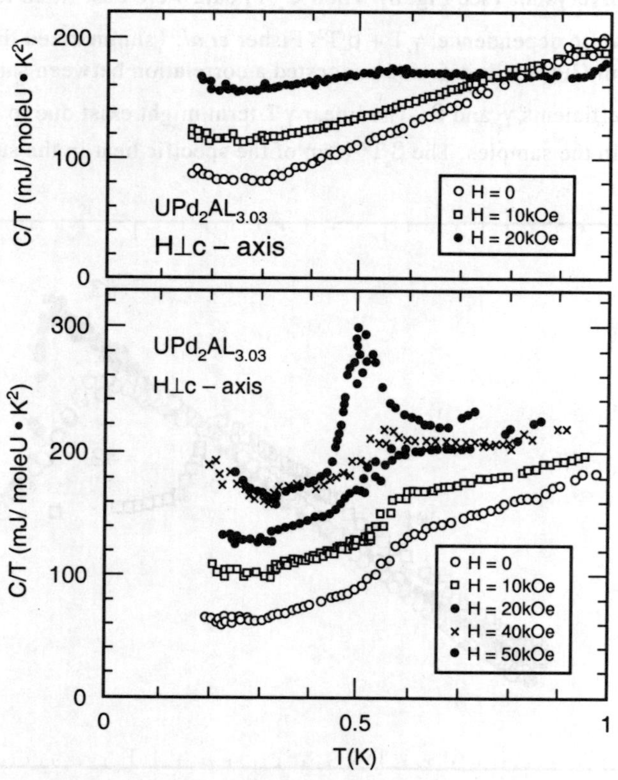

Fig. 23. Specific heat divided by temperature C/T of UPd_2Al_3 (upper frame) and $UPd_2Al_{3.03}$ (lower frame) as a function of temperature. A broad peak around 0.55 K in UPd_2Al_3 for H = 20 kOe is due to a normal-superconducting phase transition (from Sato *et al.* 1993).[37]

in URu_2Si_2 was found, which was much smaller than the values of 100 - 260 mJ/K^2mol in the compound UPt_3. Similar quadratic temperature dependence was obtained in a polycrystalline sample of URu_2Si_2: $C_s(T) = \gamma_s T + \beta_s T^2$ with $\gamma_s = 6.5$ mJ/moleK^2 and $\beta_s = 93$ mJ/moleK^3. Specific heat data of three URu_2Si_2 single crystals could be represented by a linear $\gamma_s T$ plus a quadratic $\beta_s T^2$ term (see Fig. 22). The largest $\gamma_s T$ term was observed for the unannealed crystal with the lowest T_c and the smallest $\Delta C/\gamma T_c$ ratio.

5. UPd_2Al_3

The "intrinsic" specific heat in UPd_2Al_3 (see Fig. 15) could be best described by $C_s(T) \propto T^{2.9}$ rather than an exponential law[35, 36] similar to the observations in the compound UBe_{13} by Mayer et al.[14] and Ravex et al.[15] Caspary et al.[25] confirmed the results obtained by Steglich et al.[35] and found $C_s(T) = \gamma_s T + \delta_s T^2$ with $\gamma_s = 0.024$ J/moleK^2. The single crystal study of UPd_2Al_3 in specific heat data by Sato et al.[37] showed an interesting result (see Fig. 23); namely, a quadratic term observed between 0.3 K and T_c (0.55K). At low temperatures down to 0.2 K, this behavior changed to a linear dependence (a flattening in C_s/T versus T) which might be attributed to the nuclear Zeeman effect as suggested by Sato et al. (1993).[37]

6. Conclusion

The specific heat in the superconducting state for the heavy fermion superconductors cannot be described by an exponential form predicted by BCS theory. A power-law temperature dependence could be determined in the compounds $CeCu_2Si_2$, UBe_{13} and UPd_2Al_3 with the exponents in the range of 2 to 3. All the specific heat data for UPt_3 and URu_2Si_2 did not show the exponential temperature dependence. They did not follow power-law behavior either except one measured by Ott et al.[34] The best fit would be a quadratic form with two terms: T^2-term and T-term. Table III listed the various form of temperature dependence and corresponding fitting parameters.

Table III. Temperature dependence of the specific heat in the superconducting state $C_s(T)$ of heavy fermion superconductors (fitting coefficient β_s is in unit of J/moleK^3, γ_s in J/moleK^2 and δ_s in J/moleK^4).

Compound	$C_s(T)$	Fit Parameters	Reference
CeCu$_2$Si$_2$	T^2 or T^3 *		Steglich *et al.* 1984[28]
	$T^{2.43}$		Bredl *et al.* 1983[11]
UBe$_{13}$	$T^{3.15}$		Ott *et al.* 1984[13]
	$T^{2.9}$		Mayer *et al.* 1986[14]
	$T^{2.9}$ or $\gamma_s T$ *	$\gamma_s = 0.11$	Ravex *et al.* 1987[15]
UPt$_3$	$\beta_s T^2 + \gamma_s T$	$\beta_s=1$ & $\gamma_s = 0.26$	Franse *et al.* 1985[16]
		$\beta_s=1.25$ & $\gamma_s=0.11$	Sulpice *et al.* 1986[17]
		$\beta_s=1.975$ & $\gamma_s=0$	Ott *et al.* 1987[34]
		$\beta_s=1.3$ & $\gamma_s=0.14$	Fisher *et al.* 1989[19]
		$\beta_s=1.25$ & $\gamma_s=0.165$	Fisher *et al.* 1989[19]
		$\beta_s=1.6$ & $\gamma_s=0.26$	Schuberth *et al.* 1990[20]
URu$_2$Si$_2$	$\beta_s T^2 + \gamma_s T$	$\beta_s=0.092$ & $\gamma_s=0.003$	Schlabitz *et al.* 1986[4]
		$\beta_s=0.0065$ & $\gamma_s=0.093$	Fisher *et al.* 1990[23]
		$\beta_s=0.087$ & $\gamma_s=0.013$	Hasselbach *et al.* 1991[24]
UPd$_2$Al$_3$	$T^{2.9}$		Steglich *et al.* 1992[35]
	T^2 or T *		Sato *et al.* 1993[37]
	$\delta_s T^3 + \gamma_s T$	$\gamma_s = 0.024$	Caspary *et al.* 1993[25]

* There were two temperature dependencies: the former T-dependence was obtained at temperature just below T_c and the latter one was at very low temperatures ($T \ll T_c$).

The excess of $C_s(T)$ at low temperatures compared with the exponential behavior predicted by the BCS theory indicated the heavy fermion superconductors are unconventional, which is consistent with a departure of the $\Delta C/\gamma T_c$ ratios from the BCS value of 1.43.

Ott *et al.*[13] argued that the approximate T^3-dependence in the specific heat in the superconducting state was due to triplet superconductivity in UBe$_{13}$. However, the arguments of Bredl *et al.*[11] used to describe the power-law behavior in CeCu$_2$Si$_2$ was for gapless superconductivity. As pointed out by Volovik and Gor'kov[33] a T^3 power-law for the superconducting electronic specific heat might occur in a singlet (BCS -type) superconductor, as well as for p-wave superconductors. The quadratic temperature dependence observed in UPt_3, URu_2Si_2 and UPd_2Al_3 might be explained in the following manner. First, a non-vanishing linear temperature term, $\gamma_s T$, was mainly due to impurities in the samples (recall that broad transitions were observed in these compounds).[19] Second, the T^2 power-law could be obtained for a polar-like d-wave superconducting state.[38]

D. Entropy Balance

At low temperatures, much smaller than the Fermi temperature (energy) and the Debye temperature, the specific heat of a heavy fermion superconductor consists of three terms: one is the electronic contribution (linear T), a second is due to phonons (T^3-term) and a third is due to spin fluctuation contributions. Thus one can write: $C_n(T) = \gamma_n T + \beta_n T^3 + \delta T^3 \ln T$ where the subscript n denotes normal state to distinguish it from the superconducting state. The details for the properties of the specific heat in the normal state in heavy fermion systems will be discussed in a separate section (Heat Capacity). At low temperatures (just above T_c), the linear temperature term plays an important role and $\gamma_n = C_n(T)/T$ is almost a constant. The specific heat in the superconducting state can be described approximately by a temperature dependence of a power-law or quadratic form.

The entropy at T_c derived from the superconducting state data is defined as $S_s(T_c) = \int_o^{T_c} (C_s/T) \, dT$, which is the area under the measured $C_s(T)$ curve in the superconducting state. The entropy at T_c extrapolated from the normal state data is given by $S_n(T_c) = \int_o^{T_c} (C_n^{extrap}/T) \, dT$. C_n^{extrap}/T is normally determined by extrapolating $\gamma_n = C_n(T)/T$ below T_c to $T = 0$ assuming a constant γ_n below T_c. For a second order phase transition of superconductivity, $S_n(T_c) = S_s(T_c)$, which is the so-called entropy balance. The observations of entropy discrepancy were claimed in the heavy fermion superconductors.[16, 39, 14, 22] In most papers, however, the values of entropy, $S_n(T_c)$ and $S_s(T_c)$, were not given. In Table IV, some data of $S_n(T_c)$, $S_s(T_c)$ and entropy discrepancy in UPt_3 were given based on the fitting parameters given by Fisher et $al.$[19] The entropy discrepancy is defined as a percentage difference between $S_s(T_c)$ and $S_n(T_c)$.

It was first notified by Franse et $al.$[16] that there was an entropy discrepancy at T_c in the compound UPt_3. The C_s/T behavior was almost linear in temperature[16] with a slope of 960 mJ/K^3mole and an intercept of 260 mJ/K^2mole.[19] The entropy discrepancy is estimated to be 6.6% with $S_s(T_c) = 181$ mJ/Kmole and $S_n(T_c) = 169$ mJ/Kmole (see Table IV).

In the review article, Stewart[26] speculated that an entropy discrepancy would exist in the UBe_{13} and $CeCu_2Si_2$ systems as well. The entropy discrepancy was estimated to be about 10% in UBe_{13}. The data were taken from the original specific heat measurement by Ott *et al.*[13] This estimation was again based on a constant γ_n. For a second order phase transition of superconductivity, the entropy balance could be reached by a 20% increase of γ_n below T_c.[26] The increase of γ_n with decreasing temperature observed in a $CeCu_2Si_2$ non-superconducting single crystal was used to support the assumption of an increase γ_n for the superconductors below T_c in this heavy fermion system.

Schlabitz *et al.*[16] found an entropy mismatch in the compound URu_2Si_2 by comparing the entropy associated with the measured $C_s(T)/T$ data and the extrapolation with the one obtained from the normal state results. The specific heat data showed a much larger S_n than S_n, which was different from the observations in the compound UPt_3. We estimated the entropy discrepancy of 30% from the figure 12: $S_s(T_c) = 61$ mJ/Kmol and $S_n(T_c) = 87$ mJ/K²mol.

Table IV. Entropy and parameters derived from the specific heat measurements in UPt_3.

T_c	γ_n	γ_s	β_s	S_n	S_s	$\dfrac{S_s - S_n}{S_s}$	Reference
K	mJ/K²mol	mJ/K²mol	mJ/K³mol	mJ/Kmol	mJ/Kmol	(%)	
0.37	426	265	875	158	158	0.20	Fisher *et al.* 1989[19]
0.40	422	260	960	169	181	6.64	Franse *et al.* 1985[16]
0.40	430	165	1340	172	173	0.69	Fisher *et al.* 1989[19]
0.47	434	140	1300	204	209	2.58	Fisher *et al.* 1989[19]
0.50	461	80	1560	230	235	1.91	Fisher *et al.* 1989[19]
0.54	426	110	1250	230	242	4.80	Sulpice *et al.* 1986[17]
0.59	450	56	1430	266	282	5.83	Revax *et al.* 1987[15]

Steglich *et al.*[39] and Mayer *et al.*[14] both suggested two possibilities to guarantee the entropy discrepancy. First, the $C_s(T)$ should contain some other degrees of freedom (e.g., magnetic) rather than a simple electronic contribution. A smaller value of specific heat at T=0 is expected and the limit of the lowest temperature in the specific heat experiments needs to be extended. Second, the extrapolation of γ_n below T_c was underestimated by assuming a constant γ_n and an upturn of γ_n below T_c had to be anticipated. The field dependence specific heat measurements of Mayer *et al.* were used to support the possibility of the increase γ_n from T_c to T=0.

Fisher *et al.*[19] listed fitting parameters γ_n, γ_s and β_s derived from various specific heat measurements of UPt_3 samples where $C_s/T = \gamma_s + \beta_s T$ and $C_n/T = \gamma_n$ (see Table III). The superconducting entropy at T_c can be calculated from the definition as $S_s(T_c) = \int_o^{T_c} (C_s/T)$ dT $= \gamma_s T_c + f(1,2) \beta_s T_c^2$. The normal state entropy at T_c is $S_n(T_c) = \int_o^{T_c} (C_n^{extrap}/T)$ dT $= \gamma_n T_c$ assuming γ_n is temperature independent at $T < T_c$. In Table IV, the entropy discrepancy is shown to be 0.2 - 6.6 % in the compound UPt_3. Compared with large experimental uncertainties in γ_s and β_s (more than 100%) obtained in various UPt_3 samples (see Table IV), the entropy difference between the superconducting and normal states is noticeably smaller, within several percent.

The large discrepancy observed in a URu_2Si_2 sample[22] could be due to a broad superconducting transition and a large uncertainty in the fitting parameters derived from the specific heat data.

III THERMAL CONDUCTIVITY

A. Introduction

Thermal conductivity studies in heavy fermion superconductors[15-16, 28, 34, 39-46] suggested that the mechanism leading to superconductivity and characteristics of the superconducting state showed an unconventional nature. These findings were consistent with other thermal and transport properties observed experimentally (see Specific Heat and Ultrasonic Attenuation). In the BCS theory, the superconducting transition is accompanied by the formation of a gap in the electronic energy excitation spectrum, which has no zeros over the entire Fermi surface. This assumption results in exponential temperature dependencies of many physical quantities such as the specific heat, ultrasonic attenuation, nuclear spin relaxation and thermal conductivity at sufficient low temperature below the superconducting transition. The non-exponential temperature dependence observed in thermal conductivity measurements in the superconducting state reveals that the gap formation assumption by the BCS theory is altered such that some intrinsic gap zeros exist on the Fermi surface. The location and characteristics of the gap zeros reflect on the symmetry of the order parameter in the superconducting state. Therefore, certain types of gap zeros results in various temperature dependencies at $T \ll T_c$ for all experiments involving the thermal excitation of quasiparticles near the nodes of the superconducting order parameter. As shown in Table V, exponential temperature dependencies were displayed by the isotropic gap structure with the assumption that quasiparticles had no zeros over entire Fermi surface. In the non-isotropic gap structures (axial or polar), power-law temperature dependencies were obtained by assuming that the gap vanished at either points or lines on the (spherical) Fermi surface.

Table V. Theoretical temperature dependencies for measurements of specific heat, sound attenuation, nuclear magnetic relaxation time and thermal conductivity in the superconducting state at low temperatures, $T \ll T_c$, for various pairing mechanisms (Δ is the superconducting order parameter).[38, 47-49]

Gap Structure	Isotropic	Axial (points)	Polar (lines)
Specific Heat, $C_s(T)$	$\exp(-\Delta/T)$	$(T/\Delta)^3$	$(T/\Delta)^2$
Sound Attenuation, $a_s(T)$	$\exp(-\Delta/T)$	$(T/\Delta)^4$	$(T/\Delta)^2$
NMR Relaxation, T_{ls}^{-1}	$\exp(-\Delta/T)$	$(T/\Delta)^5$	$(T/\Delta)^3$
Thermal Conductivity, $\kappa_s(T)$	$\exp(-\Delta/T)$	$(T/\Delta)^3$	$(T/\Delta)^2$

B. Temperature Dependence

1. $CeCu_2Si_2$

Thermal conductivity studies in the $CeCu_2Si_2$ system were reported by the Steglich group.[28, 39] These researchers presented the first thermal conductivity data in a $CeCu_{2.02}Si_{1.98}$ polycrystal. As shown in Fig. 17, the thermal conductivity ratio κ_s/κ_n was plotted logarithmically as a function of T_c/T. The quantities κ_s and κ_n represent the thermal conductivities in the superconducting and normal states, respectively. The significant deviation of κ_s/κ_n from the BCS prediction is shown in Fig. 17. Steglich et al. (1984) claimed that no simple power law was found for $\kappa_s(T)$. Fig.24 shows the thermal conductivity, $\kappa_s(T)$, for the same sample of $CeCu_{2.02}Si_{1.98}$. As displayed in the inset of Fig. 24, the temperature dependence of $\kappa_s(T)/T$ below 0.2 K appeared linear and a finite intercept was obtained. The best fit for $\kappa_s(T)$ at temperatures below 0.25 K was a quadratic equation:

$$\kappa_s(T) = \gamma T + \beta T^2$$

where $\gamma = 0.28$ mW/K^{2}cm and $\beta = 1.8$ mW/K^{3}cm. It was pointed out by Steglich et al. that a substantial phonon contribution was expected below T_c. Therefore, the

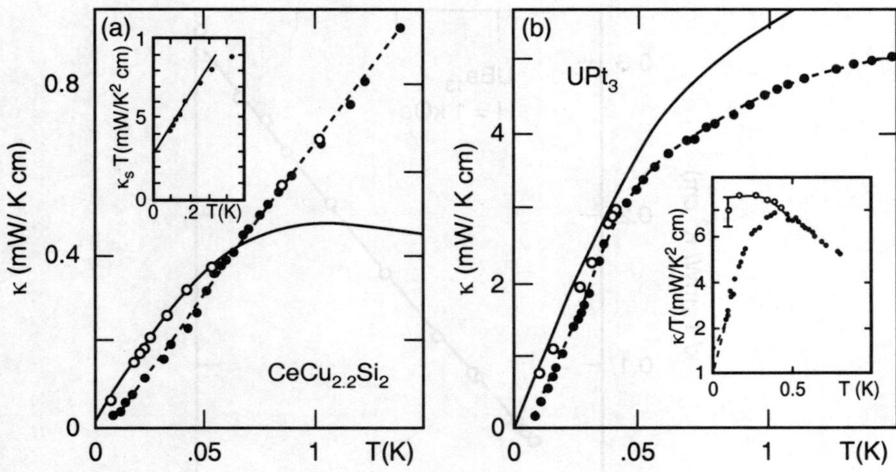

Fig. 24. Thermal conductivity κ for polycrystalline samples of $CeCu_{2.02}Si_{1.98}$ (a) and UPt_3 (b). Inset shows low-T data in plots κ/T vs. T. Full circles: B = 0; open circles: B = 2.5 T (from Steglich *et al.* 1985).[39]

electronic contribution below T_c could be obtained only when the phonon contribution was successfully subtracted from $\kappa_s(T)$. Compared with the specific heat data, a nearly T^3-term was observed in the low temperature data.

Jaccard *et al.*[44] listed coefficients of the thermal conductivity for three heavy fermion superconductors: $CeCu_2Si_2$, UBe_{13} and UPt_3. The coefficients listed for the compound $CeCu_2Si_2$ differed significantly from those found in the same compound by Steglich *et al.*[39] However, no reference, sample information or experimental data were given in the paper.[44]

2. UBe_{13}

Thermal conductivity measurements on a single crystal of UBe_{13} are shown in Fig. 25. A linear line in $\kappa_s(T)/T$ for H=1 kOe and $0.1 < T < 0.8$ K was observed: $\kappa_s(T) = \beta T^2$ with $\beta = 0.38$ mW/K^{3}cm. In comparison, specific heat experiments[13-15] displayed a power-law temperature dependence with exponents close to 3. The T^2-dependence of $\kappa_s(T)$ was interpreted as a purely electronic contribution. That could be described by the thermal excitations of a superconductor with strong gap anisotropy.

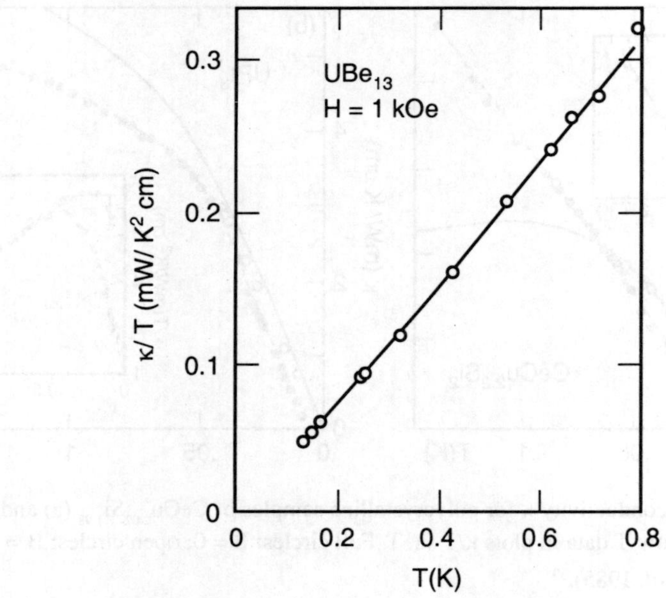

Fig.25. In the superconducting phase of UBe$_{13}$, κ/T as a function of T (from Jaccard et al. 1985).[41]

Low temperature experiments of the thermal conductivity in UBe$_{13}$ were performed by Ravex et al.21 Jaccard and Flouquet[42] and Ott et al.[34] Ravex et al. measured the thermal conductivity, $\kappa_s(T)$, on a polycrystalline sample of UBe$_{13}$ at temperature down to 53 mK (see Fig. 26). No single fit could be made over the entire temperature range below T$_c$ (T$_c$ = 0.845 K). However, the low temperature data below 100 mK showed a linear temperature dependence: $\kappa_s(T) = \gamma T$ with γ = 0.03 mW/K^{2}cm, different from earlier results by Jaccard et al.[40-41] Jaccard and Flouquet[42] used the data reported by Ravex et al.[41] to conclude that the early claims of T^3 or T^2 power-law dependence of the specific heat and thermal conductivity appeared erroneous. On the contrary, the $\kappa_s(T)$ data measured by Ott et al. down to 50 mK deviated from a linear temperature dependence below 100 mK as reported by Ravex et al. and Jaccard and Flouquet (see Fig. 27). Ott et al. pointed out that their data, as shown in Fig. 27, were close to a T^2-dependence at low temperatures. Clearly, sample quality can play a critical role in achieving the various temperature dependencies of these experimental quantities.

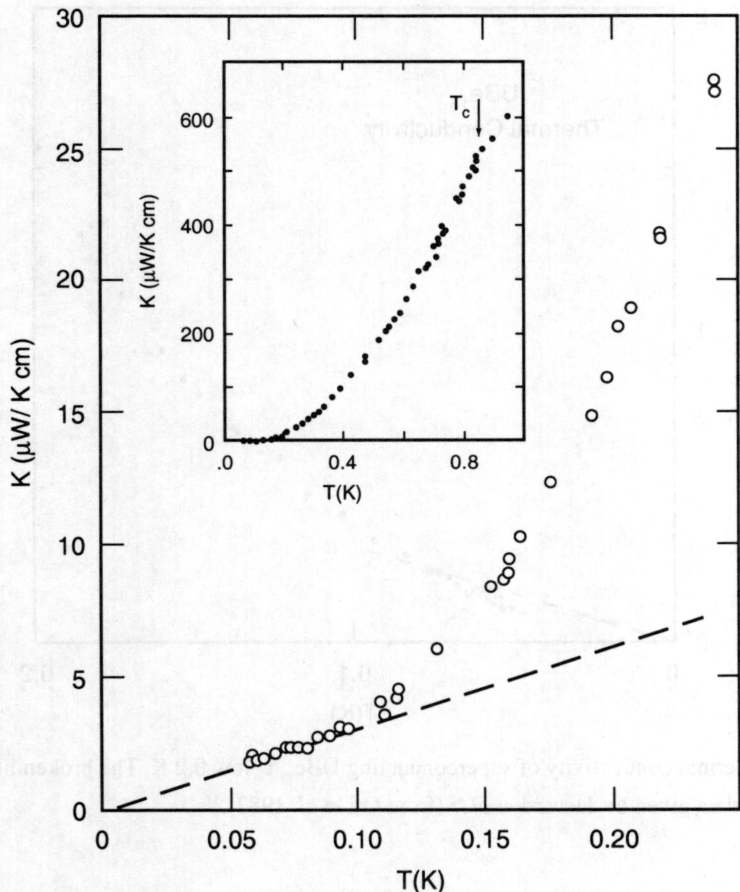

Fig. 26. Thermal conductivity κ of UBe$_{13}$. The dashed line represents the linear temperature term κ = γT. The inset shows the data below 1 K (from Ravex *et al.* 1987).[15]

3. UPt$_3$

Thermal conductivity experiments were reported on both polycrystalline[16, 17, 34, 37, 50, 51] and single crystal samples of UPt$_3$. Fig. 28 shows the temperature dependence of the thermal conductivity, $\kappa_s(T)$, of polycrystalline UPt$_3$. At temperatures below T_c, $\kappa_s(T)$ follows a power-law behavior[16, 50] with an exponent between 1 and 2. As indicated by the dashed line in Fig. 28, a nearly linear temperature dependence was found at very low temperatures between 0.1 and 0.2 K: $\kappa_s(T) = \beta T^2$ with $\beta = 32$ mW/K^{3}cm. As T → 0,

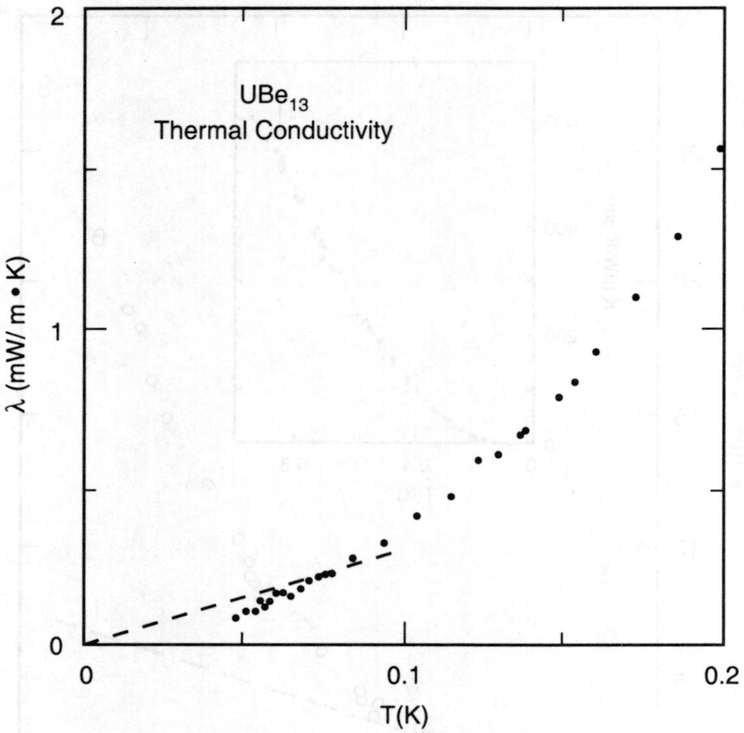

Fig. 27. Thermal conductivity of superconducting UBe$_{13}$ below 0.2 K. The broken line represents data given by Jaccard *et al.*[42] (from Ott *et al.* 1987).[34]

$\kappa_s(T)/T$ approached zero. The coefficient b was larger than the value observed in UBe$_{13}$ by Jaccard *et al.*[41] by two orders of magnitude. Franse *et al.*[16] and Steglich *et al.*[39] explained that the observed power-law dependence of the thermal conductivity was dominated by the electronic contribution and was consistent with the contribution of the thermal excitations in a superconductor with strongly anisotropic pairing. Comparison studies between CeCu$_2$Si$_2$ and UPt$_3$ and theoretical predictions for the electronic contribution revealed that phonon contributions below T$_c$ were only a few percent of the total heat transport[18,39] (see Fig. 24). This result was opposite to the early speculation[28] for CeCu$_2$Si$_2$.

The unusually high thermal conductivity in a polycrystalline sample of UPt$_3$ was confirmed.[17,51] (1985b) with a strong T^2 term. $\kappa_s(T)$ data between 0.035 and 0.2 K (see Fig. 29) could be fitted by $\kappa_s(T) = \gamma T + \beta T^2$, with $\gamma = 0.55$ mW/K^{2}cm and $\beta = 20$ mW/K^{3}cm.

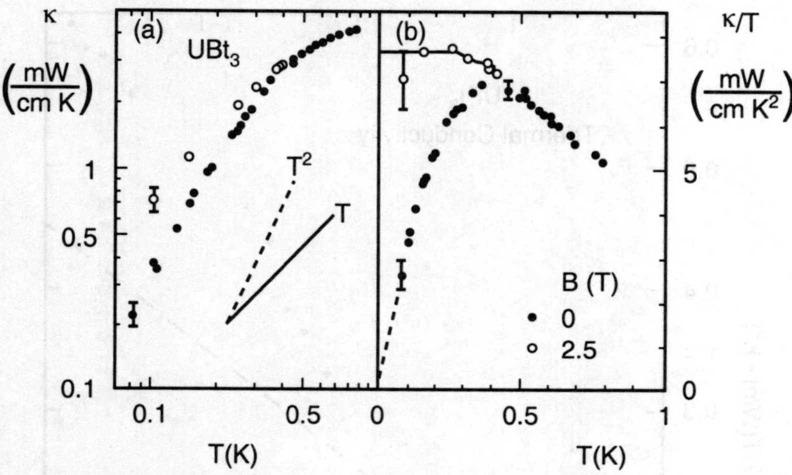

Fig. 28. Temperature dependence of the thermal conductivity of polycrystalline UPt$_3$ in a double-logarithmic plot (a), and as κ/T vs. T (b) (from Franse *et al*. 1985).[16]

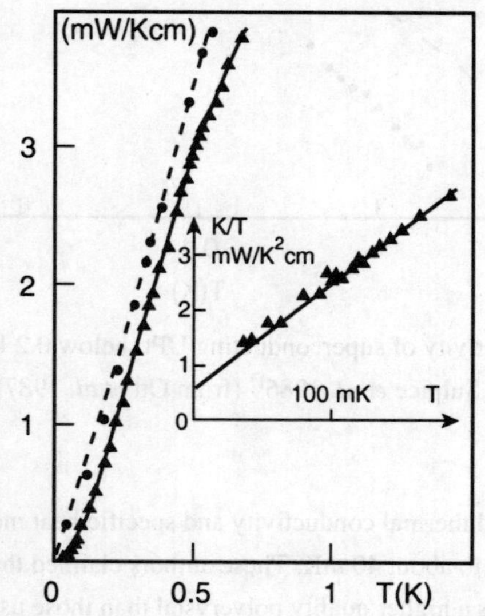

Fig. 29. Thermal conductivity of UPt$_3$ at H = 1 and 70 kOe (from Jaccard *et al*. 1985).[51]

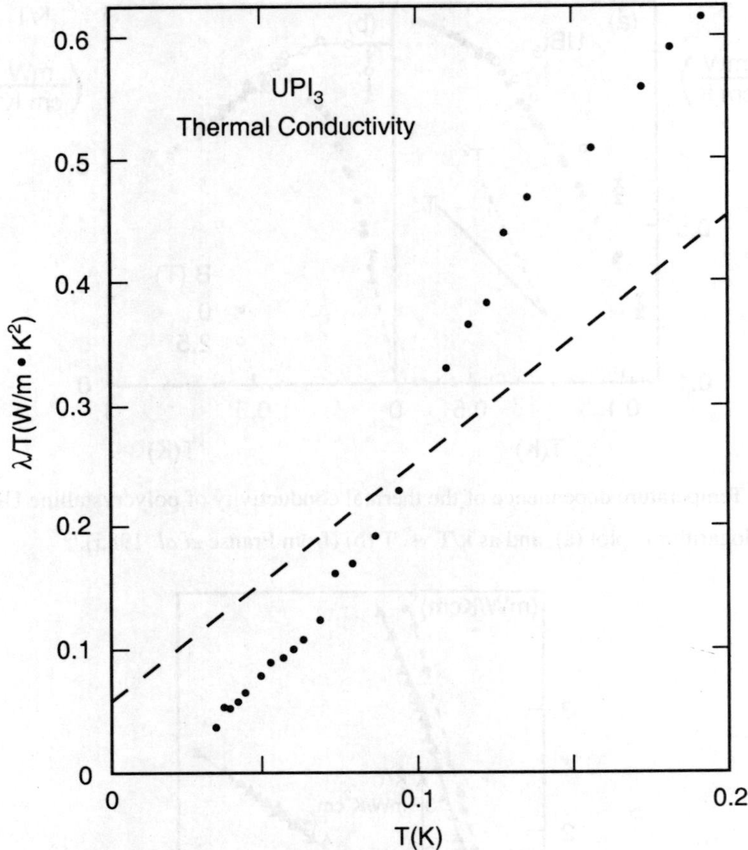

Fig. 30. Thermal conductivity of superconducting UPt₃ below 0.2 K. The broken line represents data given by Sulpice *et al.* 1986[17] (from Ott *et al.* 1987).[34]

Ott *et al.*[34] published thermal conductivity and specific heat measurements at very low temperatures, down to about 40 mK. These authors claimed that the UPt₃ sample used in these experiments was a higher quality polycrystal than those used in previous investigations because this sample had been used successfully for de Haas-van Alphen studies. Fig. 30 shows thermal conductivity data for UPt₃ below 0.2 K (note: λ in Ott's paper is our κ_s). Ott *et al.* concluded that no simple power-law could be fitted over an extended temperature range and the temperature dependence of $\kappa_s(T)$ included a term stronger than T^2.

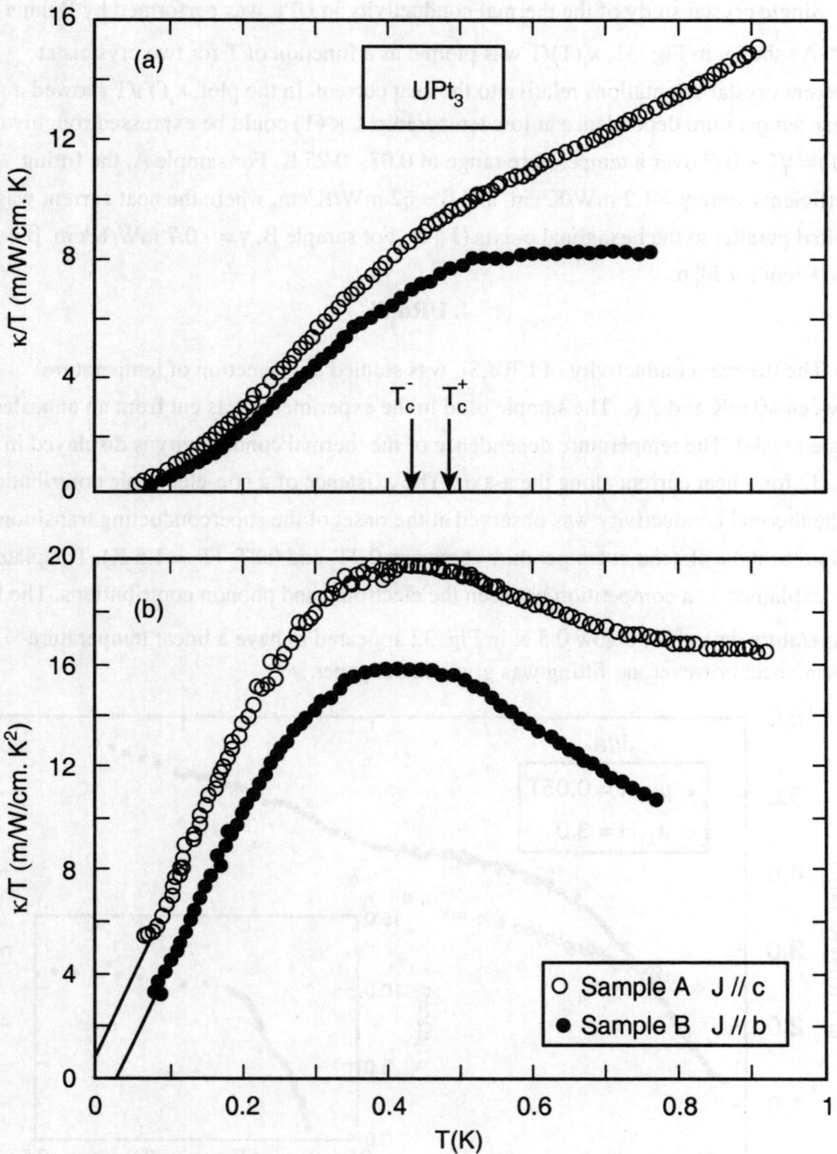

Fig. 31. (a) Thermal conductivity κ(T) of UPt$_3$ as a function of temperature, for sample A with heat current J parallel to the c-axis and for sample B with J ∥ b. The two critical temperatures T_c^- and T_c^+ correspond to the two transitions observed in thermodynamic properties. (b) Same as in (a), with the data plotted as κ/T vs. T. The solid lines are linear fits to the low-temperature data (from Behnia *et al.* 1991).[44]

Single crystal study of the thermal conductivity in UPt$_3$ was performed by Behnia *et al.*[44] As shown in Fig. 31, $\kappa_s(T)/T$ was plotted as a function of T for two crystals at different crystal orientations relative to the heat current. In the plot, $\kappa_s(T)/T$ showed a linear temperature dependence at low temperatures. $\kappa_s(T)$ could be expressed roughly as $\kappa_s(T) = \gamma T + \beta T^2$ over a temperature range of 0.07 - 0.25 K. For sample A, the fitting coefficients were $\gamma = 1.2$ mW/K^{2}cm, and $\beta = 62$ mW/K^{3}cm, where the heat current was applied parallel to the hexagonal c-axis (J $\parallel$ c). For sample B, $\gamma = -0.7$ mW/K^{3}cm, $\beta = 55$ mW/K^{2}cm for J $\parallel$ b.

4. URu$_2$Si$_2$

The thermal conductivity of URu$_2$Si$_2$ was studied as a function of temperature[53] between 40 mK and 2 K. The sample used in the experiments was cut from an annealed single crystal. The temperature dependence of the thermal conductivity is displayed in Fig. 32 for a heat current along the a-axis. The existence of a non-electronic contribution to the thermal conductivity was observed at the onset of the superconducting transition. $\kappa_s(T)$ reached a plateau at temperatures between 0.7T$_c$ and 0.8T$_c$ (T$_c$ = 1.5 K). The plateau was explained as a competition between the electronic and phonon contributions. The low temperature data $\kappa_s(T)$ below 0.5 K in Fig. 32 appeared to have a linear temperature dependence; however, no fitting was given in the paper.

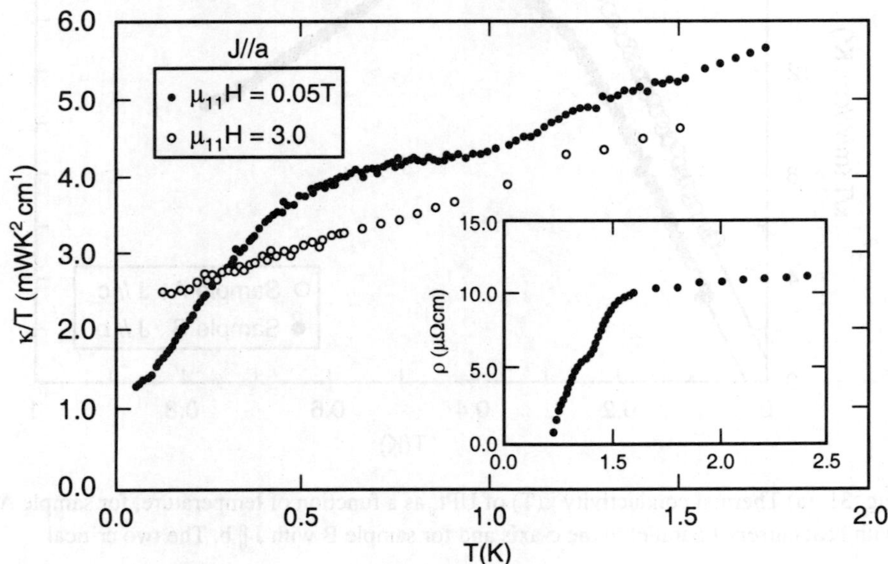

Fig. 32. Thermal conductivity of URu$_2$Si$_2$ as a function of magnetic field at T = 1 K for three different orientations of a magnetic field (from Behnia *et al.* 1992).[53]

5. Conclusion

The essential finding in the studies of the temperature dependence of the thermal conductivity of heavy fermion superconductors (see Table VI) was that $\kappa_s(T)$ did not follow an exponential temperature dependence at low temperatures as predicted by the BCS theory. Jaccard *et al.* and Franse *et al.* observed a power-law temperature dependence of $\kappa_s(T)$ with an exponent of 2 in a UBe_{13} crystal and in a UPt_3 polycrystalline sample. On the other hand, Steglich *et al.*, Jaccard *et al.*, Sulpice *et al.* and Behnia *et al.* reported a T^2 - dependence of $\kappa_s(T)$ with an additional T term in a $CeCu_2Si_2$ polycrystal and both polycrystalline and single crystalline samples of UPt_3. These power-law or quadratic temperature dependencies of $\kappa_s(T)$ indicate the presence of zeros in the gap function.[38, 54, 48., 49] Ott *et al.*[34] and Behnia *et al.*[53] did not find a simple exponential or power law in $\kappa_s(T)$ down to 40 mK in a high quality polycrystal of UPt_3 and a single crystal of URu_2Si_2, respectively.

The phonon contribution to the thermal conductivity in the superconducting state was negligible[18, 39] in the compounds $CeCu_2Si_2$ and UPt_3. However, the thermal conductivity in the superconducting state might be influenced strongly by the contributions[53] due to phonons or magnons in the compound URu_2Si_2.

Table VI. Coefficients γ and β of the thermal conductivity $\kappa_s(T)$ in the superconducting state: $\kappa_s(T) = \gamma T + \beta T^2$ for $T < T_c$.

System	γ mW/K²cm	β mW/K³cm	T_c K	Ref.
$CeCu_2Si_2$	0.28	1.8	0.65	Steglich *et al.* 1985[39]
	0.7	2.8	0.56	Jaccard & Flouquet 1987[42]
UBe_{13}		0.38	0.854	Jaccard *et al.* 1985[41]
	0.03		0.845	Ravex *et al.* 1987[15]
UPt_3		32		Franse *et al.* 1985[16]
	0.55	20	0.52	Jaccard *et al.* 1985[51]
	12	62		Behnia *et al.* 1991[44]
	-0.7	55		Behnia *et al.* 1991[44]

C. $\kappa_s(T)$ Analysis

A comparison of the temperature dependence of the specific heat and the thermal conductivity showed interesting results. In $CeCu_2Si_2$, $C_s(T)$ had a power law temperature dependence, while two different temperature terms were found in $\kappa_s(T)$: γT and βT^2. The compound UPt_3 displayed a large T^2-term in both $C_s(T)$ and $\kappa_s(T)$. If an additional linear temperature dependence was assumed to be due to the impurities in the samples, then the T^2 power-law could be taken as evidence for a polar-like superconducting state in UPt_3.[38, 54] A discrepancy was found in UBe_{13} where $C_s(T)$ showed on a T^3-dependence, but $\kappa_s(T)$ had a T^α-dependence with $\alpha = 1$ or 2.

Grewe and Steglich[50] studied the UPt_3 data measured by Ott et al.[34] (see Fig.30). and obtained a linear coefficient $\gamma = 2$ mW/K^{2}cm by extrapolating to $T = 0$ from the linear section above 0.13 K in the plot of $\kappa_s(T)/T$. This linear coefficient was used by Grewe and Steglich to establish a picture underlying the thermal transport in the superconducting UPt_3: the superconducting transition at 0.5 K involved only part of the Fermi surface and left another part with light electrons unchanged. The light electrons made a considerable contribution to $\kappa(T)$ as measured by γT, which then decreased strongly below 0.13 K (Grewe and Steglich 1991). Well below $T = 0.13$ K, Grewe and Steglich explained that almost the whole Fermi surface was involved in an anisotropic superconducting state as inferred from an apparent pure T^2 dependence with no linear term in $\kappa_s(T)$.

D. Anisotropy and Field Dependence

The anisotropy of the thermal conductivity was studied by Behnia et al.[44] in UPt_3 single crystals. The normalized $\kappa_s(T)$ was about the same for different heat directions (J $\parallel$ c and J $\parallel$ b). Since two crystals were used in the experiments, the authors were not able to draw a conclusion whether there was a small anisotropy at low temperatures, although a small deviation was observed in the plot. If the concentration of impurities did not have a dramatic effect on the temperature dependence, it was suggested that an isotropic distribution of the gap might exist in UPt_3. However, this conclusion was in conflict with the assumption of strong gap anisotropy (i.e. polar-like gap function) used to explain the T^2 power-law in $\kappa_s(T)$.

Lussier et al.[55] reported the anisotropy of the thermal conductivity on a single crystal of UPt_3 with a current parallel and perpendicular to the hexagonal axis. From Fig. 33, where κ/T is normalized at $T_c = 0.5$ K, it is seen that the anisotropy is independent of temperature above T_c, and $\kappa_c/\kappa_b = 2.8$ from 0.5 to 0.8 K within 2% (κ_c and κ_b are normalized thermal conductivity coefficients at J $\parallel$ c and J $\parallel$ b, respectively).

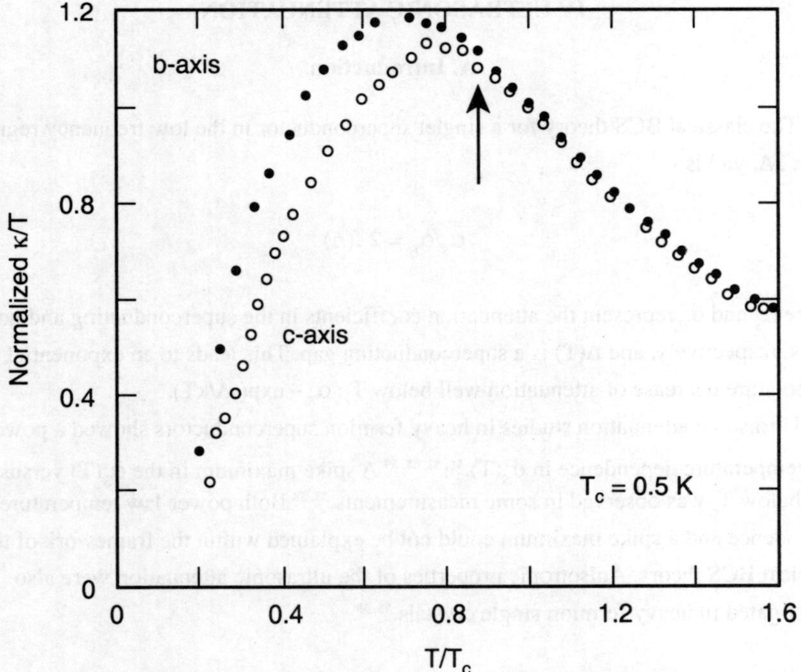

Fig. 33. Thermal conductivity divided by temperature vs. reduced temperature for the two directions of heat current. κ_b and κ_c are both normalized to unity at T = 0.5 K. Note the additional anisotropy appearing below T_c^- (arrow) (from Lussier *et al.* 1994).[55]

Behnia *et al.*[21, 44, 45] investigated the field dependence and anisotropy of the thermal conductivity in UPt$_3$ crystals. The difference of $\kappa_s(H)$ due to the orientations of the field, current and crystalline axes was less than 15%. At low fields, $\kappa_s(H)$ was larger for H ∥ J than that for H ⊥ J and independent of the orientation of the crystal. When the field was higher than 1/2 H$_{c2}$, $\kappa_s(H)$ was found to depend on the relative orientation of the field and the crystalline axes and was insensitive to the direction of the current.

IV ULTRASONIC ATTENUATION

A. Introduction

The classical BCS theory for a singlet superconductor in the low frequency regime, $\omega \ll 2\Delta$, yields

$$\alpha_S/\alpha_N = 2\, f(\Delta)$$

where α_S and α_N represent the attenuation coefficients in the superconducting and normal states, respectively, and $\Delta(T)$ is a superconducting gap. This leads to an exponential temperature decrease of attenuation well below T_c: $\alpha_S \sim \exp(-\Delta/kT)$.

Ultrasonic attenuation studies in heavy fermion superconductors showed a power-law temperature dependence in $\alpha_S(T)$.[56, 57, 58, 59] A spike maximum in the $\alpha_S(T)$ versus T plot below T_c was observed in some measurements.[57, 58] Both power-law temperature dependence and a spike maximum could not be explained within the framework of the classical BCS theory. Anisotropic properties of the ultrasonic attenuation were also investigated in heavy fermion single crystals.[58, 59]

B. Temperature Dependence

Sound velocity measurements were made in UPt$_3$ by Bishop *et al.*[56] The sample used in the experiment was a UPt$_3$ single crystal with T_c of 0.48 K. The experiment was performed on a pulse spectrometer of conventional design and an ultrasound pulse of the longitudinal polarization was propagated along the c-axis. The frequency range was between 50 and 600 MHz with a low temperature limit of about 0.06 K. The results shown in Fig. 34 yielded a T^2 power-law for the ultrasonic attenuation at low temperatures (i.e. linear line in $\alpha_S(T)$ versus T^2), which was inconsistent with the exponential temperature dependence predicted by the BCS theory. The T^2 temperature dependency was in good agreement with the theoretical assumption that UPt$_3$ was an anisotropic (triplet) superconductor in a polar-like state, where the gap vanished along a line on the Fermi surface.[56, 18, 47]

Sound propagation in the heavy fermion superconductor UBe$_{13}$ was studied by Golding *et al.*[57] at frequencies from 0.9 to 2.4 GHz and temperatures down to 0.01 K. The sample was a single crystal with a superconducting transition onset of 0.9 K. A longitudinal sound pulse was used with the sound propagation oriented along the c-axis. The attenuation coefficient below 1/2 T_c was inconsistent with an exponential temperature dependence but was well characterized by a T^2-dependence as shown in Fig. 35.

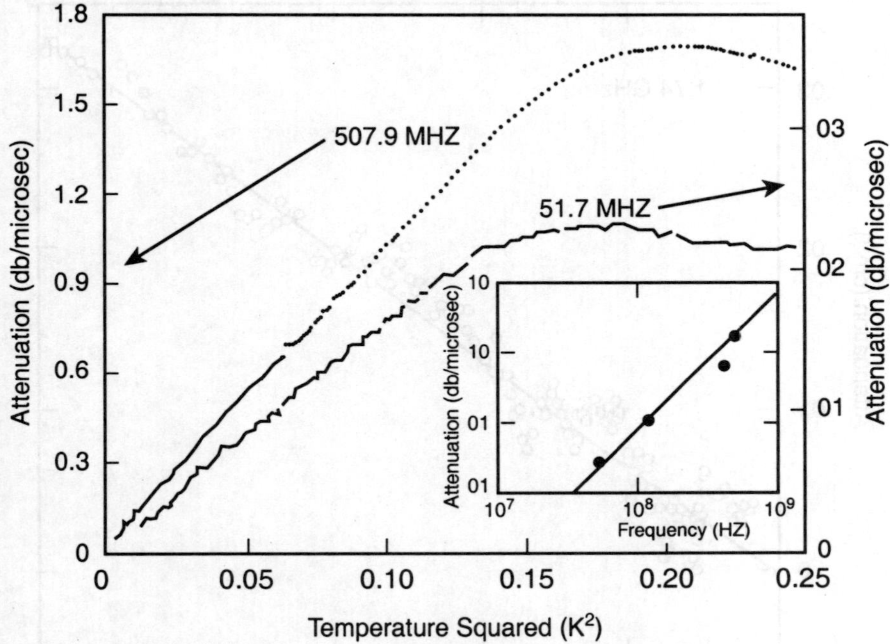

Fig. 34. Ultrasonic attenuation of UPt$_3$ at two different frequencies plotted vs. T^2. The inset shows the frequency dependence. The solid line has an ω^2 frequency dependence (from Bishop *et al*. 1984).[56]

Müller *et al*.[58] investigated the ultrasonic attenuation in a superconducting UPt$_3$ single crystal where both the longitudinal and transverse ultrasound pulse were propagated along the c-axis. A T^3 term was observed in the longitudinal ultrasonic sound attenuation as shown in Fig. 36, which deviated strongly from the BCS theory. The ultrasonic attenuation data α_S/α_N showed a T^2 dependence over a wide temperature range $0.1 < T/T_c < 0.6$, which was interpreted as due to an anisotropic polar-like (triplet) superconducting states.[60]

Single crystal studies of the transverse sound were performed by Shivaram *et al*. in UPt$_3$.[59] For the polarization in the basal plane (i.e. the transverse ultrasound), the attenuation varied linearly with temperature over a large temperature range between T$_c$ and 35 mK (see Fig. 37). This observed linear temperature dependence of α_s(T) was different from previous results by Bishop *et al*.[56] or Müller *et al*.[58]

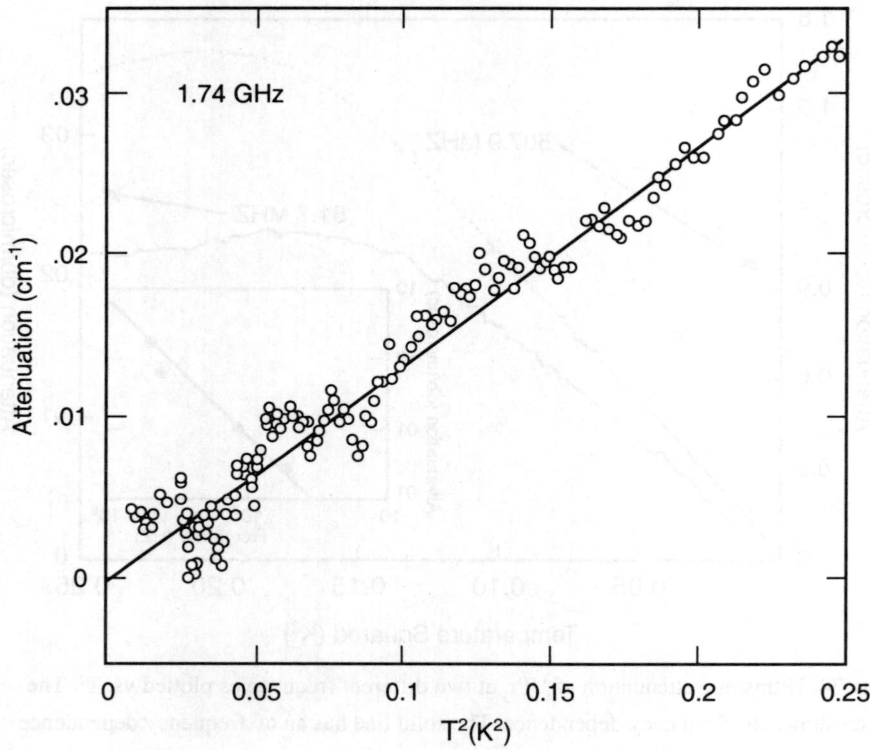

Fig. 35. Temperature dependence of the attenuation from 100 to 500 mK of UBe_{13}. The data are consistent with a T^2 power law (from Golding *et al.* 1985).[57]

Thalmeier *et al.*[61] reported attenuation results for the various elastic modes in UPt_3 in the superconducting state. The longitudinal modes (c_{11} and c_{33}) exhibited an approximately T^3 temperature dependence for the attenuation from 0.1 K to T_c. The attenuation for transverse modes showed T^3 and T dependencies for c_{44} and c_{66}, respectively.

Ultrasonic attenuation experiments were mainly studied in heavy fermion superconducting single crystals of UPt_3 and UBe_{13} (see Table VII). Power-law temperature dependencies were observed in the ultrasonic attenuation and different exponents of the power-law were found. This power-law temperature dependence strongly indicated deviations from an exponential decrease of $\alpha_s(T)$, which was predicted by the classical BCS theory. This might suggest that the gap function vanishes for some region on the Fermi surface and thus that, at the very least, the gap function depends strongly on the direction of propagation (see discussion in Anisotropy). Another possibility is that the gap

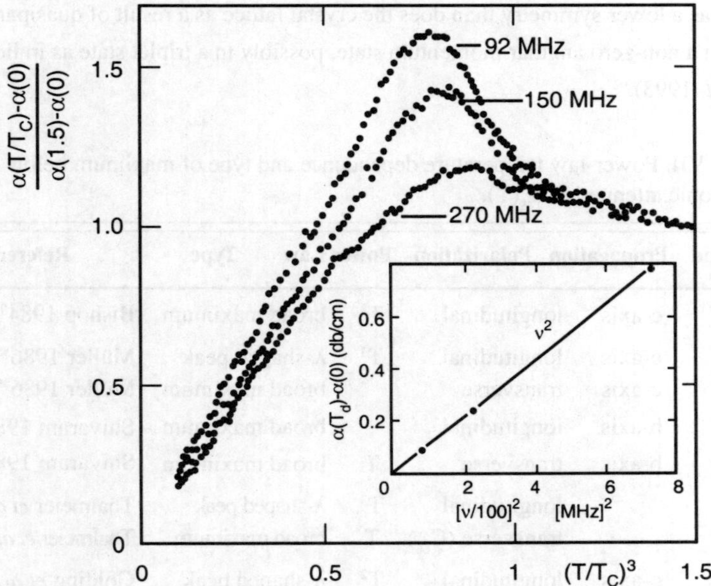

Fig. 36. Relative change of the longitudinal attenuation coefficient of a UPt$_3$ single crystal as a function of $(T/T_c)^3$ for three different frequencies (from Müller *et al.* 1986).[58]

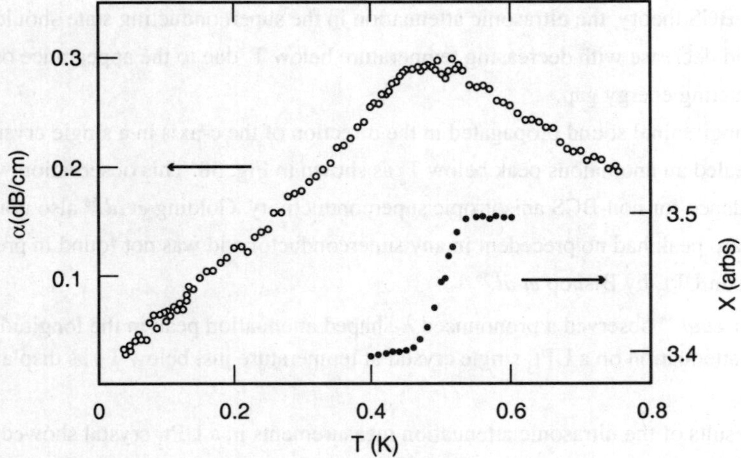

Fig. 37. Attenuation of transverse sound at 94 MHz of UPt$_3$ with $\hat{q} \parallel \hat{b}$ and $\hat{e} \parallel \hat{a}$ as measured from its extrapolated zero-temperature value. The ac susceptibility (right scale) is simultaneously measured through the superconducting transition. Note the linear dependence of the attenuation down to the lowest temperatures measured (from Shivaram *et al.* 1986).[59]

function has a lower symmetry than does the crystal lattice as a result of quasiparticles that pair in a non-zero angular momentum state, possibly in a triplet state as in liquid ^{3}He (Hess *et al.* 1993).[62]

Table VII. Power-law temperature dependence and type of maximum below T_c in the ultrasonic attenuation $\alpha_s(T)$.

Compound	Propagation	Polarization	Power Law	Type	Reference
UPt$_3$	c-axis	longitudinal	T^2	broad maximum	Bishop 1984[56]
	c-axis	longitudinal	T^3	λ-shaped peak	Müller 1986[58]
	c-axis	transverse		broad maximum	Müller 1986[58]
	b-axis	longitudinal		broad maximum	Shivaram 1986[59]
	b-axis	transverse	T	broad maximum	Shivaram 1986[59]
		longitudinal	T^3	λ-shaped peak	Thalmeier *et al.* 1992[61]
		transverse (C_{66})	T	broad maximum	Thalmeier *et al.* 1992[61]
UBe$_{13}$	c-axis	longitudinal	T^2	λ-shaped peak	Golding *et al.* 1985[56]

C. $\alpha_s(T)$ Maximum

In the BCS theory, the ultrasonic attenuation in the superconducting state should have a rapid decrease with decreasing temperature below T_c due to the appearance of the superconducting energy gap.

The longitudinal sound propagated in the direction of the c-axis in a single crystal of UBe$_{13}$ revealed an anomalous peak below T_c as shown in Fig. 38. This observation was a strong evidence for non-BCS anisotropic superconductivity. Golding *et al.*[56] also pointed out this sharp peak had no precedent in any superconductor and was not found in previous studies in UPt$_3$ by Bishop *et al.*[56]

Müller *et al.*[58] observed a pronounced λ-shaped attenuation peak in the longitudinal ultrasonic attenuation on a UPt$_3$ single crystal at temperature just below T_c, as displayed in Fig. 39.

The results of the ultrasonic attenuation measurements in a UPt$_3$ crystal showed that maxima of attenuation appeared for both polarizations at temperature slightly below T_c (Shivaram *et al.* 1986b) (see Fig. 40).[59]

Similar anomalous behavior in attenuation below the superconducting transition temperature was observed in URu$_2$Si$_2$.[63]

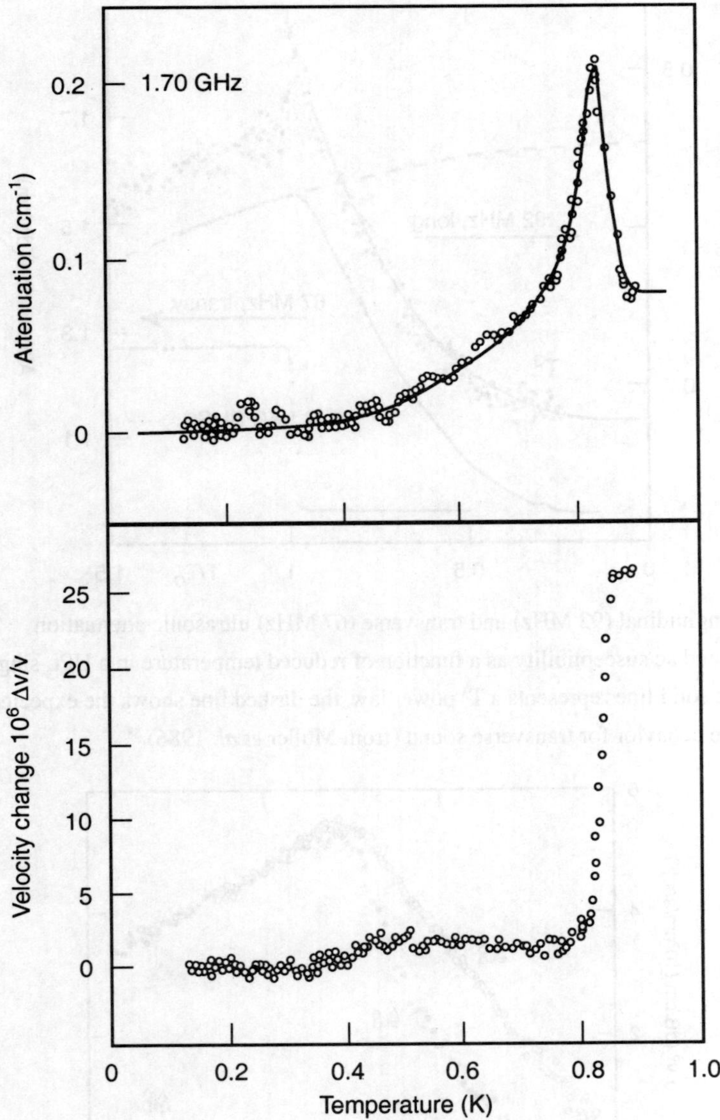

Fig. 38. Acoustic attenuation and velocity changes in UBe$_{13}$ below the superconducting transition T$_c$ = 0.9 K. In contrast to all other superconductors, the attenuation in UBe$_{13}$ exhibits a peak rather than a sharp decrease below T$_c$, The velocity discontinuity is inversely proportional to the specific heat jump at T$_c$. The low-frequency ac susceptibility indicates a transition temperature indistinguishable from that given by the velocity (from Golding *et al.* 1985).[56]

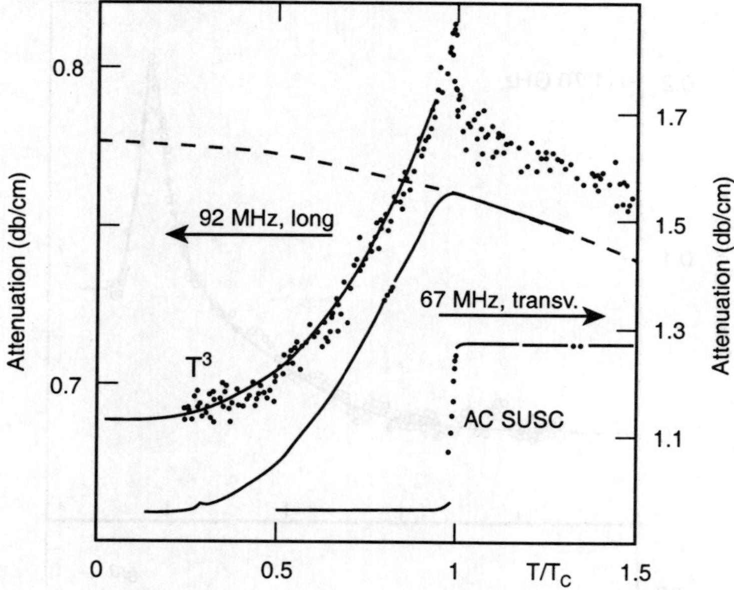

Fig. 39. Longitudinal (92 MHz) and transverse (67 MHz) ultrasonic attenuation coefficient and ac susceptibility as a function of reduced temperature in a UPt$_3$ single crystal. The solid line represents a T^3 power law, the dashed line shows the expected normal state behavior for transverse sound (from Müller *et al.* 1986).[58]

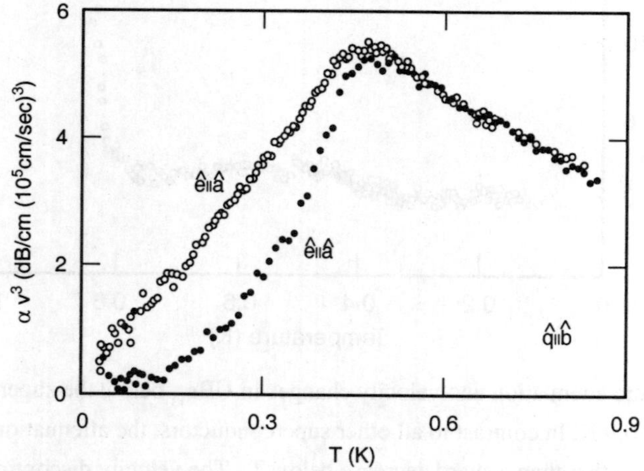

Fig.40. αv^3 (proportional to the viscosity) vs. temperature at 132 MHz in a UPt$_3$ crystal. The values for the two polarizations are identical in the normal state, but there is an enhanced viscosity in the superconductor for ê ∥ â (from Shivaram *et al.* 1986).[59]

All the ultrasonic attenuation experiments showed a maximum in the plot of $\alpha_s(T)$ versus T at temperatures at or below T_c (see Table VII). These results are in contrast to the BCS theory which predicts a rapid exponential drop with decreasing temperature in $\alpha_s(T)$ below T_c. Two kinds of maxima were observed[58] in the ultrasonic attenuation measurements. One was a broad maximum observed in the transverse sound propagated along the c-axis of UPt_3, and both the transverse and longitudinal waves propagated perpendicular to the c-axis of UPt_3. The other maximum was a spike peak with a λ-shape found in the longitudinal ultrasound in UBe_{13} (Golding *et al.* 1985).[56] Contradicting results were obtained in the longitudinal ultrasonic attenuation and propagation in the c-axis of UPt_3, where a flat maximum was observed by Bishop *et al.* (1984) and a λ-shaped peak was found by Müller *et al.*

The excitation of a collective mode of sound waves was used to explain the large attenuation peak observed in UBe_{13} by Golding *et al.* Shivaram *et al.* argued that the pronounced longitudinal attenuation in UPt_3 was regarded as a "correlation hump" of the density-density correlation function caused by the condensation of the heavy fermions into the superconducting states.

D. Anisotropy

Müller *et al.* studied anisotropy of the ultrasonic attenuation in a UPt_3 crystal for different polarizations (i.e. the longitudinal and transverse sound waves) and the same propagation along the c-axis. Both the temperature dependencies and the maxima below T_c showed anisotropic effects. At temperature just below T_c, a pronounced λ-shaped peak was observed in the longitudinal ultrasonic attenuation, while a rather broad maximum appeared in the transverse sound (see Fig. 39). A T^3 term was observed only in the longitudinal ultrasonic sound attenuation (see Fig. 36).

The results for two polarizations propagated along the b-axis (see Fig. 40) showed a noticeable anisotropy in their temperature dependencies.[59] An enhanced attenuation of the transverse sound was seen in the superconducting state, while no distinct behavior was observed in normal state. This was interpreted to indicate that UPt_3 is a polar superconductor with a line of nodes of the gap perpendicular to the hexagonal c-axis.[64] The maxima of attenuation appeared for both polarizations at temperature slightly below T_c and no significant difference in the shapes was displayed between two polarizations (see Fig. 40). The temperature dependence of $\alpha_s(T)$ showed a nearly linear behavior, but only in the transverse sound (see Fig. 40).

The attenuation was characterized by a decrease in the superconducting state with decreasing temperature. The anisotropic properties of the ultrasonic attenuation were investigated. Below T_c, a spike with a λ-shape in $\alpha_s(T)$ was observed only in the longitudinal sound along the c-axis propagation in the compound UPt_3. A power-law T^3-dependence was observed in the longitudinal sound propagated along the c-axis and a linear T-dependence was found in the transverse sound propagated along the b-axis.

V UPPER CRITICAL FIELD

A. Introduction

Upper critical fields, $H_{c2}(T)$, give important information about the parameters characterizing the normal Fermi-liquid and the superconducting states.[65] $H_{c2}(T)$ is determined by diamagnetic pair-breaking near T_c. The possible presence of Pauli limiting at lower temperatures, i.e., the breaking of spin alignments preferred by the superconducting state, allows for conclusions concerning the symmetry type of the superconducting order parameter. The slope of $H_{c2}(T)$ at T_c, dH_{c2}/dT, is influenced by the quasiparticle-band structure of the clean material and by impurities.

The heavy fermion superconductors are type II superconductors with κ as large as 50 or more. These rank among the largest κ values found in conventional superconductors, for which the large value of κ occurs as a result of scattering from impurities. In contrast, the heavy fermion substances are pure materials. The temperature dependence of $H_{c2}(T)$ for the heavy fermion systems is generally unusual. The large slopes of $H_{c2}(T)$ at T_c observed in heavy fermion superconductors might be due to the existence of a large effective mass in these compounds.

B. Experimental Methods

When bulk superconductivity was first discovered in the compound $CeCu_2Si_2$ by Steglich et al.,[1] a large value of the slope of the upper critical field, dH_{c2}/dT, was reported to be approximately - 10 T/K. This slope was obtained as an average change of magnetic fields over the change of transition temperatures. The transition temperatures were determined at three different magnetic fields of 0, 0.5 and 1T from independent experiments of resistivity, ac susceptibility and specific heat.

Later, the temperature dependent upper critical field, $H_{c2}(T)$, was determined more accurately from the superconducting transition curves (more data points) at numerous magnetic fields obtained in various experiments including both dc and ac resistivity, ac susceptibility, calorimetry, and others. Fig. 41 shows transport measurements for a UBe_{13}

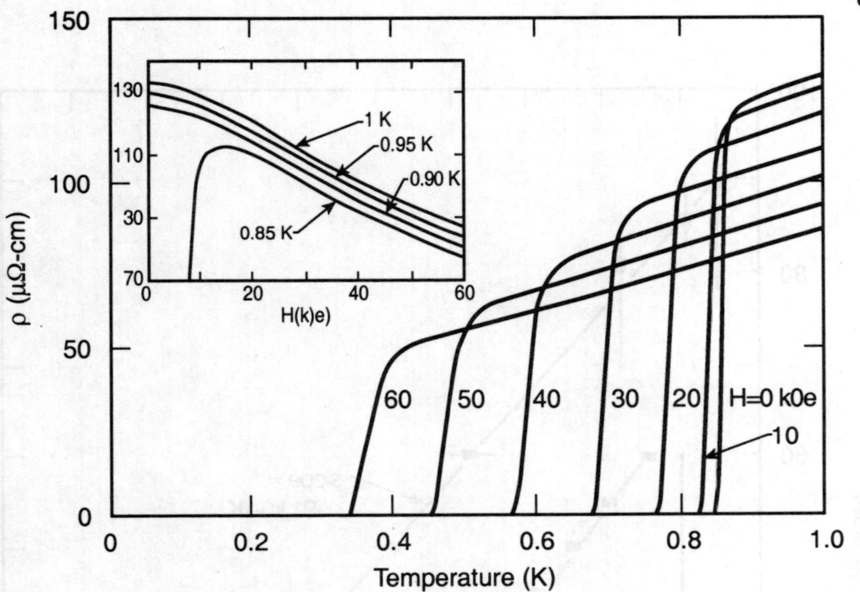

Fig. 41. Selected electrical resistivity ρ vs. temperature data for a single crystal specimen of UBe_{13} in various applied magnetic fields between 0 and 60 kOe. Shown in the inset are ρ vs. H isotherms between 0 and 60 kOe at 0.85, 0.90, 0.95, and 1.00 K. The lines are smooth curves that have been drawn through the data points (from Maple *et al.* 1985).[66]

single crystal where each transition curve was measured as electrical resistivity versus temperature at a given magnetic field. For each magnetic field, H, the transition temperature, T_c, was determined from the midpoint of the resistive transition curve. Then, the upper critical field, $H_{c2}(T)$, was plotted as H versus T_c as shown in Fig. 42.

Similarly, the $H_{c2}(T)$ curve can be determined from transition curves in ac-susceptibility experiments. Inset of Fig.43 shows selected three ac-susceptibility superconducting transitions measured parallel to the c-axis in magnetic fields of 0, 0.52 and 0.81 T for a single crystal of URu_2Si_2. The superconducting transition temperature was determined as the 50% point of the transition in the ac-susceptibility as a function of field. Therefore, the upper critical field versus temperature is plotted in Fig. 43 for field applied both parallel and perpendicular to the c-axis.

Most upper critical field curves were obtained from the resistively measured transition curves. The discrepancy between resistively and inductively measured $H_{c2}(T)$ curves in $CeCu_2Si_2$ is seen in Fig. 44. The difference between two $H_{c2}(T)$ curves measured by resistivity ρ and ac susceptibility χ_{ac} is mainly due to the inhomogeneities in the

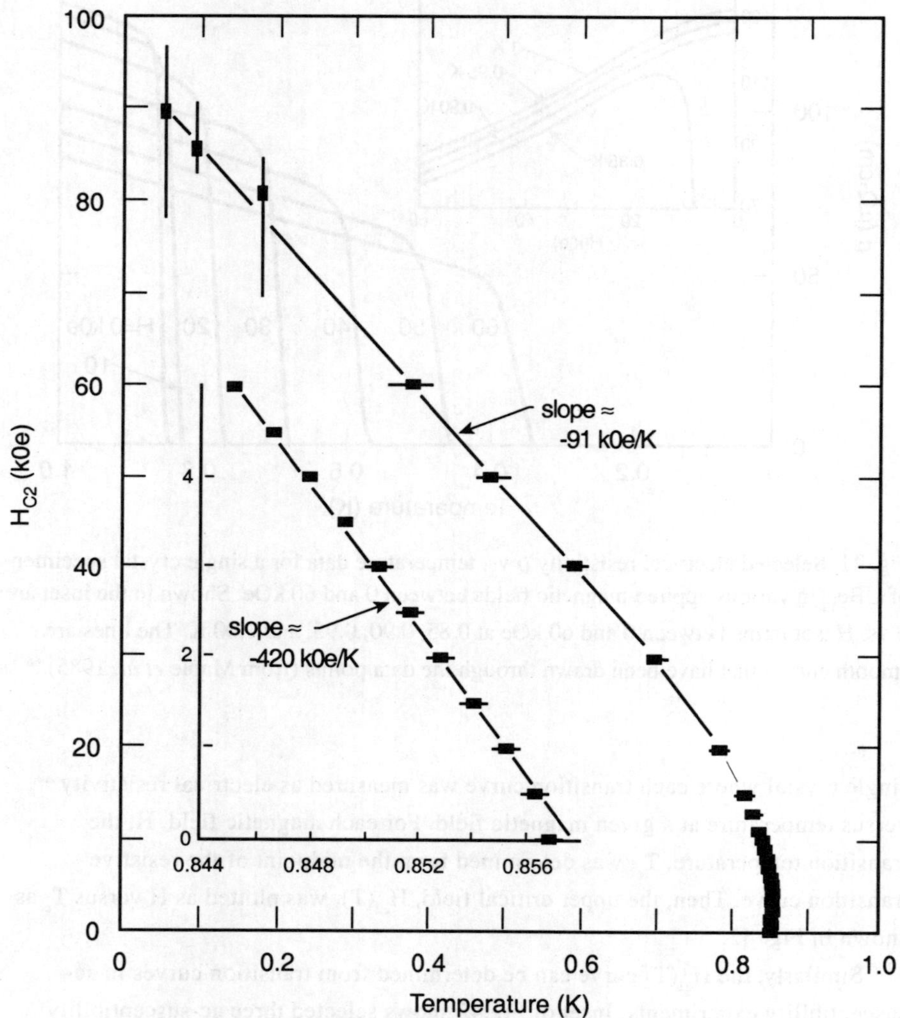

Fig. 42. Upper critical magnetic field H_{c2} vs. temperature T for a single crystal specimen of UBe$_{13}$. Shown in the inset are H_{c2} vs. T data in the vicinity of the zero-field T_c of 0.857 K. The lines are a guide to the eye. Horizontal and vertical bars indicate experimental uncertainties (from Maple *et al.* 1985).[66]

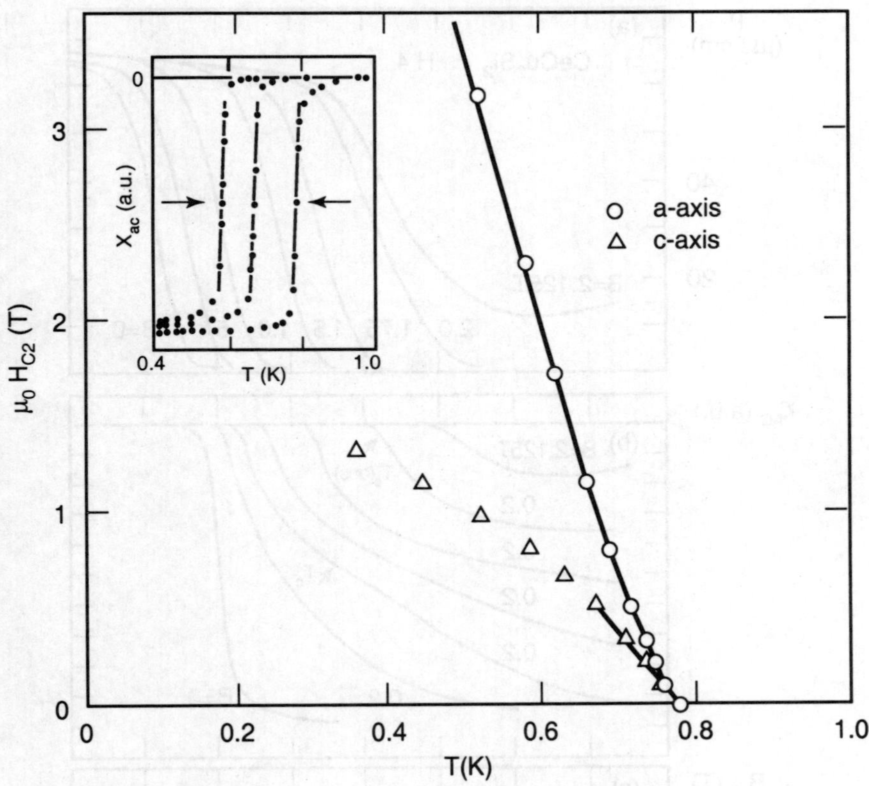

Fig. 43. Upper critical field $\mu_o H_{c2}$ of URu_2Si_2 vs. temperature parallel to the a and c axes. The inset shows three ac-susceptibility superconducting transitions measured parallel to the c-axis in applied magnetic fields of 0, 0.52, and 0.81 T (from Palstra et al.).[21]

superconductor. $T_c(\chi_{ac})$ is the average transition temperature due to whole sample volume. $T_c(\rho)$, on the other hand, is determined as the first portion of sample become superconducting. Therefore, the transition curves measured from transport method are more sharp, so the determination of T_c from resistive data is simpler and more reliable than ac-susceptibility or other measurements. Contrary to the case of $CeCu_2Si_2$, the inductively determined transitions for UBe_{13} remained sharp even in high fields. The difference among $H_{c2}(T)$ curves determined from resistive, inductive and calorimetric measurements is small. The shapes of the $H_{c2}(T)$ curves are almost the same as shown in Fig. 45.

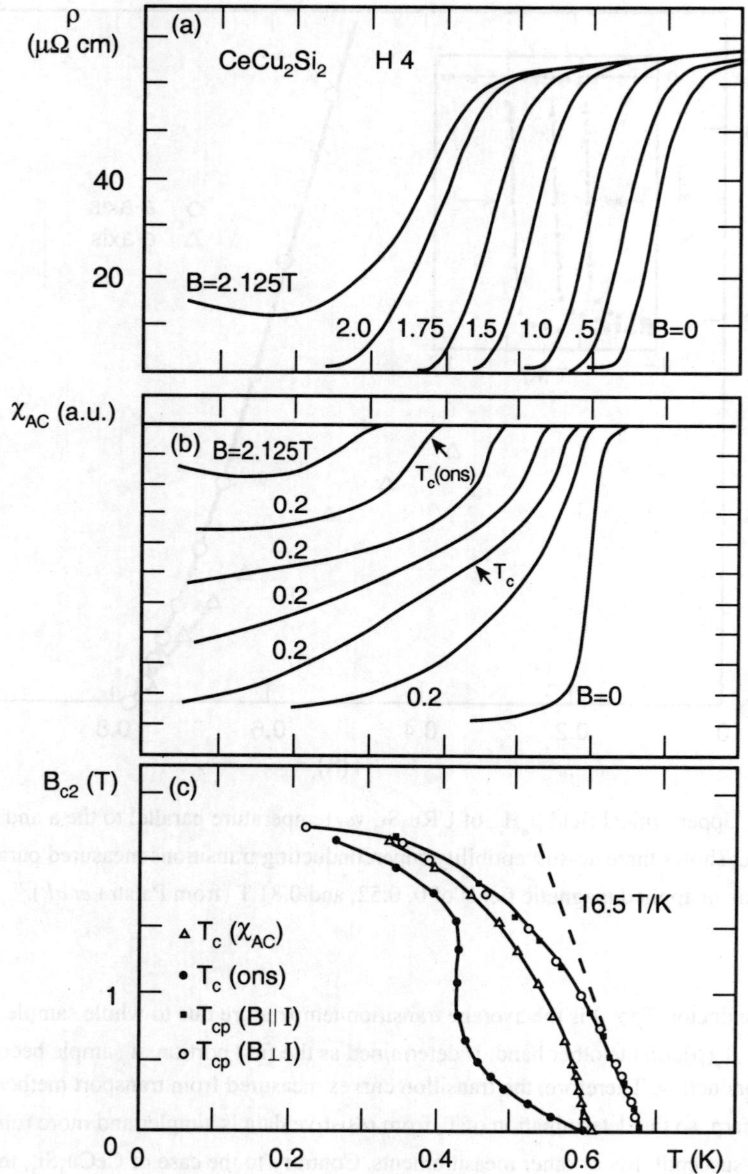

Fig.44. (a) Resistive transitions of the unannealed CeCu$_2$Si$_2$ polycrystal #4. (b) Inductive transitions of the same sample. (c) Upper critical field B$_{c2}$(T), determined from the transitions in (a) (■) and (b) (△, ●), as well as from a resistivity experiment with I ⊥ B (○). (from Rauschschwalbe).[67]

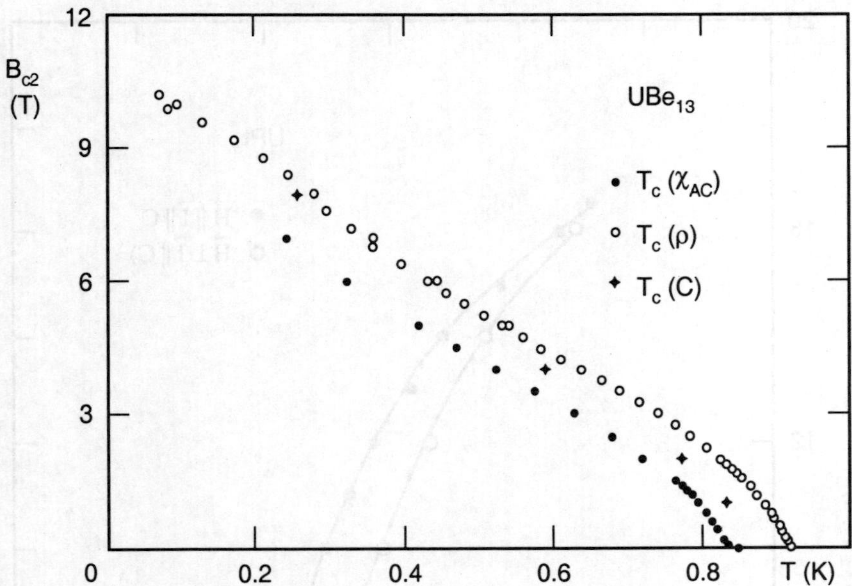

Fig. 45. Upper critical field B_{c2}(T), determined for UBe_{13} from measurements of the resistivity ($T_c(\rho)$), of the ac susceptibility ($T_c(\chi_{ac})$), and of the specific heat ($T_c(C)$), respectively (from Rauchschwalbe).[67]

Upper critical field studies yield two important qualities: the slope of the upper critical field at the superconducting transition, $-dH_{c2}/dT$, and the zero temperature upper critical field, $H_{c2}(0)$.

Generally, the upper critical field H_{c2}(T) is linearly temperature dependent near T_c in conventional superconductors. The slope of the upper critical field at the transition temperature, dH_{c2}/dT, is determined from the linear portion of H_{c2}(T) near T_c. As shown in the inset of Fig. 41, a very smooth linear relationship of H_{c2} versus T near T_c is observed between 1 and 0.4 T with a slope of $-dH_{c2}/dT= 4.2 \pm 2$ T/K. However, some H_{c2}(T) data observed in heavy fermion superconductors indicate a large curvature near T_c. The estimation of the slope at T_c, therefore, becomes inaccurate. Fig. 46, for example, demonstrates a curvature behavior in H_{c2}(T) near T_c in the case of field perpendicular to the current. The cited value of the slope is found by different workers to vary from 1 to 4 T/K.

$H_{c2}(0)$ is defined as the upper critical field at zero temperature. Experimentally, $H_{c2}(0)$ is determined by the extrapolation of the upper critical field H_{c2}(T) as $T \rightarrow 0$.

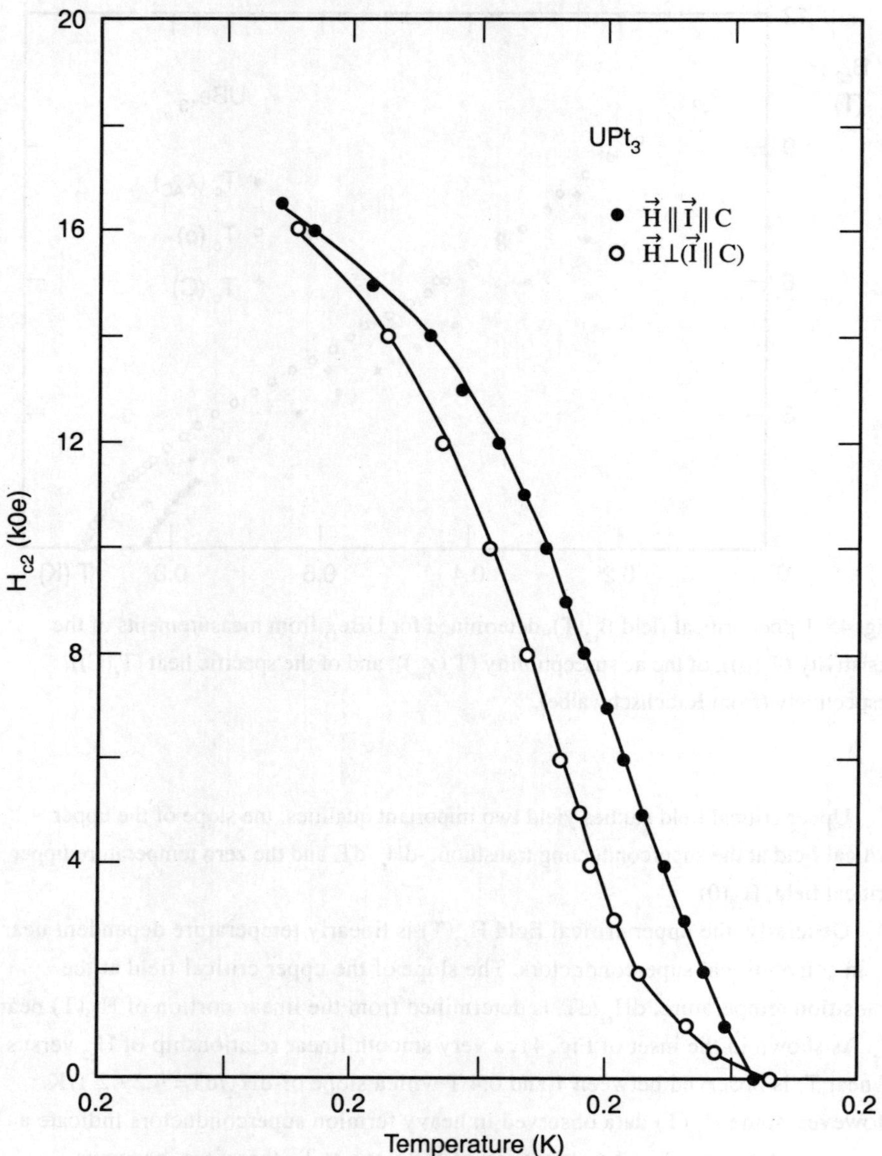

Fig. 46. Upper critical field vs. temperature data for UPt$_3$ with applied field H parallel or perpendicular to the current I (from Chen *et al.*).[68]

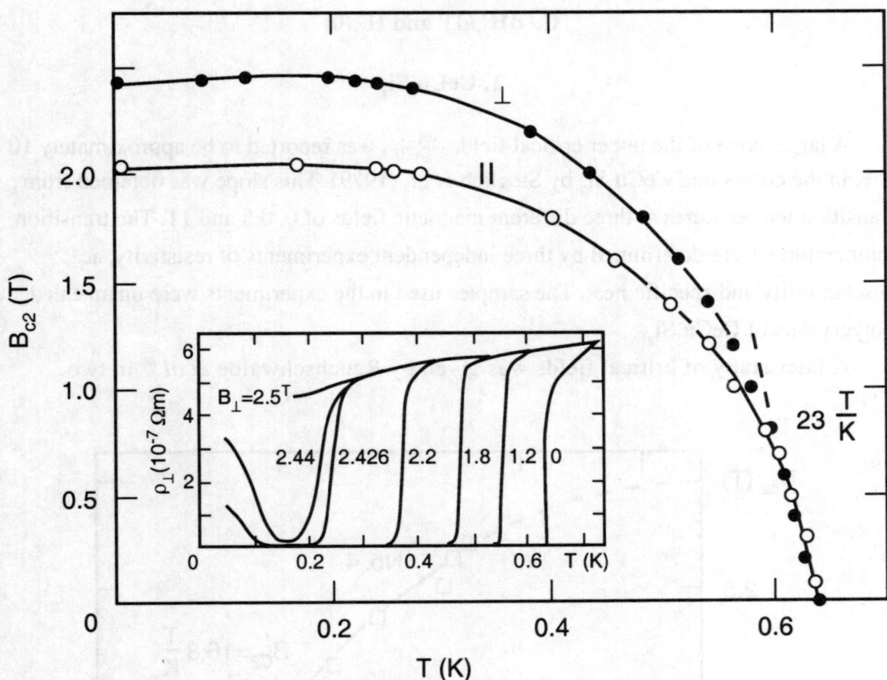

Fig. 47. B_{c2} vs. T, as obtained from the midpoints of ρ vs. T curves taken at different external fields for a $CeCu_2Si_2$ single crystal. Field B and current I are aligned to each other, either parallel ($B_{c2\parallel},\rho_\parallel$) to[1] or perpendicular ($B_{c2\perp},\rho_\perp$) to Ce planes. Inset shows $\rho_\perp$ vs. T at differing fields for this crystal (from Assmus *et al.*).[12]

As displayed in Fig. 47, the $H_{c2}(0)$ values of 2.4 and 2.0 T are simply obtained by extrapolation of low temperature $H_{c2}(T)$ data to T=0 for the two field configurations. However, the extrapolation method becomes difficult when the upper critical field is too high to measure in some experiments. For example, the lack of high field data in a URu_2Si_2 crystal at field perpendicular to the c-axis as shown in Fig. 43 made the estimation of $H_{c2}(0)$ impossible.

The next section presents the upper critical field studies in the order of heavy fermion superconducting compounds: $CeCu_2Si_2$, UBe_{13}, UPt_3, URu_2Si_2 and UPd_2Al_3.

C. dH_{c2}/dT and $H_{c2}(0)$

1. CeCu$_2$Si$_2$

A large slope of the upper critical field, $-\frac{dH_{c2}}{dT}\big|_{Tc}$, was reported to be approximately 10 T/K in the compound CeCu$_2$Si$_2$ by Steglich *et al.* (1979). This slope was obtained from transition temperatures at three different magnetic fields of 0, 0.5 and 1T. The transition temperatures were determined by three independent experiments of resistivity, ac susceptibility and specific heat. The samples used in the experiments were unannealed polycrystals of CeCu$_2$Si$_2$.

A later study of critical fields was given by Rauchschwalbe *et al.*[69] in two

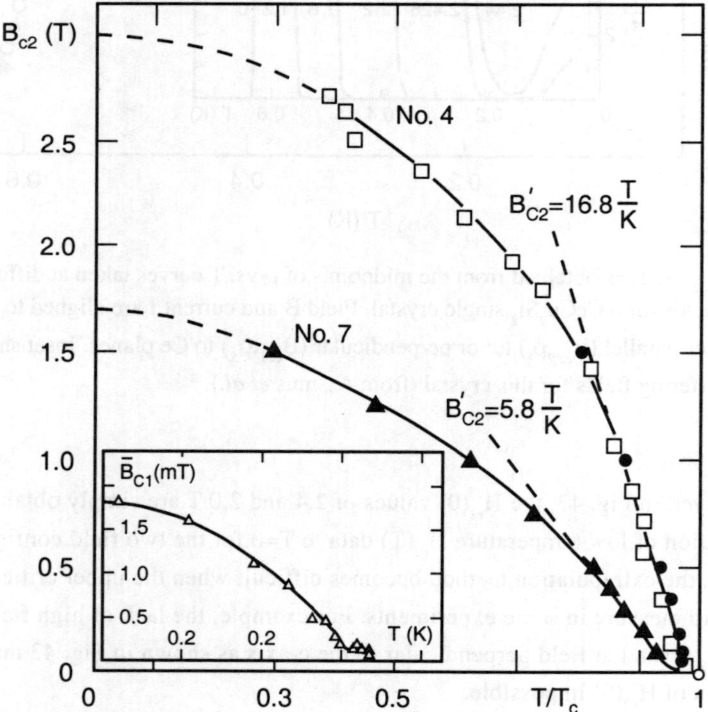

Fig. 48. Upper critical field, B$_{c2}$, of CeCu$_2$Si$_2$ as a function of the reduced temperature, T/T$_c$. While T$_c$ is the transition temperature at B$_c$ = 0 as measured, T$_{co}$ is defined by extrapolation of linear B$_{c2}$(T) dependence to B = 0. Data were obtained from ac susceptibility (triangles: No. 7, T$_{co}$ = 0.64 K; squares: No. 4, T$_{co}$ = 0.66 K) or specific heat (circles: No. 4, T$_{co}$ = 0.56 K) (from Rauchschwalbe *et al.*).[69]

annealed $CeCu_2Si_2$ polycrystals, where one sample was relatively pure and another was less clean. The $H_{c2}(T)$ data were determined from the transition curves obtained from both inductive and calorimetric measurements. At low fields near T_c, an upward curvature was found in the H_{c2} versus T plot as shown in Fig. 48. This curvature was attributed to inhomogeneities in the samples. The slopes were estimated from a linear region of $H_{c2}(T)$ close to T_c. The $-dH_{c2}/dT$ values were 5.8 and 16.8 T/K in the relatively pure and less clean samples, respectively. $H_{c2}(0)$ was found from the extrapolation of the upper critical field $H_{c2}(T)$ as $T \rightarrow 0$ and the corresponding values of 1.7 T and 3 T were obtained.

Assmus *et al.*[12] first published single crystal data of an as-grown (unannealed) $CeCu_{2.6}Si_2$ crystal with 30% excess Cu. The $H_{c2}(T)$ was obtained from transport measurements. The value of $-dH_{c2}/dT$ was found to be 23 T/K (see Fig. 47) for both field orientations (parallel and perpendicular) to the crystalline c-axis. This slope was larger than the values reported in previous studies on polycrystalline samples.[1, 69] The $H_{c2}(0)$ values were obtained in the single crystal of $CeCu_{2.6}Si_2$ with considerable anisotropy: $H_{c2}(0)$ = 2.4 and 2.0 T for field parallel and perpendicular to the c-axis, respectively.

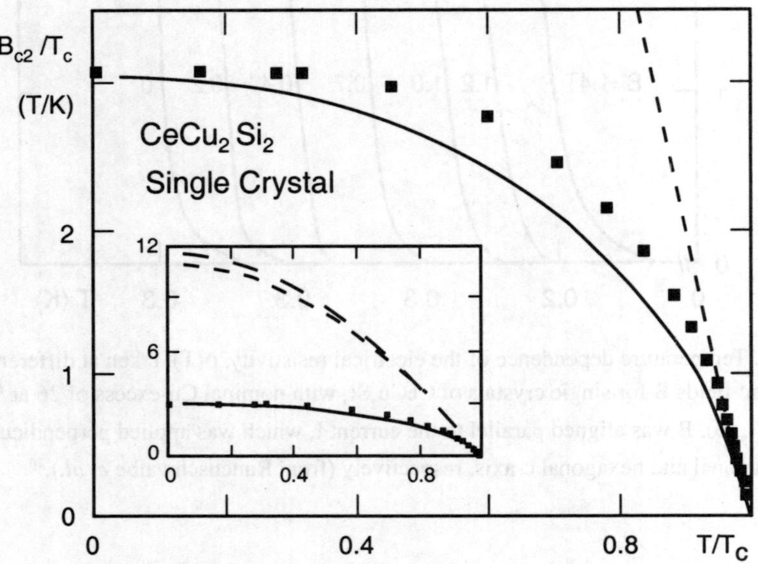

Fig. 49. B_{c2}/T_c vs. T/T_c for a $CeCu_2Si_2$ single crystal (T_c = 0.65 K). Inset shows same data on larger scale (from Rauchschwalbe *et al.*).[70]

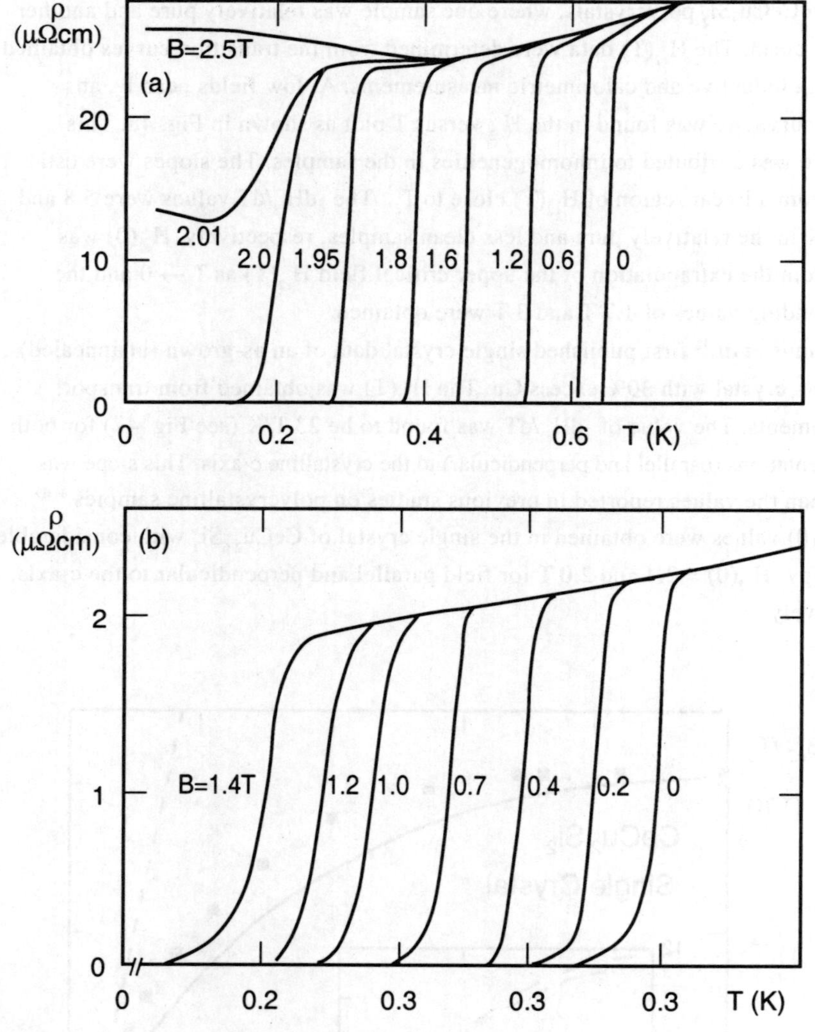

Fig. 50. Temperature dependence of the electrical resistivity, ρ(T), taken at different magnetic fields B for single crystals of $CeCu_2Si_2$ with nominal Cu-excess of 26 at % (a) and UPt_3 (b). B was aligned parallel to the current I, which was applied perpendicular to the tetragonal and hexagonal c-axis, respectively (from Rauchschwalbe *et al.*).[70]

A similar slope of 21 T/K was obtained in an annealed $CeCu_{2.53}Si_2$ crystal[70] with 26% excess Cu at H ⊥ c (see Fig. 49). The upper critical field was determined from a temperature dependence of the electrical resistivity at different magnetic fields (see Fig. 50). The $H_{c2}(0)$ value was found to be 2.0 T for $H \perp c$, which was the same as the result obtained in an unannealed crystal at same field orientation by Assmus et al.[12]

A more thorough study of the upper critical field on seven polycrystalline and five single crystalline samples of $CeCu_2Si_2$ was published by Rauchschwalbe.[67, 71] In polycrystals, values of $-dH_{c2}/dT$ between 11 and 16.5 T/K were listed, which included the previous studies[12, 69, 70] excluding one value of 5.8 T/K obtained from a purer sample (Rauchschwalbe et al. 1982).[69] Single crystals yielded relatively higher values of $-dH_{c2}/dT$ between 13.5 and 23 T/K. The difference in the slopes between polycrystalline and single crystalline samples could be explained by the inhomogeneities in polycrystals causing broadened superconducting transitions and thus averaged (lower) upper critical fields. The $H_{c2}(0)$ values were summarized to be between 1.7 - 2.4 T in both single crystals and polycrystals with no observable difference in $H_{c2}(0)$ between single crystals and polycrystals.

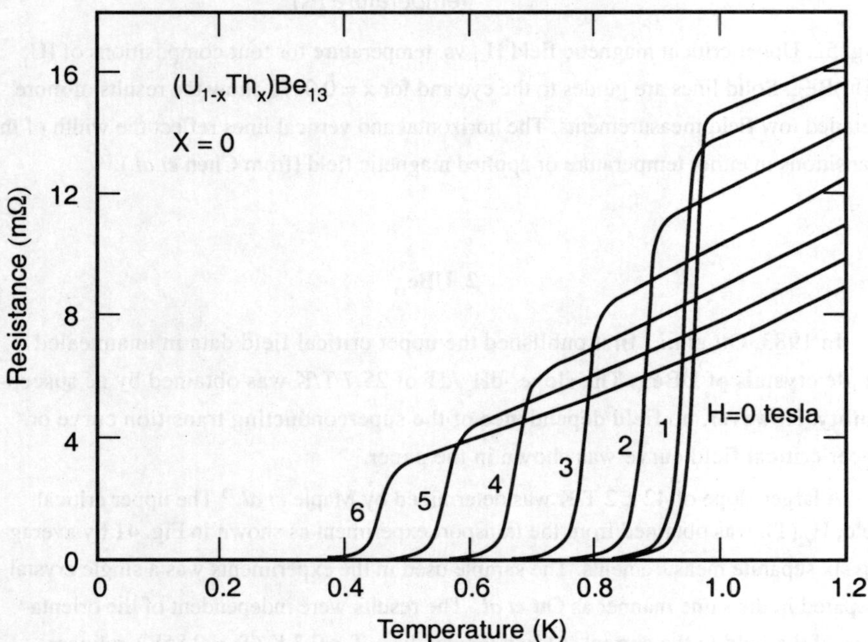

Fig. 51. ac electrical resistance vs. temperature in various applied magnetic fields in T for a polycrystalline sample of UBe_{13} (from Chen et al.).[72]

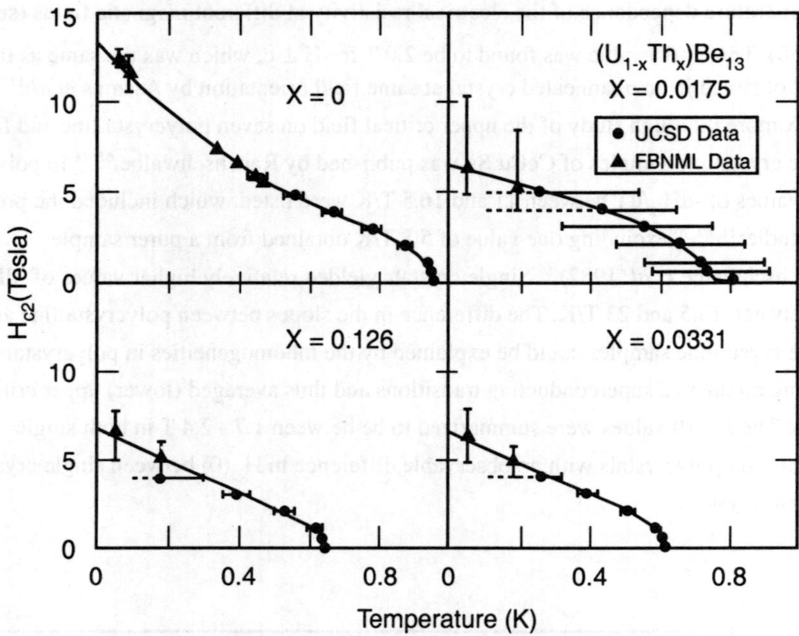

Fig. 52. Upper critical magnetic field H_{c2} vs. temperature for four compositions of $(U_{1-x}Th_x)Be_{13}$. Solid lines are guides to the eye and for x = 0.0175 show the results of more detailed low field measurements. The horizontal and vertical lines reflect the width of the transitions in either temperature or applied magnetic field (from Chen *et al.*).[72]

2. UBe$_{13}$

In 1983, Ott *et al.*[3] first published the upper critical field data in unannealed single crystals of UBe$_{13}$. The slope -dH_{c2}/dT of 25.7 T/K was obtained by ac susceptibility. However, no field dependence of the superconducting transition curve or upper critical field curve was shown in the paper.

A larger slope of 42 ± 2 T/K was determined by Maple *et al.*[73] The upper critical field, H_{c2}(T), was obtained from the transport experiment as shown in Fig. 41 by averaging six separate measurements. The sample used in the experiments was a single crystal prepared in the same manner as Ott *et al.*. The results were independent of the orientations of the field to the current. At low temperature, T < 0.7 K (T_c = 0.85K), a linear temperature dependence of the upper critical field was found with a smaller slope of dH_{c2}/dT = 9.1 T/K (see Fig. 42). There was no indication of saturation of H_{c2}(T) down to

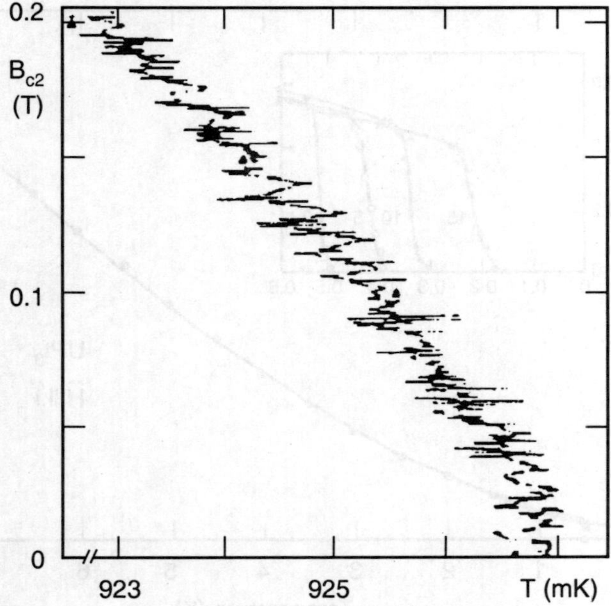

Fig. 53. Continuously recorded $B_{c2}(T)$ curve for UBe_{13} in the direct vicinity of T_c. Triangles are the results of the discrete experiments (from Rauchschwalbe).[67]

the low temperature limit 50 mK with $H_{c2} = 9$ T. By extrapolating the linear portion of the $H_{c2}(T)$ data obtained between 50 mK and 0.7 K, we estimate $H_{c2}(0)$ to be about 9.5 T.

Chen et. al.[72] reported upper critical field studies in a polycrystalline sample of the compound UBe_{13}. The ac electrical resistance versus temperature was measured in various magnetic fields and the upper critical field was defined from the transition curves, using the traditional 50% method (see Fig. 51). A slope $-\frac{dH_{c2}}{dT}|_{T_c}$ of 35 T/K was found. Due to the non-linear behavior of $H_{c2}(T)$ at low temperatures (see Fig. 52), the $H_{c2}(0)$ data could not be determined. The low temperature limit for the experiment was 90 mK and H_{c2} was found to be 12 T at this temperature.

A similar critical field study for a UBe_{13} polycrystalline sample, was published by Rauchschwalbe (1987d). In Fig. 45, three upper critical field curves were displayed from resistive, inductive and calorimetric measurements. Since the superconducting transitions for the compound UBe_{13} remained sharp even in high fields, the discrepancy among $H_{c2}(T)$ curves was small and the shapes of the $H_{c2}(T)$ curves were almost the same. Values for $-\frac{dH_{c2}}{dT}|_{T_c}$ were estimated as 25 T/K and 23 T/K as determined by resistivity and ac

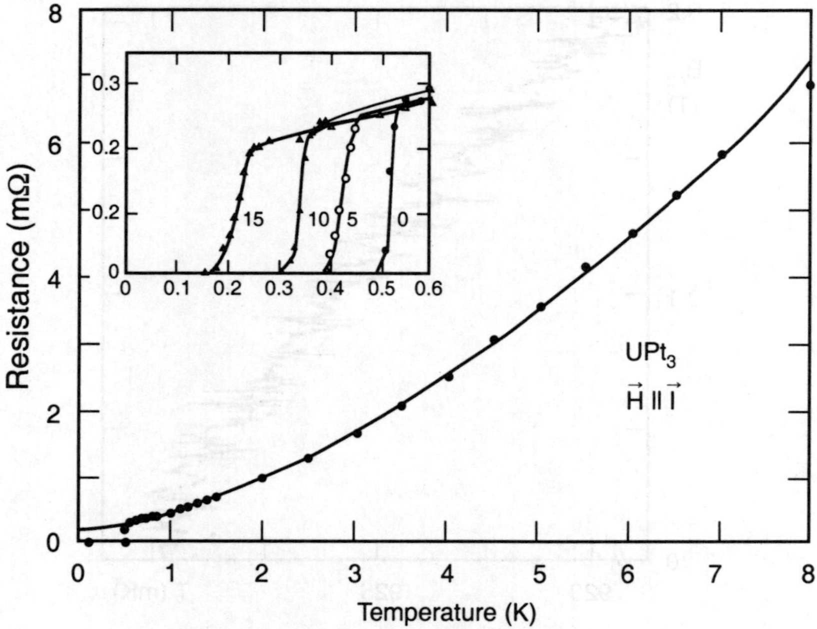

Fig. 54. Electrical resistance R vs. temperature of UPt$_3$ in zero applied magnetic field. The solid line represents a linear fit. Shown in the inset are resistive superconducting transition data in magnetic fields H of 0, 5, 10, and 15 kOe with H parallel to the current I (from Chen *et al.*).[68]

susceptibility, respectively, and H$_{c2}$(T) saturated at 10.2 T as T approached to zero. A continuously recorded H$_{c2}$(T) method was discussed in details by Rauchschwalbe.[68] This method indicated -dH$_{c2}$/dT of the compound UBe$_{13}$ could be as large as 100 T/K or larger (see Fig. 53).

3. UPt$_3$

When superconductivity was discovered in the compound UPt$_3$, the critical field was measured resistivity with the field parallel to the c-axis of the crystal. -dH$_{c2}$/dT was found[26] to be about 2 T/K, which was smaller than those obtained in the CeCu$_2$Si$_2$ and UBe$_{13}$ systems.

Later, Chen *et al.*[68] measured the electrical resistance as a function of temperature (see Fig. 54) in single crystals of UPt$_3$ prepared in the same matter as Stewart *et al.*[3]

Upper critical field versus temperature is shown in Fig. 46 for H perpendicular and parallel to I. The slopes of -dH_{c2}/dT were 6.3 T/K and 4 T/K for H $\parallel$ I and H $\perp$ I, respectively.[74, 75] An $H_{c2}(0)$ value of 1.7 T was estimated in the crystal of UPt_3 by the extrapolating of the low temperature $H_{c2}(T)$ data to T=0.

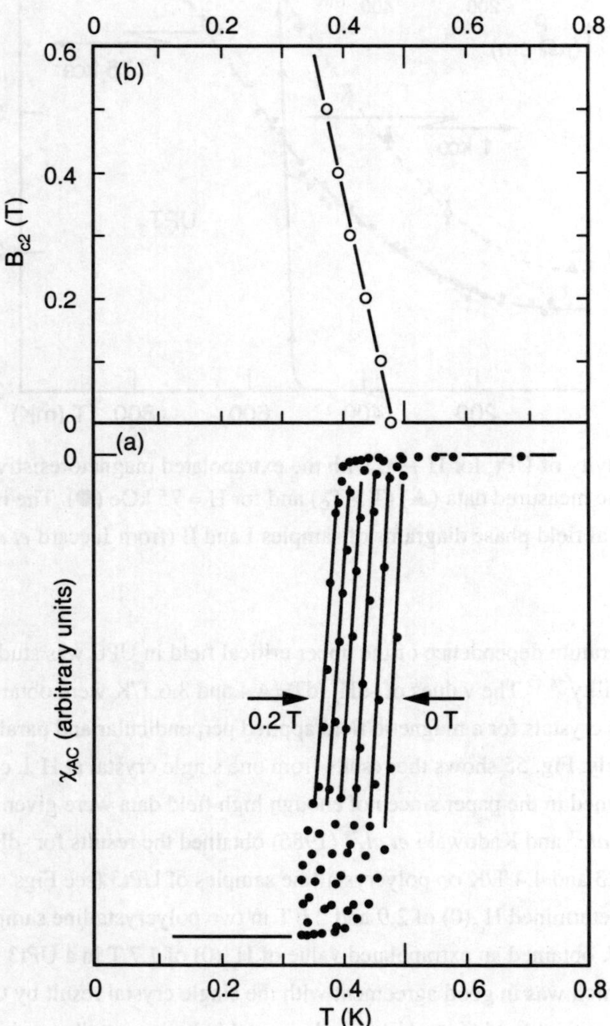

Fig. 55. (a) ac susceptibility (in arbitrary units) in zero magnetic field and in fields of 0.1, 0.2, 0.3, 0.4 and 0.5 T (directed along the a-axis). (b) Upper critical field as a function of temperature; T_c is defined at the 50% point (indicated by arrows in (a)) (from de Visser *et al.*).[76]

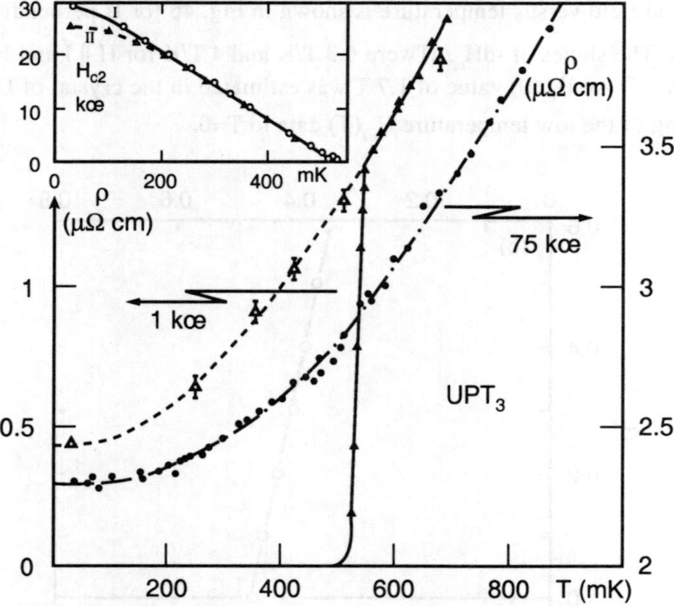

Fig. 56. Resistivity of UPt$_3$ for H → 0 with the extrapolated magnetoresistivity data (△) (T < T$_c$) and the measured data (▲) (T > T$_c$) and for H = 75 kOe (●). The inset shows the upper critical field phase diagrams of samples I and II (from Jaccard *et al*.).[51]

The temperature dependence of the upper critical field in UPt$_3$ was studied by means of ac susceptibility.[76, 77] The values of -dH$_{c2}$/dT= 4.4 and 3.6 T/K were obtained on two different single crystals for a magnetic field applied perpendicular and parallel to the c-axis, respectively. Fig. 55 shows the results from one single crystal at H ⊥ c. No H$_{c2}$(0) data were obtained in the paper since not enough high field data were given.

Jaccard *et al*.[51] and Kadowaki *et al*.[78] (1986) obtained the results for -dH$_{c2}$/dT with the values of 6.3 and 4.4 T/K on polycrystalline samples of UPt3 (see Figs. 56 and 57). Jaccard *et al*. determined H$_{c2}$(0) of 2.9 and 2.6 T in two polycrystalline samples of UPt3. Kadowaki *et al*. obtained an extrapolated value of H$_{c2}$(0) of 1.7 T in a UPt3 polycrystalline sample, which was in good agreement with the single crystal result by Chen *et al*.

The upper critical fields for a UPt$_3$ single crystal in both crystallographic directions of H ∥ c and H ⊥ c are presented in Fig. 58. The results of -dH$_{c2}$/dT were 6 and 4 T/K for H ∥ c and H ⊥ c, respectively. H$_{c2}$(0) of 1.5 T was given by Rauchschwalbe.[67]

The study of the angular dependence of the upper critical field in UPt$_3$ was performed by Shivaram *et al*.[75] Two single crystals from the same batch were used in this work.

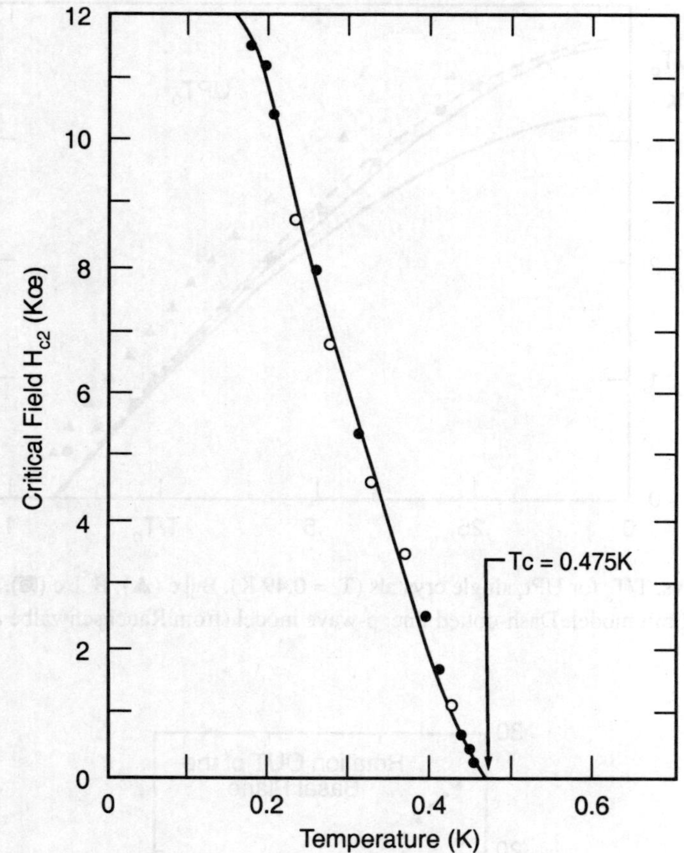

Fig. 57. Temperature dependence of the superconducting upper critical field H$_{c2}$ in UPt$_3$ polycrystal. A dc current density of 27 A/cm^2 was used (from Kadowaki *et al.*).[78]

Fig. 59 presents critical field data for the field parallel and perpendicular to the basal plane. Values of the slope -dH$_{c2}$/dT were found to be 7.7 and 4.5 T/K for the field parallel and perpendicular to the c-axis of the crystal, respectively. No anisotropy was observed for sample rotation in the basal plane, which was consistent with the hexagonal symmetry of UPt$_3$. The H$_{c2}$(0) values were found in the range of 2.1 - 2.8 T for two field orientations in two samples.

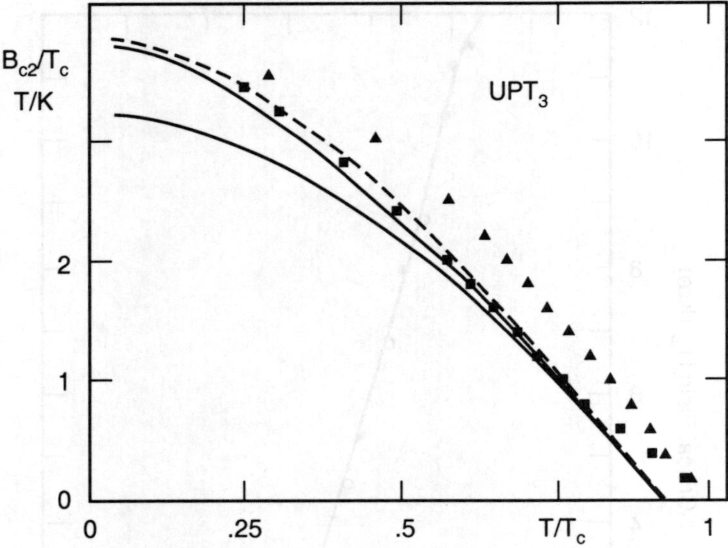

Fig. 58. B_{c2} vs. T/T_c for UPt$_3$ single crystals (T_c = 0.49 K). B ∥ c (▲); B ⊥ c (■). Solid lines: dirty-limit model. Dash-dotted line: p-wave model (from Rauchschwalbe *et al.*).[70]

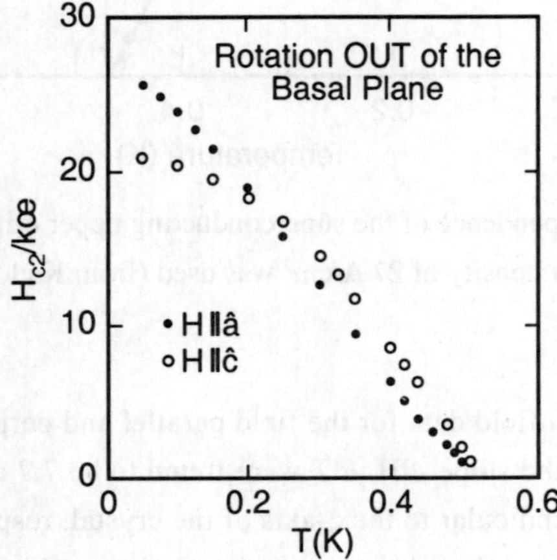

Fig. 59. The upper critical field, H$_{c2}$, in (filled circles) and perpendicular to (open circles) the basal plane of UPt$_3$ single crystals. The current is along $\hat{b}$ and always perpendicular to H. The curves for intermediate angles also cross at T ~ 200 mK (from Shivaram *et al.*).[59]

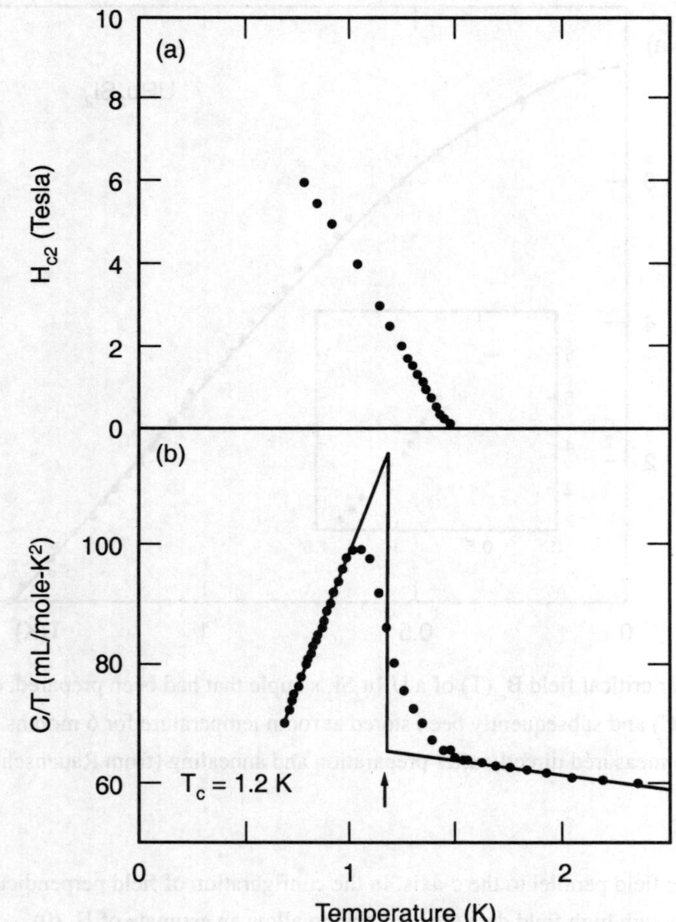

Fig. 60. (a) Upper critical field H$_{c2}$ vs. temperature for URu$_2$Si$_2$. (b) Specific heat C divided by temperature T vs. T for URu$_2$Si$_2$ (from Maple *et al.*).[66]

4. URu$_2$Si$_2$

Palstra *et al.*[21] first published the upper critical field data for a URu$_2$Si$_2$ crystal (see Fig. 43). The slope of 4 T/K was determined by transition curves measured by ac susceptibility in both directions for the applied field parallel and perpendicular to the c-axis. The H$_{c2}$(0) values were not given in the paper. From their data, we estimate H$_{c2}$(0)

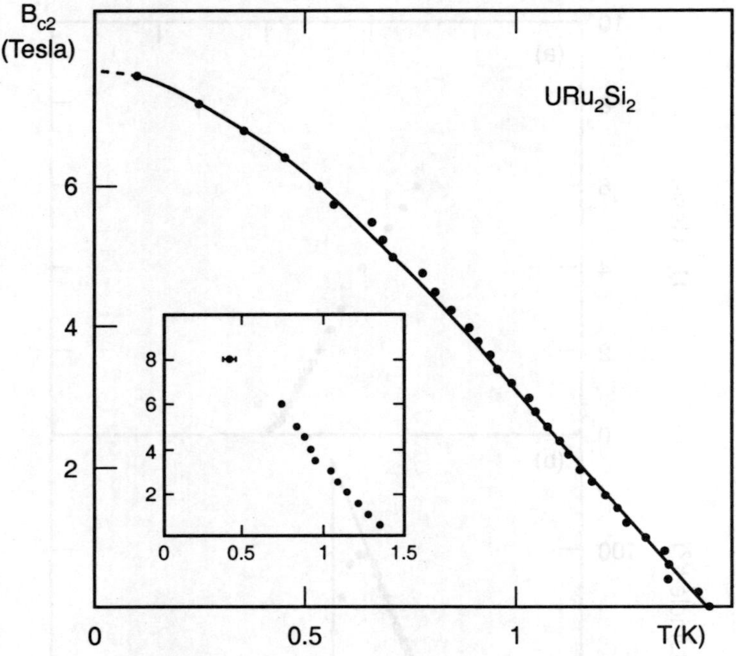

Fig. 61. Upper critical field $B_{c2}(T)$ of a URu_2Si_2 sample that had been prepared, annealed (24 h at 850°C) and subsequently been stored at room temperature for 6 months. The inset shows $B_{c2}(T)$ measured directly after preparation and annealing (from Rauchschwalbe).[67]

~1.5 T for the field parallel to the c-axis. In the configuration of field perpendicular to the c-axis, not enough high field data are available to allow an estimate of $H_{c2}(0)$.

A relatively large slope $-dH_{c2}/dT$ was obtained in an annealed polycrystal of URu_2Si_2 by Maple et al.[66] The upper critical field curve was determined from resistively measured superconducting transitions. The $H_{c2}(T)$ data show a linear temperature dependence above T_c, as displayed in Fig. 60, which yields an initial slope of 9.2 T/K.

Fig. 61 shows the upper critical field determined resistively on a URu_2Si_2 polycrystalline sample. The $H_{c2}(T)$ began with a slope $-dH_{c2}/dT$ of 7.6 T/K and reached $H_{c2}(0)$ of 8 T.[67, 22]

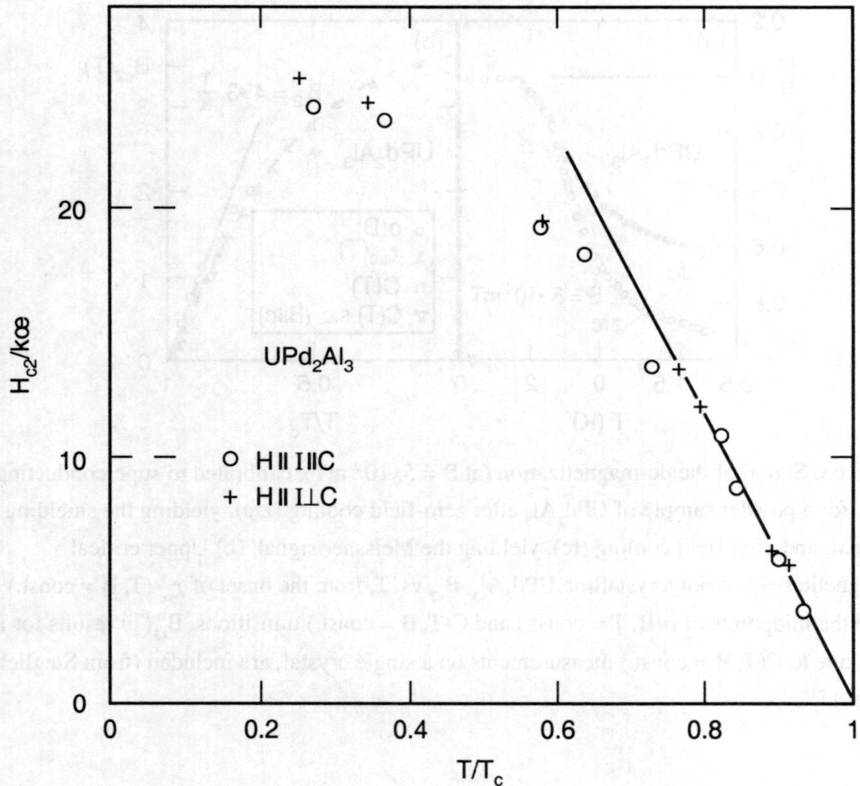

Fig. 62. Upper critical magnetic field H_{c2} for H ∥ c and H ⊥ c as a function of the normalized temperature by T_c, obtained from the magnetoresistivity measurements in UPd_2Al_3 single crystal (from Sato et al.).[37]

5. UPd_2Al_3

Sato et al.[37] reported upper critical field study on unannealed single crystals of UPd_2Al_3 from magnetoresistivity measurements. They obtained a value of $-dH_{c2}/dT$ of 3.6 T/K for the magnetic field parallel and perpendicular to the c-axis in two different crystals (see Fig. 62). $H_{c2}(0)$ was not given in the paper. By simple extrapolation of $H_{c2}(T)$ to T=0 in Fig. 62, we estimate $H_{c2}(0)$ to be 2.5 T. Geibel et al.[6] obtained a slope of 4.3 T/K in a polycrystalline UPd_2Al_3 from specific heat and resistivity measurements (see Fig. 63).

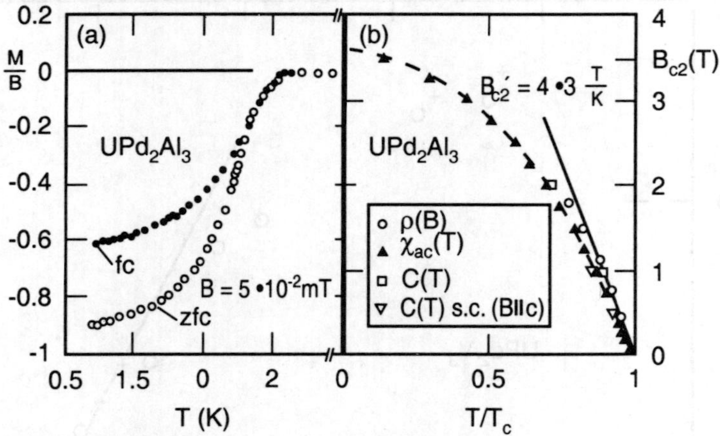

Fig. 63. Signal of the dc-magnetization (at $B = 5 \times 10^{-2}$ mT), calibrated to superconducting Cd, for a powder sample of UPd_2Al_3 after zero-field cooling (zfc), yielding the shielding signal, and after field cooling (fc), yielding the Meissner signal. (b) Upper critical magnetic field of polycrystalline UPd_2Al_3, B_{c2} vs. T, from the onset of χ_{ac} (T, B = const.) and the midpoints of ρ(B, T = const.) and C(T, B = const.) transitions. B_{c2}(T) results for B $\parallel$ c, due to C(T, B = const.) measurements on a single crystal, are included (from Steglich et al.).[79]

6. Conclusion

a. -dH$_{c2}$/dT

As shown in Table VIII, large slopes of the upper critical field, $-dH_{c2}/dT$, are observed in heavy fermion superconductors. The UBe_{13} system has an anomalously large slope up to ~100 T/K, possibly exceeded only by that for some of the high-T_c superconductors. The slopes are in the range 2 - 7.6 T/K in the UPt_3, URu_2Si_2 and UPd_2Al_3 families. Large variations of the slope were found in the compounds $CeCu_2Si_2$ and UBe_{13}. In $CeCu_2Si_2$, this variation may be due to the sample dependence. The reason for large variations in the compound UBe_{13} may simply be the fact that the slope is too large to estimate accurately.

b. H$_{c2}$(0)

Table VIII lists the results of the upper critical field at zero temperature, H_{c2}(0), obtained in heavy fermion superconductors. The UBe_{13} system has a large H_{c2}(0) value of

10 T obtained from both single crystals and polycrystals, which was consistent with the result of a large slope at T_c. The $H_{c2}(0)$ values are in the range 1.3 - 3 T in both polycrystalline and single crystalline samples of the compounds $CeCu_2Si_2$ and UPt_3, and single crystals of the compounds URu_2Si_2 and UPd_2Al_3. Large variations of $H_{c2}(0)$ are found in URu_2Si_2 between single crystals and polycrystals, which might be due to the broad superconducting transitions.

Table VIII. The slope of upper critical field, $-dH_{c2}/dT$, and the upper critical field at zero temperature, $H_{c2}(0)$, in heavy fermion superconductors.

Compound	Sample	$-\frac{dH_{c2}}{dT}\|_{T_c}$ (T/K)	$H_{c2}(0)$	Reference
$CeCu_2Si_2$		10		Steglich *et al.* 1979[1]
	2 polycrystals	5.8, 16.8	1.7, 3	Rauchschwalbe *et al.* 1982[69]
	crystal	23	2.4, 2.0	Assmus *et al.* 1984[12]
	crystal	21	2.0	Rauchschwalbe *et al.* 1985[70]
	polycrystals	11 - 16.5	1.3 - 2.25	Rauchschwalbe 1987[67]
	crystals	13.5 - 23	1.93 - 2.45	Rauchschwalbe 1987[67]
UBe_{13}	crystal	25.7		Ott *et al.* 1983[2]
	crystal	42	9.5	Maple *et al.* 1985[73]
	polycrystal	35		Chen *et al.* 1985[72]
	polycrystal	23 - > 100	10.2	Rauchschwalbe 1987[67]
UPt_3	crystal	2		Stewart *et al.* 1984[3]
	crystal	6.3, 4	1.7	Chen *et al.* 1984[68]
	crystal	4.4		de Visser *et al.* 1984[76]
	polycrystal	6.3	2.9, 2.6	Jaccard *et al.* 1985[51]
	polycrystal	4.4	1.7	Kadowaki *et al.* 1986[78]
	crystal	5.6, 4	1.5	Rauchschwalbe *et al.* 1985[70]
	crystal	7.7, 4.5	2.1 - 2.81	Shivaram *et al.* 1986[59]
URu_2Si_2	crystal	4	1.5	Palstra *et al.* 1985[21]
	polycrystal	9.2		Maple *et al.* 1986[66]
	polycrystal	7.6	8	Rauchschwalbe 1987[67]
UPd_2Al_3	crystal	3.6	2.5	Sato *et al.* 1992[37]
	polycrystal	4.3		Steglich *et al.* 1991[79]

D. Anisotropy

Assmus *et al.*[12] observed no anisotropy for the slope of the upper critical field in a single crystal of $CeCu_2Si_2$. dH_{c2}/dT did not depend on the orientation of the crystal with respect to the magnetic field within an experimental uncertainty of 1 T/K (see Fig. 47). However, a small anisotropy $H_{c2}(0)$ was found: $H_{c2}(0) = 2.4$ and 2.0 T for fields parallel and perpendicular to the c-axis, respectively. In both cases, the current through the sample was parallel to field (H $\parallel$ I). Rauchschwalbe[67] reported the upper critical field studies on various single crystals of $CeCu_2Si_2$. $H_{c2}(0)$ showed a small anisotropy: $H_{c2}(0)$ at H $\parallel$ c was greater than that at H $\perp$ c by 20%. $H_{c2}(T)$ did not depend on the direction of the measuring current in $CeCu_2Si_2$. The measurements with the current parallel and perpendicular to the field gave identical results as shown in Fig. 44.

Chen et al.[68] measured the resistive superconducting transition for a current along the c-axis (I $\parallel$ c). UPt_3 showed a distinct anisotropy in its H_{c2} versus T/T_c behavior (see Fig. 46). This anisotropy became much smaller at high fields and the two $H_{c2}(T)$ curves almost crossed at the lowest temperature of measurement, i.e. 0.12 K, with $H_{c2}(0) \sim 1.7$ T. In the configuration of H parallel to I and c, $H_{c2}(T)$ showed a linear temperature dependence in a large temperature range. This linear portion of $H_{c2}(T)$ had a slope $-dH_{c2}/dT$ of 6.3 T/K. For H $\perp$ I, $H_{c2}(T)$ was found to have a large curvature above T_c and $-dH_{c2}/dT$ was estimated to be close to 4 T/K. The anisotropy effect in H_{c2} observed in UPt_3 by Chen et al. was the opposite of that found by Assmus et al. in $CeCu_2Si_2$. The upper critical field study by Rauchschwalbe et al.[70] in UPt_3 indicated a marked anisotropic behavior (see Fig. 58). For the field parallel to the c-axis, $H_{c2}(T)$ had a linear temperature dependence over a large range of temperature between T_c and $0.6\ T_c$ with a slope $-dH_{c2}/dT$ of 5.6 T/K. In the configuration of H $\perp$ c, a discontinuous slope of the upper critical field was observed around 0.4 T and reached a value at T_c of 4 T/K. Fig.59 shows the upper critical field measurements in a UPt_3 crystal for field parallel and perpendicular to the c-axis, $H_{c2\parallel}$ and $H_{c2\perp}$, respectively. An anomalous crossover in the anisotropy of $H_{c2}(T)$ was found with decreasing temperature. For temperatures near T_c, $H_{c2\parallel}/H_{c2\perp}$ was about 1.3 and dropped to ~ 0.8 as T approached 0. This was consistent with a published value:[80] $H_{c2\parallel}/H_{c2\perp} = 1.6$ at $T = 0.932\ T_c$.

Palstra et al.[21] studied the upper critical field data for a URu_2Si_2 crystal (see Fig. 43). No information of current direction was given in the publication. The slope of 4 T/K was obtained in both directions for field parallel and perpendicular to the c-axis. At lower temperature, the anisotropy of $H_{c2}(T)$ was enhanced significantly with decreasing temperature: $H_{c2\parallel}(0.5K) = 1$ T and $H_{c2\perp}(0.5K) = 3.1$ T. Obviously, the value of $H_{c2}(0)$ with the applied field parallel to the c-axis would be much smaller than that of the perpendicular field configuration. This anisotropy effect of $H_{c2}(0)$ is opposite to $CeCu_2Si_2$ and UPt_3 systems.

Table IX lists anisotropy the upper critical field data of $CeCu_2Si_2$, UPt_3 and URu_2Si_2 single crystals. Compared with $CeCu_2Si_2$, UPt_3 shows a large anisotropy 30-40% in the slopes of $-dH_{c2}/dT$ for the field parallel and perpendicular to the c-axis. Both $CeCu_2Si_2$ and UPt_3 show none or small anisotropy in $H_{c2}(0)$. Within the uncertainty of the experiments, $H_{c2}(T)$ is independent to current directions in $CeCu_2Si_2$ and UPt_3. In contrast to $CeCu_2Si_2$ and UPt_3, URu_2Si_2 has no anisotropy in the slope of the upper critical field at T_c, but marked difference of $H_{c2}(T)$ at temperatures lower than T_c.

The large anisotropy at T_c in H_{c2} in UPt_3 was used by Varma[47] to argue that UPt_3 is a p-wave superconductor, in agreement with arguments for p-wave superconductivity in UPt_3 put forward by Stewart et al.[3] when they discovered superconductivity coexisting with strong evidence (the $T^3\ln T$ term in the heat capacity) for spin fluctuations.

The reduced slope dH_{c2}/dT in UPt_3 for the magnetic field applied parallel to the basal plane ($H \perp c$) may be understood from the assumption of non-isotropic gap structures with some intrinsic zeros on the Fermi surface.[38, 47] Assuming a line of nodes exists in the basal plane, a large screening current can be formed for the field applied parallel to the hexagonal c-axis without pair breaking. By contrast, these nodes are not effective for the field applied in the basal plane. As a result, small screening current leads to a rapid decrease in the condensation energy, indicated by a reduced critical field slope.[52]

Table IX. Anisotropy $H_{c2}(T)$ of heavy fermion superconductors.

Compound	Configuration	$-\frac{dH_{c2}}{dT}\big|_{T_c}$ (T/K)	$H_{c2}(0)$ (T)	Reference
$CeCu_2Si_2$	$H \parallel c$, $H \parallel I$	23	2.4	Assmus et al. 1984[12]
	$H \perp c$, $H \parallel I$	23	2.0	
UPt_3	$H \parallel c$, $I \parallel c$	6.3	1.7	
	$H \perp c$, $I \parallel c$	4	1.7	Chen et al. 1984[68]
UPt_3	$H \parallel c$, $H \parallel I$	5.6	1.5	Rauchschwalbe et al. 1985[70]
	$H \perp c$, $H \parallel I$	4	1.5	
UPt_3	$H \parallel c$, $I \parallel c$	77.6		
	$H \perp c$, $I \parallel c$	45.1	28.1	Shivaram et al. 1986[59]
	$H \parallel c$, $I \perp c$	77.2	21.1	
	$H \perp c$, $I \perp c$	45.9	25.9	
URu_2Si_2	$H \parallel c$	4	~ 1.5	Palstra et al. 1985[21]
	$H \perp c$	4	$\gg 3$	

E. $H_{c2}(T)$ Maximum

Fig. 47 displays a flat maximum of $H_{c2}(T)$ near 0.2 K for the field applied both parallel and perpendicular to the c-axis of an as-grown $CeCu_2Si_2$ crystal. The low temperature limit in the resistivity experiments was 0.02 K. Reentrant behavior was claimed since superconductivity occurred in an intermediate temperature range around 0.2 K. The amplitude of this maximum was $H_{c2}(0.2K) - H_{c2}(0) = 20 - 50$ mT, which was about 1% of $H_{c2}(0)$. The reentrant behavior was not observed (see Fig. 48) in polycrystalline samples of the compound $CeCu_2Si_2$.[69] This absence of reentrance might be explained that the lowest temperature of 0.2 K used in the inductive and calorimetry measurements was not low enough or the small inhomogeneities in the sample smeared out the effect. However, single crystal studies in an annealed $CeCu_2Si_2$ crystal did not show a clear maximum in H_{c2} versus T plot between T_c and < 0.04 K (see Fig. 49). Rauchschwalbe[67] concluded that the maximum in $H_{c2}(T)$ occurred mainly in stoichiometric polycrystals and in the superconducting single crystals with moderate Cu excess. In single crystals with a very large Cu excess ($CeCu_{2.8}Si_2$), this effect might also be disappeared (Rauchschwalbe 1987d). $H_{c2}(T)$ maximum is not observed in some samples of the compound $CeCu_2Si_2$. This can be simply due to the uncertainty in the determination of the upper critical field (>1 %) as a midpoint of broad transition curve.

F. Superconducting Parameters

The phase diagram of a conventional superconductor is determined by the ratio of the penetration depth to the coherence length. The penetration depth, λ, is defined as the characteristic length scale over which the magnetic field varies within the material. The coherence length, ξ, is the characteristic length to measure the variation of the order parameter. These superconducting parameters can be determined by upper critical field studies.

Generally, there are two mechanisms of pair-breaking: orbital pair breaking and Pauli limit. Both pair-breaking mechanisms lead to the destruction of the superconductivity at a critical field. Orbital pair breaking takes place at small magnetic field. Therefore, it determines the initial slope of the upper critical field at T_c. One can obtain the orbital critical field as $T \to 0$, $H_{c2}{}^*(0)$, from an extreme-type-II electron-spin effect theory of Werthamer, Helfand and Holhenberg:[81]

$$H_{c2}{}^*(0) = 0.693 \left(-\frac{dH_{c2}}{dT}\Big|_{T_c} \right) T_c \tag{1}$$

In the absence of other pair-breaking effects, $H_{c2}(0) = H_{c2}*(0)$. The Pauli limit, on the other hand, is due to the influence of the magnetic field on the electrons in the superconducting state. The Pauli paramagnetic limiting field at $T = 0$ K, H_{po}, in the absence of spin-orbit scattering for a BCS superconductor was given by the relation[82] (Clogston 1962)

$$H_{po}(0) = 1.84 \, T_c \qquad (2)$$

The superconducting coherence length at $T = 0$ K, ξ_o, can be determined from the equation

$$H_{c2}*(0) = \frac{\Phi_o}{2\pi\xi_o^2} \qquad (3)$$

where $\Phi_o = 2.07 \times 10^{-7}$ Oe cm^2 is the flux quantum.

For a type II superconductor, $\kappa = \lambda/\xi > \sqrt{2}$. The Ginzburg-Landau parameter κ_{GL} can be deduced from the ratio of the upper and lower critical fields (H_{c2} and H_{c1}) such as

$$H_{c2}/H_{c1} = \frac{2\kappa^2}{\ln\kappa} \qquad (4)$$

The penetration depth λ can be determined from a knowledge of ξ and κ obtained from experiments. This determination yields information about quasiparticle masses and about the temperature dependent magnitude of the superconducting order parameter.

The critical field study by Rauchschwalbe et al.[69] yielded some superconducting parameters that characterize the novel superconducting state of $CeCu_2Si_2$. The transition temperature, T_c, was measured by ac susceptibility and found to be 0.64 and 0.66 K in two $CeCu_2Si_2$ polycrystalline samples: the former sample was "very clean" and latter one was "less clean". The slopes at T_c were 5.8 and 16.8 T/K for two samples, respectively. By assuming some strong interaction between conduction electrons[69, 83] the superconducting parameters for the "very clean" and "less clean" samples, respectively: the BCS coherence length, $\xi_o = 19, 3$ nm; London penetration depth as $T \rightarrow 0$, $\lambda_L(0) = 200, 300$ nm; and Ginzburg-Landau parameter, $\kappa_{GL} = 22, 100$. Rauchschwalbe et al. concluded that ξ_o was comparable with the mean free path of the quasiparticles $\ell = 12$ nm in the "very clean" sample. The "less clean" sample represented the "dirty limit" (ℓ was not given for this sample in this paper). The value of $H_{c2}*(0)$ was found as 2.6 T using eq.1, which was 50 % higher than the value of 1.7 T obtained from extrapolation of measured $H_{c2}(T)$ at low temperatures. It was argued that the smaller value of measured $H_{c2}(0)$ was caused by the presence of other pair-breaking mechanisms such as Pauli paramagnetic limiting or exchange scattering from paramagnetic impurities.

The upper critical field study in single crystals of the compound UPt_3 was carried out by transport experiments.[68] The slopes of $-dH_{c2}/dT$ were 6.3 T/K and 4 T/K for H || I and H $\perp$ I, respectively. The $H_{c2}(0)$ value of 1.7 T was estimated by the extrapolation of the low temperature $H_{c2}(T)$ data to T=0. Similarly, $H_{c2}*(0)$ was estimated to be 2.2 T using eq.1. Equation 3 yielded $\xi_o = 12$ nm. For $T_c = 0.52$ K, $H_{po}(0)$ was estimated to be 0.957 T using eq.2. $H_{c2}(0)$ exceeded $H_{po}(0)$ by a factor of two, which was explained by the spin-orbit scattering.

de Visser $et\ al.$[76] estimated the superconducting parameters in UPt_3 by using the relations given by Orlando $et\ al.$[69] They found that the BCS coherence length $\xi_o = 20$ nm, the mean free path $\ell = 36$ nm, the London penetration depth $\lambda_L(0) = 360$ nm, and the Ginzburg-Landau parameter $\kappa_{GL} = 23$. Therefore, de Visser $et\ al.$ concluded that their samples were in neither the dirty limit ($\ell << \xi_o$) nor the clean limit ($\ell >> \xi_o$) and UPt_3 was indeed a type-II superconductor ($\kappa >> 1$).

Jaccard $et\ al.$[51] measured $H_{c2}(0)$ of 2.9 and 2.6 T in two polycrystalline samples of UPt_3. The superconducting coherence length was calculated by using eq.3 and substituting orbital critical field $H_{c2}*(0)$ by the measured $H_{c2}(0)$: $x_o = 11.4$ nm. The electronic mean free path ℓ was 36 nm. These authors claimed that the UPt_3 might be the first clean heavy fermion superconductor. Jaccard $et\ al.$ also measured the lower critical field $H_{c1} = 60$ Oe at 100 mK and calculated a k value of 26 by using eq.4. Maple $et\ al.$[73] analyzed the upper critical field data for a UBe_{13} crystal in a dirty-limit approximation. The slope $-dH_{c2}/dT$ was 42 T/K and $H_{c2}(50\ mK) = 9$ T determined from experiments. The zero-temperature orbital critical field $H_{c2}*(0)$, given by eq.1, was 25 T. Using this $H_{c2}*(0)$ value, authors estimated the coherence length at T=0, ξ_o, by means of the dirty limit expression:[84]

$$H_{c2}*(0) = \frac{\Phi_o}{4.54\xi_o\ell_{tr}}\left(\ell_{tr} << \xi_o\right) \qquad (5)$$

where ℓ_{tr} is the transport mean-free path and can be calculated by

$$\ell_{tr} = \frac{1.27 \times 10^4}{\rho(Z/\Omega)^{2/3}} \qquad (6)$$

where ρ is the resistivity near T_c in units of Ωcm, Z is the number of conduction electrons per unit cell and Ω is the unit cell volume in cm^3. Given the lattice parameter of UBe$_{13}$, 1.0254 nm, and Z=24, eqs.5 and 4 yield ℓ_{tr} =1.29 nm and ξ_o = 14.2 nm. The paramagnetic limiting field at T = 0 K, H$_{po}$ = 1.58 T, in the absence of spin-orbit scattering was estimated for T$_c$ = 0.857 K using eq.2. The H$_{c2}$(0) exceeded H$_{po}$(0) by 16 times. Maple et $al.$ used the same argument that the additional spin-orbit scattering was presumably response for it.

Kadowaki et $al.$[78] analyzed the upper critical field data of the compound UPt$_3$ by using the equation for strong coupling superconductivity. The superconducting coherence length was calculated to be ξ_o = 18.7 nm. Another publication on a UPt$_3$ crystal by Shivaram et $al.$[75] showed ℓ = 180 nm deduced from ρ_o and ξ_o = 11 nm obtained from the slopes at T$_c$. The anisotropic mass ratio $\varepsilon^2 = m_\perp/m_\parallel = 1.7$ was obtained from the angular dependence of the upper critical field measurements.

The superconducting parameters of UPd$_2$Al$_3$ and UNi$_2$Al$_3$ were estimated by Geibel et $al.$[6] and Steglich et $al.$[79] using the strong coupling theory. For the compound UPd$_2$Al$_3$, Geibel et $al.$ found the transport mean-free path ℓ_{tr} = 72 nm, the BCS coherence length ξ_o = 8.5 nm, the magnetic field penetration depth λ = 400 nm and the Ginzburg-Landau parameter κ = 50. In comparison, the penetration depth of 625 $\pm$ 125 nm was obtained in the compound UPd$_2$Al$_3$ from muon-spin-rotation (μ^+SR) spectroscopy.[85] Feyerherm et $al.$[86] obtained the London penetration depth from muon-spin-rotation measurements: λ = 480 $\pm$ 50 nm and 450 $\pm$ 50 nm for the magnetic field applied parallel and perpendicular to c-axis, respectively. In UNi$_2$Al$_3$, ℓ_{tr} = 47 nm, ξ_o = 24 nm, λ = 330 nm and κ = 14 were found by Steglich et $al.$[79] Since $\ell_{tr}/\xi_o > 1$, both compounds might belong to the superconductors with unconventional order parameters.

VI. IMPURITY EFFECT

A. Introduction

Conventional superconductors with an isotropic gap do not significantly alter their superconducting transition temperature or change their thermodynamic properties with the introduction of nonmagnetic impurities.[87] Only magnetic impurities act as depairing centers. The depairing effect due to nonmagnetic impurities on unconventional states was studied by Balian and Werthamer.[88] In addition to the suppression of T$_c$, impurities change the low temperature properties, such as specific heat, ultrasonic attenuation, NMR relaxation rate.

92

The influence of impurities on heavy fermion superconductors is particularly intriguing. Heavy fermion superconductivity is very sensitive to any kind of chemical impurities. It appears to be unimportant whether the impurity ions carry a magnetic moment or not.[89] The impurity or doping effect was studied in the case of $CeCu_2Si_2$.[90, 35, 91, 92, 93]

The investigation of impurity effects in the compounds $Ce_{1-x}M_xCu_{2.2}Si_2$ (M = La, Y, Sc, Lu, Th and Gd) shows that the specific heat discontinuity at T_c was suppressed more strongly than T_c and T_c-depression depended upon the particular dopants.[90, 35, 91-93] These results raised the possibility that the "Kondo-hole" picture may be applicable to impurity scattering in the superconducting state of heavy fermion systems.[97, 98, 99, 100, 101, 102]

A peculiar feature in the system $U_{1-x}Th_xBe_{13}$ is a double-peak structure discovered in the specific heat measurements. This result revealed a complicated phase diagram of $U_{1-x}Th_xBe_{13}$ that showed at least two and possibly more phase transitions. The lower temperature transition was argued to be associated with either an antiferromagnetic transition or a transition to another superconducting state.

B. Chemical Substitution

1. $Ce_{1-x}M_xCu_{2.2}Si_2$, M = La, Y, Sc, Lu and Th

Specific heat experiments were performed on the compound $CeCu_2Si_2$ doped with non-magnetic impurities of La, Y, Sc, Lu and Th substituting for Ce. The superconducting transition temperature of $Ce_{1-x}M_xCu_{2.2}Si_2$ as function of concentration x is plotted in Fig. 64 where M = La, Y and Sc. T_c decreased monotonically with increasing impurity concentration.[90] The rate of the T_c-depression upon concentration depended on the nature of the doping atoms. Fig. 65 shows the specific heat measurements on $Ce_{1-x}M_xCu_{2.2}Si_2$ where $M_x = Y_{0.03}$, $La_{0.03}$ and $La_{0.09}$ with undoped $CeCu_{2.2}Si_2$ for comparison. T_c decreased with increasing dopant concentration of La and Y, which was consistent with earlier observation.[90] The specific heat discontinuity, ΔC, was suppressed more strongly than the transition temperature T_c. A substantial linear temperature term of $C_s(T)$ was observed with doped samples. The coefficient of linear term γ_s was larger for the Y than for the La doping. These results were interpreted as support for the assumption that the linear $\gamma_s T$ term might exist due to impurities in the samples.[19] Fig. 66 displays $\Delta C(x)/\Delta C(0)$ versus $T_c(x)/T_c(0)$ for $Ce_{1-x}M_xCu_{2.2}Si_2$ with M = La, Y and Lu. A similar plot of the reduced specific heat discontinuity versus the reduced transition temperature is shown in Fig. 67 with M = La, Y and Th. Figs. 68 and 69 present the results of the T_c dependence on the impurity concentration.[35]

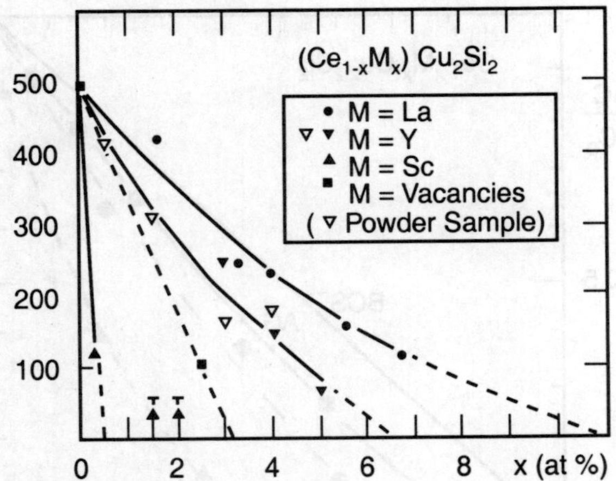

Fig.64. Transition temperature of $(Ce_{1-x} M_x)Cu_2Si_2$ as function of concentration x (vacancies). Critical concentrations $x_{cr} = x(T_c \to 0)$ deviate from those reported by Bredl et al.[11] This is caused by the additional heat treatment (from Spille et al.).[90]

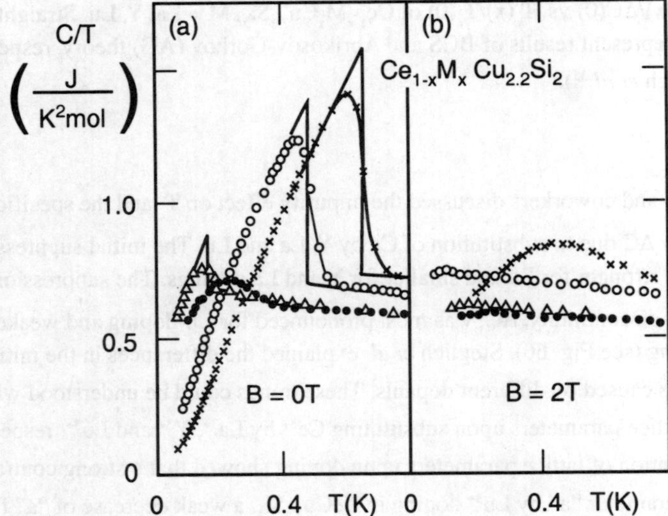

Fig. 65. Specific heat of $CeCu_{2.2}Si_2$ as well as of the quasibinary systems $Ce_{1-x}M_xCu_{2.2}Si_2$ with M = La and Y at B = 0 (a) and 2 T (b). For the clarity, units are per mole of the actual alloy, rather than per mole of Ce. Lines through the data points are guides to the eye; thin solid lines mark idealized specific heat jumps. x = 0: x; x = 0.03: O (La), ● (Y); x = 0.09: △ (La) (from Steglich et al.).[103]

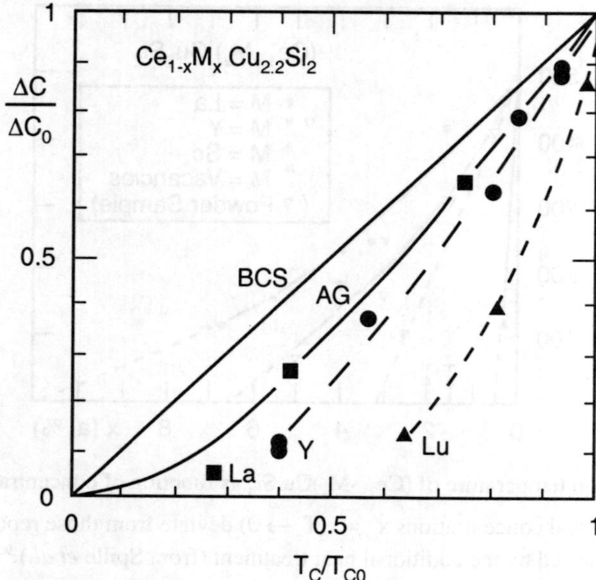

Fig. 66. $\Delta C(x)/\Delta C(0)$ vs. $T_c(x)/T_c(0)$ of Ce$_{1-x}$M$_x$Cu$_{2.2}$Si$_2$, M = La, Y, Lu. Straight line and solid curve represent results of BCS and Abrikosov-Gorkov (AG) theory, respectively (from Steglich *et al.*[35]).

Steglich and coworkers discussed the impurity effect on T_c and the specific heat discontinuity ΔC due to substitution of Ce by Y, La and Lu. The initial suppression of T_c, $-dT_c/dx$, was stronger for Lu and smaller for Y and La dopings. The suppression of the specific heat discontinuity, ΔC, was most pronounced for Lu-doping and weaker for Y and La doping (see Fig. 66). Steglich *et al.* explained the differences in the initial T_c depression as caused by different dopants. These trends could be understood with the change of lattice parameters upon substituting Ce^{3+} by La^{3+}, Y^{3+} and Lu^{3+}, respectively. The investigation of lattice parameters upon doping showed that a strong contraction of the lattice parameter "a" by Lu^{3+} doping in CeCu$_{2.2}$Si$_2$, a weak decrease of "a" for Y^{3+}, but a weak expansion in the case of La^{3+} doping. Therefore, the T_c depression was increased in the order from Lu^{3+} to Y^{3+} and La^{3+} (see Fig. 66). Ahlheim *et al.*[92] indicated the same initial drop of the transition temperature due to La, Y and Th impurities (see Fig. 69). At $x > 5$ at.%, $T_c(x)$ exhibited a positive curvature for Y-doping, a negative curvature for La and Th doping. This result was expected by the size mismatch between Ce and the respective dopant: the lattice parameters (a) were 0.4154, 0.3967 and 0.4104 nm for

LaCu$_2$Si$_2$, YCu$_2$Si$_2$ and ThCu$_2$Si$_2$, respectively, compared with 0.4105 nm in the compound CeCu$_2$Si$_2$.

Some kind of "pair-breaking" by nonmagnetic dopants was suggested by Steglich *et al.*,[35, 91] Ahlheim *et al.*,[92-93] and Grewe and Steglich.[52] However, the absolute strength of this pair-breaking action was unexpectedly weak. For example, a 1 at.% concentration of La, Y and Th doping did not change T$_c$ significantly (see Fig. 69). As shown in figs. 66 and 67, the experimental results were considerably below the straight line representing the behavior of the BCS superconductors containing paramagnetic impurities, and less pronounced than the solid curve indicating the Abrikosov-Gor'kov theory[104-105] for stable impurity moments in ordinary superconductors.[106] This observation might lead to an argument that the compound CeCu$_2$Si$_2$ was a "strongly anisotropic superconductor" exhibiting either an unconventional or a conventionally anisotropic order parameter.

Similar behavior as shown in Fig. 66 was observed in Kondo superconductors associated with moment instabilities, which was successfully described by Müller-Hartmann and Zittartz (MHZ).[107] Ce$_{1-x}$M$_x$Cu$_{2.2}$Si$_2$ with M = La, Y, and Lu could be interpreted by the "Kondo-hole" picture.[97, 98, 99, 100, 101, 102] The superconducting host

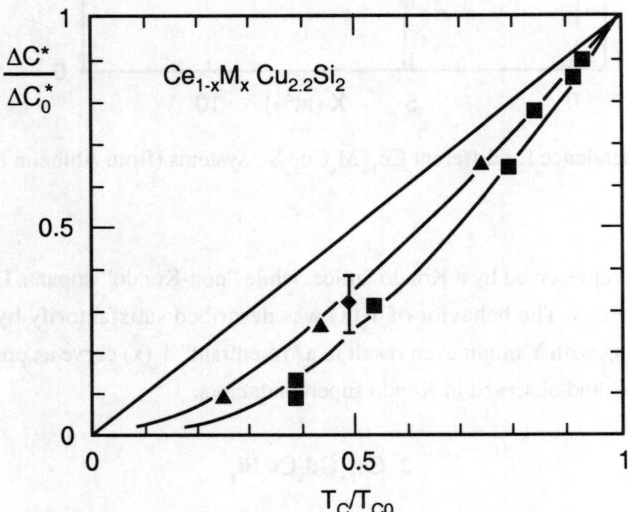

Fig. 67. Reduced specific heat jump, $\Delta C^*/\Delta C_o^* = (\Delta C (x) /\gamma (x,T_c)/(\Delta C (0)/\gamma(x=0,T_c))$ vs. reduced transition temperature $T_c(x)/T_c(0)$ for the system Ce$_{1-x}$M$_x$Cu$_{2.2}$Si$_2$, with M = Y ($\blacksquare$), M = La ($\blacktriangle$) and M = Th ($\blacklozenge$). The diagonal line shows the result for the BCS-theory, the other lines are guides to the eye. For the point of the thoriated sample an error-bar is drawn (from Ahlheim *et al.*).[92]

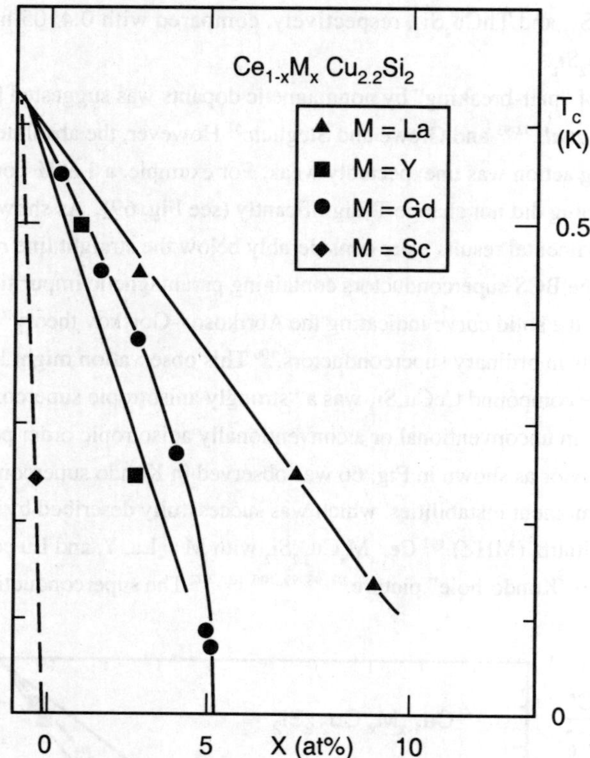

Fig. 68. T$_c$ dependence for different Ce$_{1-x}$M$_x$Cu$_{2.2}$Si$_2$ systems (from Ahlheim *et al.*).[93]

CeCu$_{2.2}$Si$_2$ was represented by a Kondo lattice, while "non-Kondo" dopants La, Y, and Lu acted as impurities. The behavior of T$_c$(x) was described satisfactorily by the MHZ theory.[35] Doping with Y might even result in a "re-entrant" T$_c$(x) curve as predicted by the MHZ theory and observed in Kondo superconductors.

2. Ce$_{1-x}$Gd$_x$Cu$_2$Si$_2$

Spille *et al.*[90] performed substitutions in CeCu$_2$Si$_2$ and found that Gd substituted for Ce depressed T$_c$ twice as fast as nonmagnetic substitutions. This result may be used as an argument against p-wave superconductivity in CeCu$_2$Si$_2$, since nonmagnetic substitutions are thought to be equally destructive to T$_c$ in a p-wave superconductor. As reflected in the depression of T$_c$ in Fig. 68, Gd impurities in CeCu$_{2.2}$Si$_2$ also cause a precipitous drop

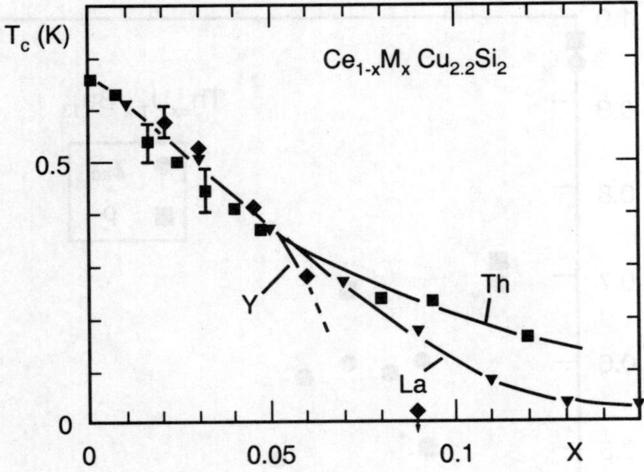

Fig. 69. Transition temperature T_c vs. concentration x of $Ce_{1-x}M_xCu_{2.2}Si_2$ with M = Y ($\blacklozenge$), M = La ($\blacktriangledown$) and M = Th ($\blacksquare$). Lines drawn are guides to the eye (from Ahlheim *et al.*).[92]

in ΔC. Ahlheim *et al.*[93] also indicated that the T_c dependence for $Ce_{1-x}Gd_xCu_{2.2}Si_2$ was in excellent agreement with the Abrikosov-Gor'kov theory. Ahlheim *et al.* used the strong ΔC depression and excellent fit to Abrikosov-Gor'kov theory as an evidence for additional paramagnetic pair breaking in the compound $CeCu_{2.2}Si_2$ due to the magnetic impurity of Gd, in contrast to diamagnetic pair breaking via Kondo holes for nonmagnetic impurities.

3. $U_{1-x}M_xBe_{13}$ M = Zr, Lu, Y, Th, Sc, Ce, La and Ba

The results of the nonmagnetic impurity studies of $U_{1-x}M_xBe_{13}$, M = Sc, Lu, Ce, Th, La, by Smith *et al.*[94, 95] are shown in Table X and Figs.70-73. From Table X, the lattice parameters for Ce and La were close to that of Th, so the T_c depressions were similar to those for Th doping. On the other hand, Sc and Ba, whose lattice parameters were the smallest and largest, respectively, resulted rather small depressions. A T_c versus x plot for $U_{1-x}Th_xBe_{13}$ (see Fig.70) indicated that the T_c determined by the resistive measurements were lower than the inductive ones. This was opposite to the common observations in ordinary superconductors where resistively determined T_c's are higher and the transitions are sharper than those obtained from magnetic measurements.

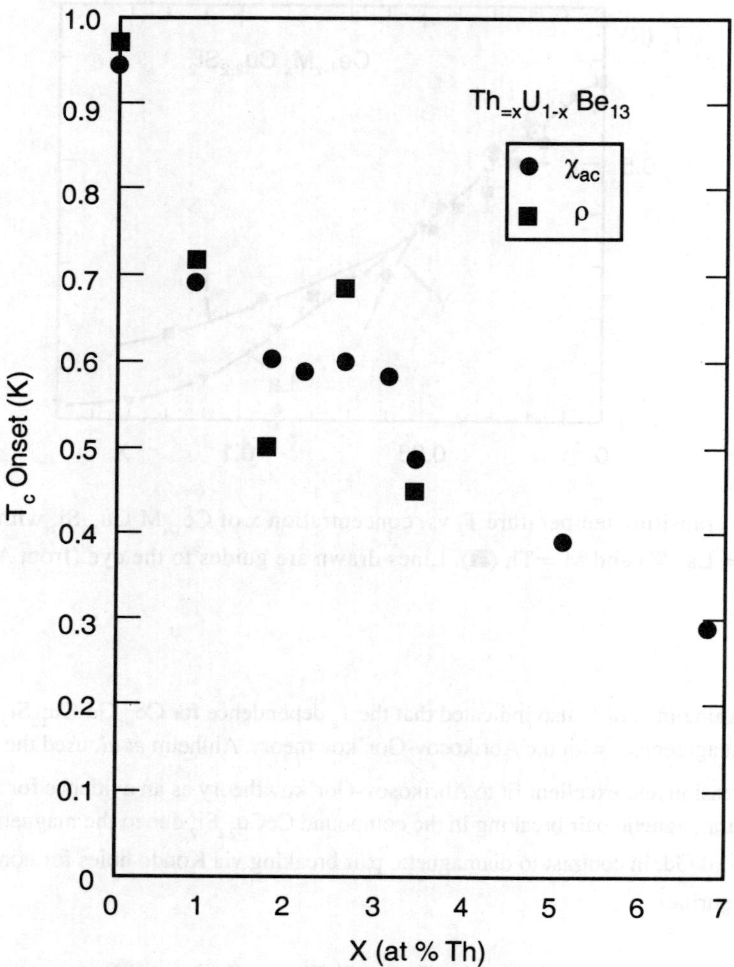

Fig.70. Superconducting transition temperatures of Th$_x$U$_{1-x}$Be$_{13}$ by ac susceptibility and resistivity measurements (from Smith *et al.*).[95]

This observation implied that lattice imperfections played an important role. The sensitivity to lattice imperfection and the almost linear x dependence of T$_c$ as shown in Fig. 70 were consistent with the prediction of p-wave superconductivity.[95, 108] Fig.71 shows the T$_c$ onset as deduced by ac susceptibility and heat capacity for Th.

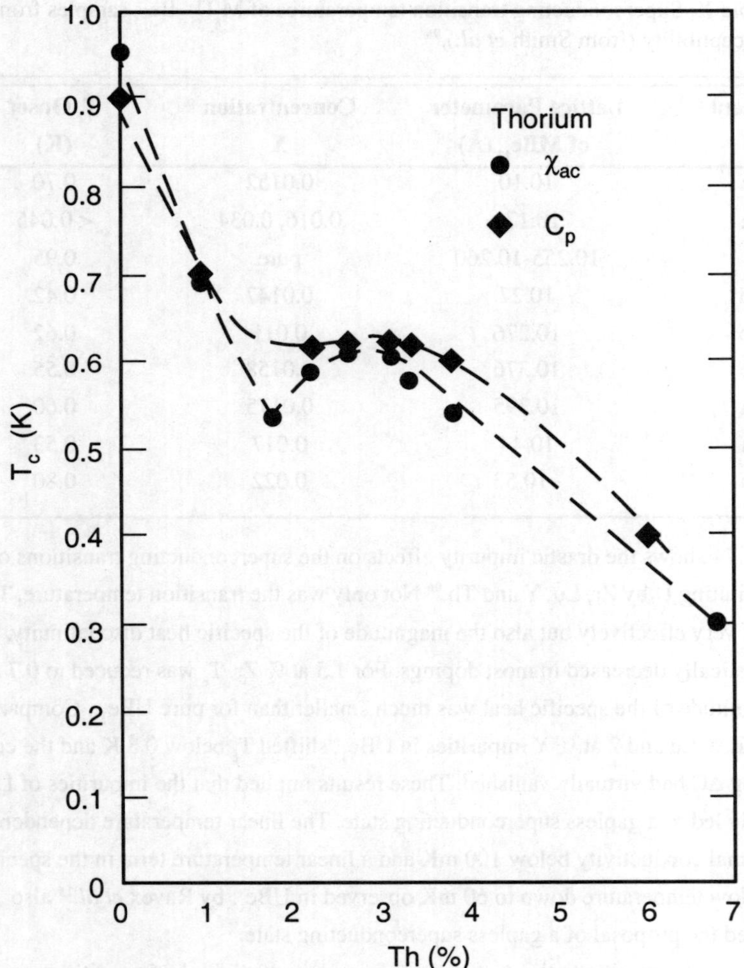

Fig. 71. Superconducting transition temperature T_c vs. Th concentration in $U_{1-x}Th_xBe_{13}$, as measured by specific beat (C_p) and ac magnetic susceptibility (χ_{ac}) measurements (from Smith *et al.*).[94]

However, no linear dependence of $T_c(x)$ could be seen, which was contradictory to the earlier results.[94] The linear x dependencies of T_c were observed for Sc, Ce and Lu impurities (see Fig.72), while no linear form appeared in Fig.73 for Y, Zr and La. Among all impurities, Lu was the most harmful one.

Table X. Superconducting transition temperatures of $M_x U_{1-x} Be_{13}$ samples from ac susceptibility (from Smith *et al*.).[95]

Element M	Lattice Parameter of MBe_{13} (Å)	Concentration X	T_c Onset (K)
Sc	10.10	0.0152	0.70
Lu	10.173	0.016, 0.034	< 0.045
U	10.255-10.260	pure	0.95
Gd	10.27	0.0147	0.42
Np	10.276	0.011	0.62
Ce	10.376	0.0158	0.55
Th	10.395	0.0175	0.60
La	10.44	0.017	0.53
Ba	~10.53	0.022	0.80

Fig.74 shows the drastic impurity effects on the superconducting transitions of UBe_{13} in substituting U by Zr, Lu, Y and Th.[96] Not only was the transition temperature, T_c, reduced very effectively but also the magnitude of the specific heat discontinuity, ΔC, was drastically decreased in most dopings. For 1.5 at.% Zr, T_c was reduced to 0.7 K and the magnitude of the specific heat was much smaller than for pure UBe_{13}. Compared with Zr, 1.6 at.% Lu and 2 at.% Y impurities in UBe_{13} shifted T_c below 0.3 K and the corresponding ΔC had virtually vanished. These results implied that the impurities of Lu and Y rapidly led to a gapless superconducting state. The linear temperature dependence of the thermal conductivity below 100 mK and a linear temperature term in the specific heat at very low temperature down to 60 mK observed in UBe_{13} by Ravex *et al*.[15] also supported the proposal of a gapless superconducting state.

Equally interesting to the destruction of superconductivity by impurities was the study of U replaced by Th in small amounts in the compound UBe_{13}.[109] As displayed in Fig.75, a sizable decrease of T_c to 0.6 K and a broading of the transition were observed for x= 0.0089 and 0.0216 of Th compared with pure UBe_{13}. For x = 0.026, 0.0308, 0.0331 and 0.0378, two specific heat anomalies around 0.4 and 0.6 K were discovered. The magnitude of the low temperature anomaly increased with increasing x and reached maximum at x = 0.0378. The double-peak feature disappeared at x = 0.0603 where T_c had dropped below 0.4 K and ΔC reduced in magnitude significantly.

The phase diagram of T_c versus x is plotted in Fig. 76. T_c is first reduced with increasing x in the same manner as observed for other impurities. In the concentration range 0.018 - 0.045, a double-peak structure was displayed as upper and lower transition

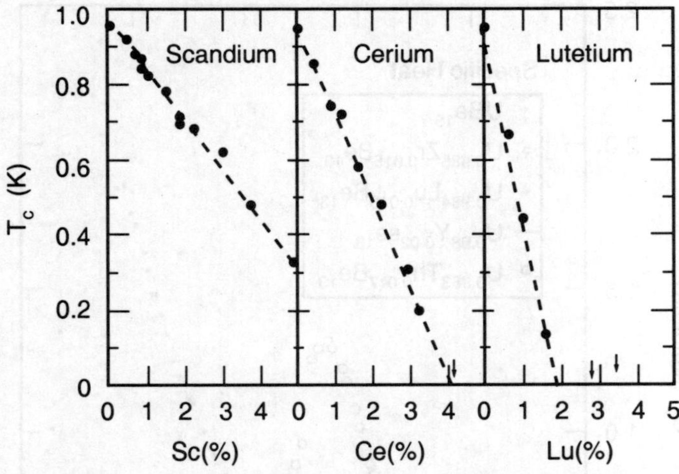

Fig. 72. Superconducting transition temperature T_c vs. impurity concentration in UBe_{13} containing Sc, Ce and Lu impurities (from Smith *et al.*).[94]

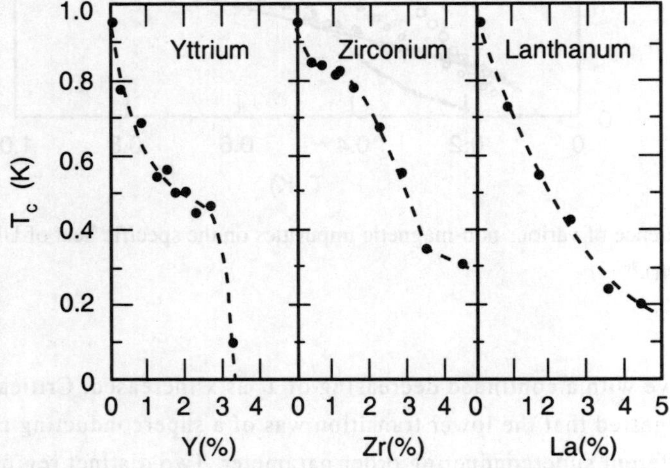

Fig. 73. Superconducting transition temperature T_c vs. impurity concentration in UBe_{13} containing Y, Zr and La impurities (from Smith *et al.*).[94]

lines. The first transition temperature with a higher T_c rose with increasing x and passed over a maximum around 0.03 and dropped again with further increasing of x. The lower transition within the superconducting state had a transition temperature of 0.4 K, independent of x. For x exceeding 0.045, the double transitions disappeared and T_c rejoined the

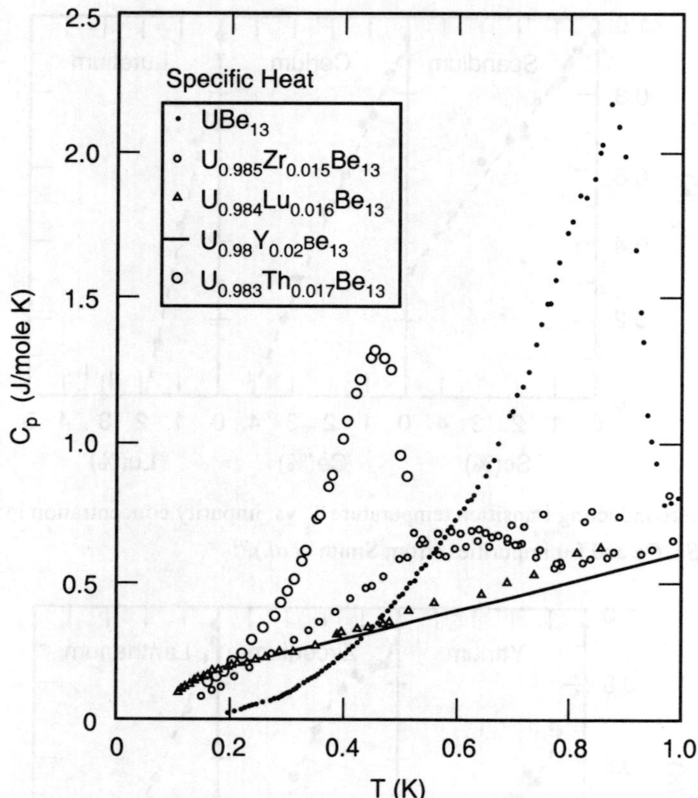

Fig. 74. Influence of various non-magnetic impurities on the specific heat of UBe$_{13}$ below 1 K (from Ott).[96]

normal curve with a continued decreasing of T_c as x increased. Critical field studies suggested that the lower transition was of a superconducting nature and led to a different superconducting order parameter. Two distinct regions of superconductivity were also claimed by the pressure study on the supercon-ducting transition temperature in $(U_{1-x}Th_x)Be_{13}$.[111] Ultrasonic attenuation and sound velocity experiments indicated a large attenuation peak and a sound velocity minimum at the lower transition (but no indication in the attenuation data for the first transition). Since both the magnitude and the shape were typical for an antiferromagnetic transition, Batlogg et al.[112] claimed that the lower transition in $U_{1-x}Th_xBe_{13}$ was a magnetic transition within the (anisotro-pic) superconducting state.

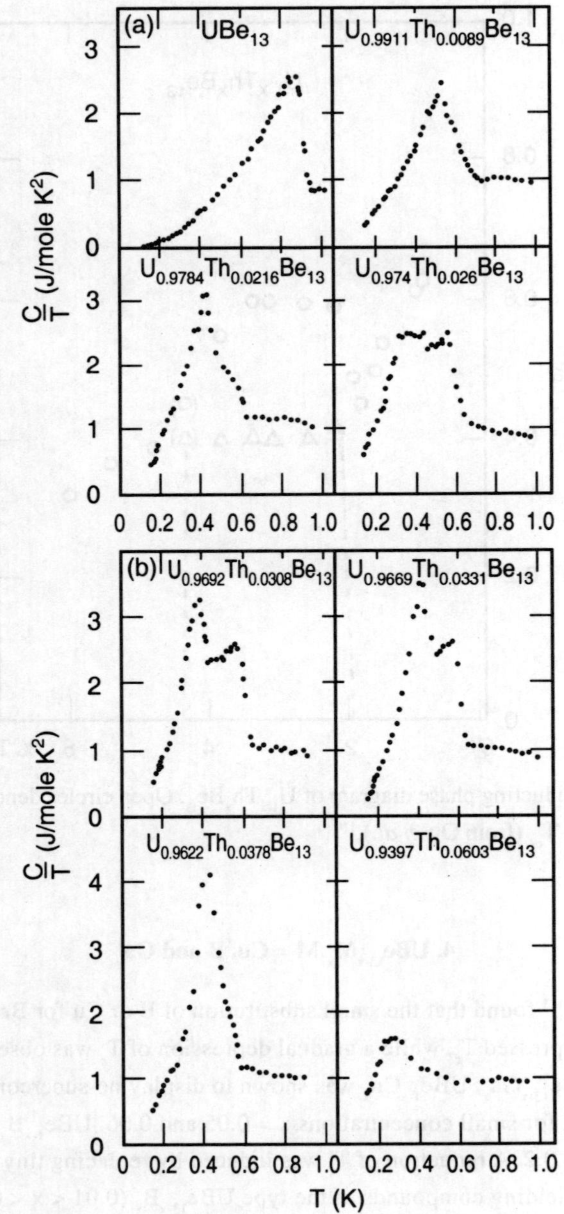

Fig. 75. Specific heat of various $U_{1-x}Th_xBe_{13}$ compounds, plotted as C_p/T vs. T for temperatures between 0.15 and 1 K (from Ott *et al.*).[109]

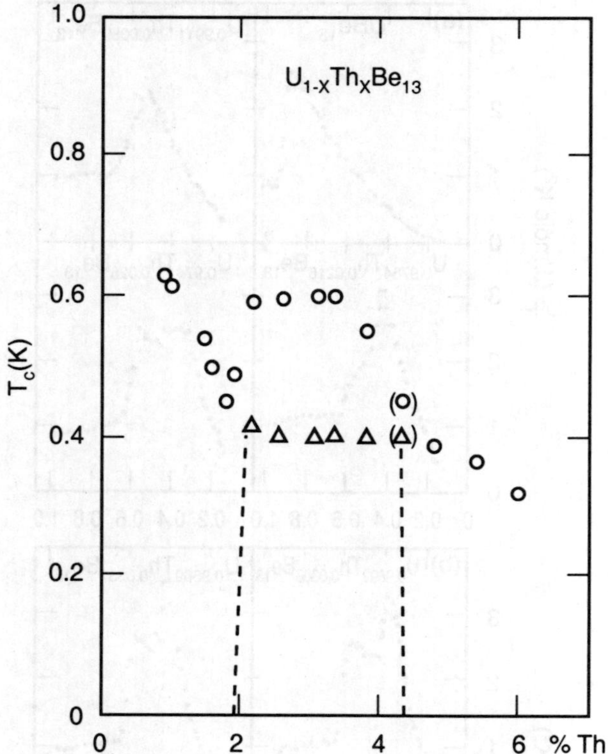

Fig. 76. Superconducting phase diagram of U$_{1-x}$Th$_x$Be$_{13}$. Open circles denote T$_{c1}$, open triangles indicate T$_{c2}$ (from Ott *et al.*).[110]

4. UBe$_{13-x}$M$_x$ M = Cu, B and Ga

Giorgi *et al.*[113] found that the small substitution of B or Cu for Be in UBe$_{13}$ dramatically suppressed T$_c$, while a gradual depression of T$_c$ was observed for Ga impurities in UBe$_{13-x}$Ga$_x$. UBe$_{1-x}$Cu$_x$ was shown to display no superconductivity down to 0.015 K for small concentrations x = 0.05 and 0.06. UBe$_{1-x}$B$_x$ had T$_c$ < 0.02 K for x =0.1 and 0.2. A reduction of T$_c$ was induced by replacing tiny amounts of Be in UBe$_{13}$ by B, yielding compounds of the type UBe$_{13-x}$B$_x$ (0.01 < x < 0.045).[89] The ΔC discontinuity at T$_c$ was surprisingly enhanced in magnitude with B impurities, as displayed in Fig. 77. This result supported the assumption that UBe$_{13}$ was a p-wave superconductor under the influence of impurity scattering.[96, 114]

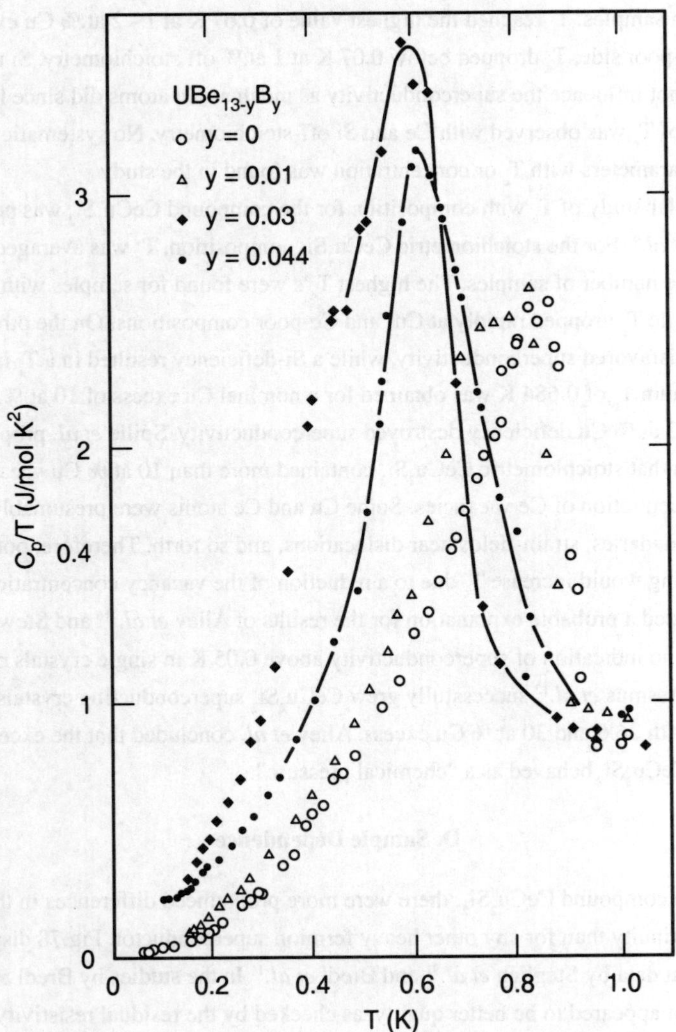

Fig. 77. C$_p$/T vs. T for superconducting UBe$_{13-y}$B$_y$ between 0.05 and I K. The solid lines are meant to guide the eye (from Ott *et al*).[89]

C. Off-Stoichiometry

Ishikawa *et al.*[115] investigated fifteen off-stoichiometric polycrystalline samples of CeCu$_2$Si$_2$. The variation of T$_c$ was particularly pronounced for the series CeCu$_{2+x}$Si$_2$. The higher superconducting transition temperatures and sharper transitions were found in copper-rich samples. T_c reached the highest value of 0.67 K at 1 - 2 at.% Cu excess. On the copper-poor side, T_c dropped below 0.07 K at 1 at.% off stoichiometry. Si and Cu atoms did not influence the superconductivity as much as Cu atoms did since less variations of T_c was observed with Ce and Si off-stoichiometry. No systematic correlation of lattice parameters with T_c or concentration was found in the study.

A similar study of T_c with composition for the compound $CeCu_2Si_2$ was performed by Spille *et al.*[90] For the stoichiometric $CeCu_2Si_2$ composition, T_c was averaged to 0.5 K from a large number of samples. The highest T_c's were found for samples with excess Cu and Ce, while T_c dropped rapidly at Cu- and Ce-poor compositions. On the other hand, excess Si disfavored superconductivity, while a Si-deficiency resulted in a T_c increase. The maximum T_c of 0.684 K was obtained for a nominal Cu excess of 10 at.%. In contrast, a 3 at.% Cu deficiency destroyed superconductivity. Spille *et al.* proposed an assumption that stoichiometric $CeCu_2Si_2$ contained more than 10 at.% Cu vacancies and a 3 at.% concentration of Ce vacancies. Some Cu and Ce atoms were presumably dispersed in grain-boundaries, strain-fields near dislocations, and so forth. Therefore, both aging and annealing would increase T_c due to a reduction of the vacancy concentration. This work provided a probable explanation for the results of Aliev *et al.*[116] and Stewart *et al.*[117] who found no indication of superconductivity above 0.05 K in single crystals of $CeCu_2Si_2$. Assmus *et al.*[12] successfully grew $CeCu_2Si_2$ superconducting crystals ($T_c = 0.65$ - 0.69 K) with 26.6 and 30 at.% Cu excess. Aliev *et al.* concluded that the excess of copper in $CeCu_2Si_2$ behaved as a "chemical pressure".

D. Sample Dependence

For the compound $CeCu_2Si_2$, there were more pronounced differences in the specific heat discontinuity than for any other heavy fermion superconductor. Fig.78 displays the specific heat data by Steglich *et al.*[91] and Bredl *et al.*[11] In the studies by Bredl *et al.*, some samples that appeared to be better quality, as checked by the residual resistivity and X-ray diffractometry, displayed no specific heat jump at all. A possible interpretation was due to gapless superconductivity:[52] the formation of a coherent and correlated Fermi-liquid state could suppress an excitation gap in the superconducting state.

The sample dependence of the specific heat discontinuity for UPt_3 as illustrated in Fig. 79 indicates the importance of lattice imperfections such as the impurities in the starting material of uranium.[67] The purer sample, which was prepared from the purer starting material, showed a higher value of T_c onset and higher and more pronounced specific heat discontinuity.[17] In the less pure sample, pair breaking apparently reduced both T_c and ΔC.[16] The data showed another interesting feature: the purer sample had a

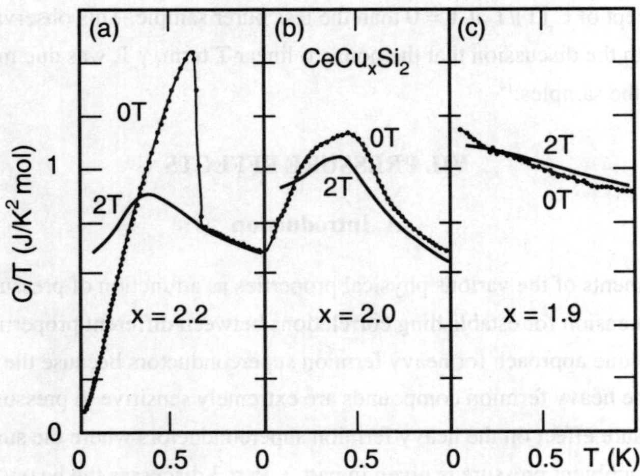

Fig. 78. Specific heat of CeCu$_x$Si$_2$ as C/T plotted against T at B= 0 and 2 T (Steglich *et al.*).[91] Data points in (b) refer to CeCu$_2$Si$_2$ sample no.10 (Bredl *et al.*).[11] Thin solid lines through B = 0 data points are guides to the eye (from Grewe and Steglich).[52]

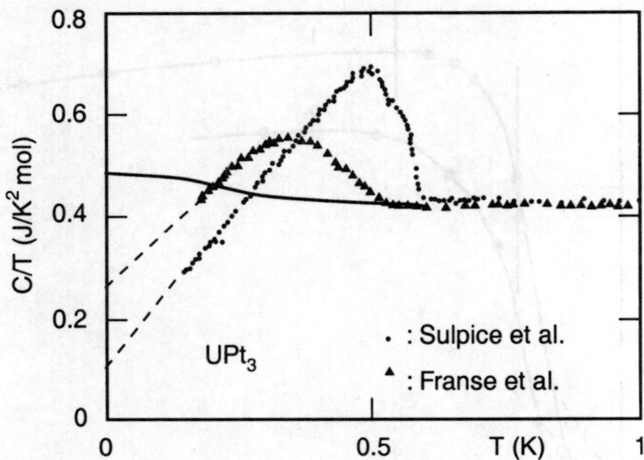

Fig. 79. Specific heat of two UPt$_3$ samples as C/T plotted against T (Franse *et al.* 1985 and Sulpice *et al.* 1986). Sample with higher T_c prepared from purer U "starting" material. Dashed lines are extrapolations of data points in the superconducting state. Solid line is a schematic extrapolation of n-state data to meet the entropy balance (from Grewe and Steglich).[52]

smaller intercept of $C_s(T)/T$ at $T = 0$ than the less purer sample. This observation was consistent with the discussion that the nonzero linear T term, $\gamma_s T$, was due mainly to impurities in the samples.[19]

VII. PRESSURE EFFECTS

A. Introduction

Measurements of the various physical properties as a function of pressure provide an additional dimension for establishing correlations between different properties. They also provide an unique approach for heavy fermion superconductors because the 4f and 5f electrons in the heavy fermion compounds are extremely sensitive to pressure.

The pressure effect on the heavy fermion superconductors where the superconductivity appears at ambient pressure is given in part 2. Part 3 discusses the heavy fermion compounds where the superconductivity is induced by the application of the high pressures.

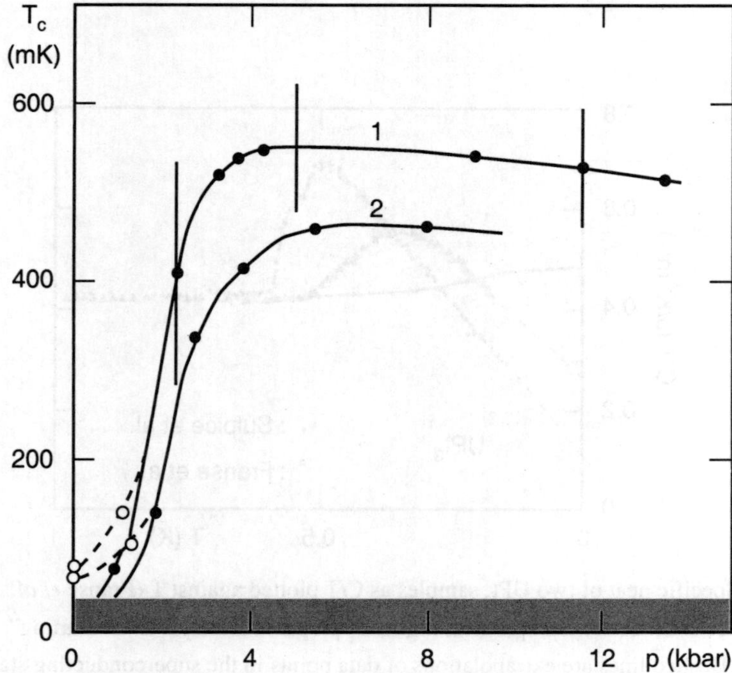

Fig. 80. The superconducting transition temperature T_c versus pressure for two $CeCu_2Si_2$ single crystals (from Aliev et al.).[125]

B. Pressure Dependent Superconducting Transition Temperature

A complicated pressure dependence of the superconducting transition temperature is observed in the compound $CeCu_2Si_2$. One interesting result is an increase of T_c with increasing pressure at low pressures up to 30 - 40 kbar. The explanation was given in terms of an enhanced Kondo coupling in $CeCu_2Si_2$ under pressure.[118-119] In contrast, the uranium heavy fermion superconductors show T_c depressions for any given range of pressures (Table XI).

Table XI. Pressure dependence of superconducting transition temperature in heavy fermion superconductors.

Compound	Sample	dT_c/dP (mK/kbar)	Reference
$CeCu_2Si_2$	crystal	187*	Aliev *et al.* 1983[120]
	polycrystal	2*	Thomas *et al.* 1993[124]
UBe_{13}	polycrystal	- 16	Chen *et al.* 1984[68]
	crystal	- 12	Chen *et al.* 1984[68]
UPt_3	crystal	- 12.6	Willis *et al.* 1984[74]
		- 15.7	Brodale *et al.* 1986[122]
	crystal whiskers	-11.3, -13.2	Behnia *et al.* 1990[46]
URu_2Si_2	polycrystal	- 97	Maple *et al.* 1986[66]
	polycrystal	- 95	McElfresh *et al.* 1987[123]
	polycrystal	- 56.2	Fisher *et al.* 1990[23]
	crystal	-35, 25	Bakker *et al.* 1992[121]
UPd_2Al_3	polycrystal	- 7	Caspary *et al.* 1993[25]

* the initial rate of dT_c/dP at low pressures.

1. $CeCu_2Si_2$

Aliev *et al.*[120, 125] studied the high pressure effect on the superconducting properties of the single crystalline samples $CeCu_2Si_2$. An increase of the superconducting transition temperature was observed in the pressure interval $0 < P < 3$ kbar (see Fig. 80). For $P > 3$ kbar, T_c gradually decreased with increasing pressure. The pressure dependence of dH_{c2}/dT is shown in Fig. 81 for the magnetic field applied both parallel and perpendicular to the direction of the current. The dH_{c2}/dT data initially increased under pressure, reached a maximum at 3 kbar and reduced at higher pressures up to 12 kbar. Aliev *et al.* (1984) explained that the increase of T_c and dH_{c2}/dT at low pressures below 3 kbar was due to the enhanced Kondo coupling in $CeCu_2Si_2$ under pressure. At $P > 3$ kbar, a continuous

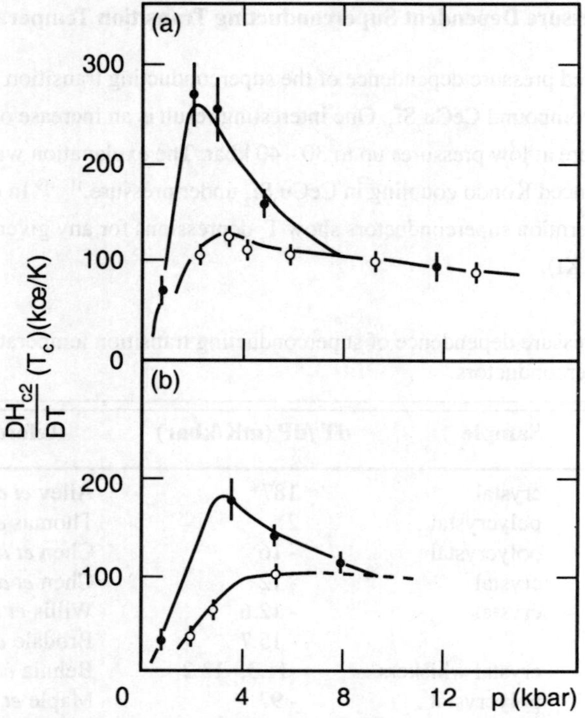

Fig. 81. The pressure dependence of the derivative of the upper critical field $dH_{c2}/dT(T=T_c)$ for two $CeCu_2Si_2$ single crystals in (a) and (b) at (O) $H \perp j$ and ($\bullet$) $H \parallel j$ (from Aliev *et al.*).[125]

transition from the Kondo lattice to the intermediate valance state resulted in a decrease of T_c and dH_{c2}/dT .

The high pressure electrical resistivity was investigated in the compound $CeCu_2Si_2$ by Bellarbi *et al.*[168] by two experimental setups. In the first one, the pressure was generated up to 40 kbar and the lowest accessible temperature was 50 mK. A second device could reach a maximum pressure of 200 kbar with a low temperature limit of 1.2 K. As displayed in Fig. 82, three pressure regions can be defined. In region I, P < 20 kbar and T_c rose slowly with increasing pressure. Region II was in the pressure range of 20 - 40 kbar, where T_c increased drastically with increasing pressure accompanied with broadened transitions. T_c reached a maximum value of 2 K at 40 kbar. At high pressures above 40 kbar (Region III), T_c reduced steadily up to 80 kbar. For P > 80 kbar, no T_c could be found above 1.2 K.

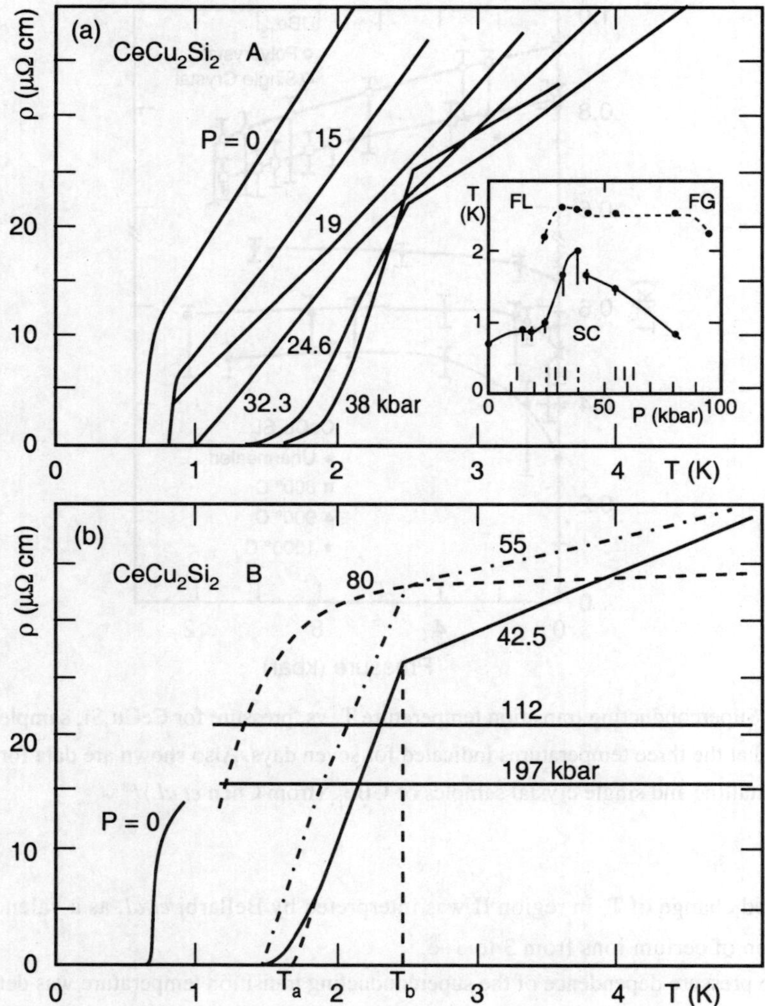

Fig. 82. (a) Low temperature resistivity and superconducting transition of $CeCu_2Si_2$ sample A as a function of pressure. (b) Low-temperature resistivity and superconducting transitions of $CeCu_2Si_2$ sample B at very high pressures. Inset of (a): Pressure dependence of the transition temperature for two samples. The apparent discontinuity at 40 kbar is a mismatch between the two different samples (from Bellarbi et al.).[118]

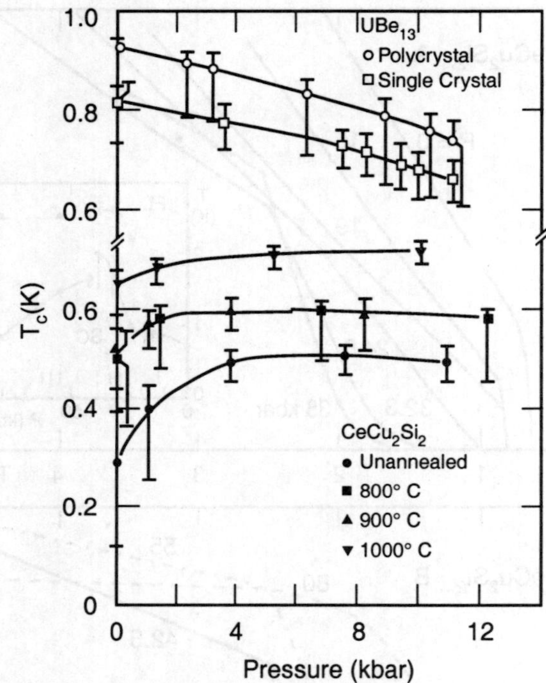

Fig. 83. Superconducting transition temperature T$_c$ vs. pressure for CeCu$_2$Si$_2$ samples annealed at the three temperatures indicated for seven days. Also shown are data for polycrystalline and single crystal samples of UBe$_{13}$ (from Chen *et al.*).[68]

The rapid change of T$_c$ in region II was interpreted by Bellarbi *et al.* as a valence transition of cerium ions from 3 to 3+δ.

The pressure dependence of the superconducting transition temperature was determined at hydrostatic pressure up to 12 kbar (Chen *et al.*).[68] Four CeCu$_2$Si$_2$ polycrystalline samples were used in the experiments. One sample was unannealed and three others were annealed at different temperatures. The highest T$_c$ was obtained in a sample annealed at 100°C. As shown in Fig. 83, T$_c$ increases rapidly with increasing pressure and reaches a plateau for all samples. The largest increase in T$_c$ was found in an unannealed sample, while a small pressure dependence of T$_c$ was observed in well-annealed samples. These results indicated that the enhancement of T$_c$ by pressure was directly related to the degree of order in the crystal lattice.

Bleckwedel and Eichler[126] performed the specific heat measurements of the heavy fermion superconductor CeCu$_2$Si$_2$ at pressures up to 5.9 kbar. The results showed T$_c$ was

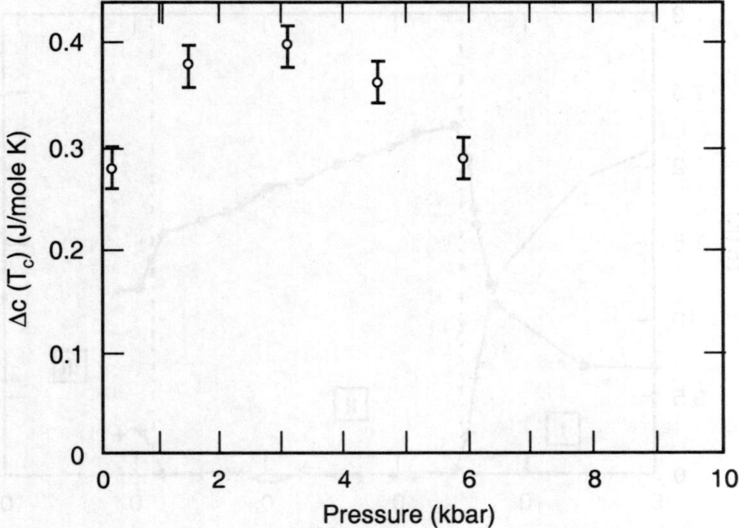

Fig. 84. Pressure dependence of the discontinuity in the specific heat at the superconducting transition in CeCu$_2$Si$_2$ (from Bleckwedel and Eichler).[126]

almost the same, but strong changes in the magnitude of the specific heat discontinuity (ΔC) at T$_c$ under the pressure were observed (see Fig. 84). The researchers suggested that the results were due to the change of the Kondo lattice structure in the quasiparticle density of states near the Fermi surface.

The measurements of T$_c$ versus P were reported in annealed polycrystals of CeCu$_2$Si$_2$ by ac susceptibility.[124] The pressure was measured with a precision better than 1 kbar. Thomas *et al.* claimed that their precise measurements (with more data points) were consistent with those obtained from Bellarbi *et al.* (1984). As displayed in Fig. 85, three regions were clearly distinct: (I) P $\leq$ 31 kbar, (II) 31 kbar $<$ P $<$ 77 kbar and (III) P $\geq$ 77 kbar. In region I, T$_c$(P) first increased slightly with an initial slope dT$_c$/dP = 2 mK/kbar. A rapid increase of T$_c$ took place for P $>$ 20 kbar and T$_c$ reached a maximum of 2.26K at 31 kbar. Region II was characterized with a linear decrease of T$_c$ with increasing P at a rate dT$_c$/dP = - 16.06 mK/kbar. Above 77 kbar (region III), T$_c$(P) dropped markedly by about 15%.

2. UBe$_{13}$

Chen *et al.*[68] presented a high pressure study in the compound UBe$_{13}$. The behavior of T$_c$ under pressure was similar for two samples: one sample was a polycrystal with a

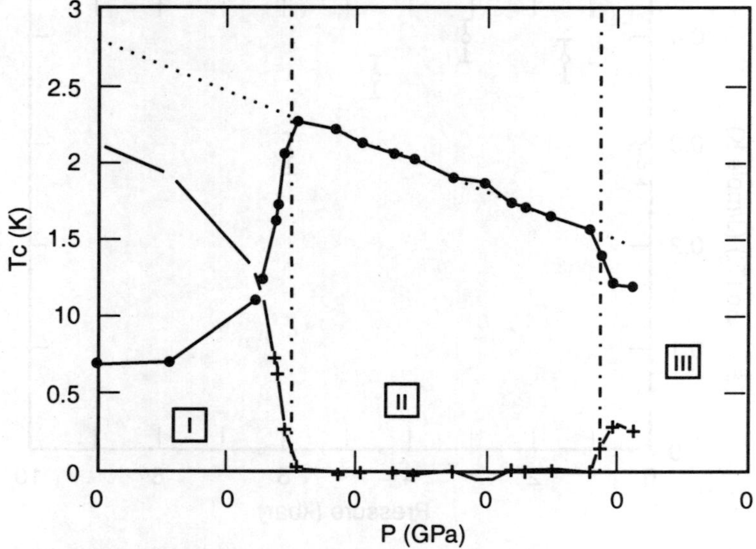

Fig. 85. Variation of superconducting T_c with pressure (circles) in annealed polycrystals of $CeCu_2Si_2$. The dashed straight line is a fit of the data between $P_1 = 3$ GPa and $P_2 = 7.7$ GPa. Crosses mark differences between the intermediate range (II) and experimental T_c values (from Thomas et at).[124]

higher T_c and another was a single crystal. The initial rates of T_c depression were 16 mK/kbar and 12 mK/kbar for the polycrystalline and single crystalline samples, respectively (see Fig. 83).

The specific heat of UBe_{13} was reported under pressures to 9.3 kbar by Phillips *et al.*[127] Fig. 86 shows the pressure dependence of the specific heat data in terms of C/T versus T. Not only was T_c depressed under the pressure, but the magnitude of the specific heat discontinuity reduced with increasing pressure as well.

3. UPt₃

Willis *et al.*[119] studied the superconducting transition temperature and the upper critical field of the compound UPt_3 at applied pressures up to 19.1 kbar. Both T_c and dH_{c2}/dT were determined by resistivity measurements. The results, obtained on a flux-grown single crystal of UPt_3 ($T_c = 0.48$ K), are shown in Fig. 87. T_c decreased as function of pressure at a rate of -12.6 mK/kbar. dH_{c2}/dT was depressed with pressure at a rate of -0.205 T/kbar. Both T_c and dH_{c2}/dT extrapo-

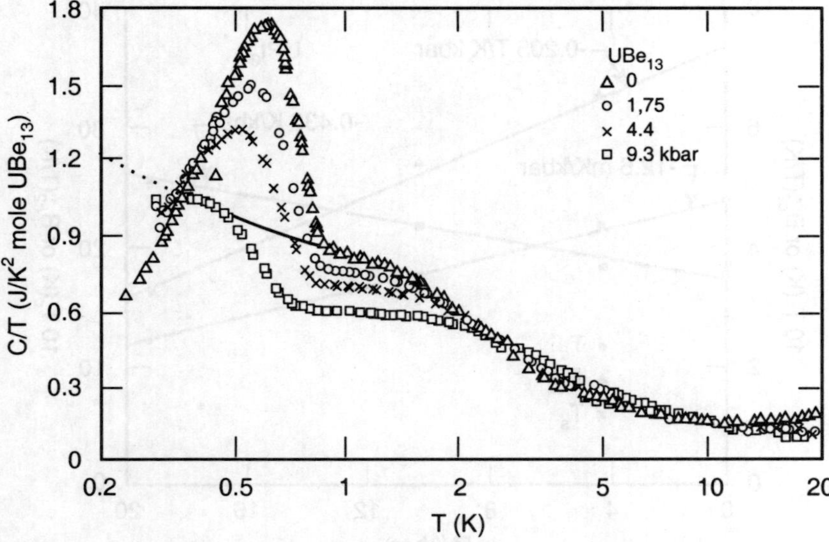

Fig. 86. The pressure dependence of C/T for UBe_{13}. The anomaly below ~ 0.9 K is associated with the transition (from Phillips *et al.*[127]).

lated to zero at 37 ± 1 kbar, corresponding to a volume change of $\Delta V/V_o$ of - 0.018 (de Visser *et al.*).[18]

Measurements of the specific heat were performed on the compound UPt_3 at three pressures 0, 3.8 and 8.9 kbar (Brodale *et al.*).[122] The superconducting transition temperatures at these three pressures were 0.50 , 0.44 and 0.36 K respectively. T_c was determined as the onset temperature of superconductivity. The average value of dT_c/dP was estimated to be - 15.7 mK/kbar, which was 26% greater than that deduced from resistivity measurements (Willis *et al.*).[119] Brodale *et al.* and Phillips *et al.* pointed out a positive correlation between the spin-fluctuation and superconductivity because the suppression of spin-fluctuation effects observed in normal state specific heat data and the relatively strong decrease in the superconducting transition temperature with increasing pressure. This correlation was not expected for a BCS superconductor. The experimental data measured by Brodale *et al.* were also analyzed by Pethick *et al.*[128] based on Fermi-liquid theory and led to the conclusion of p-wave superconductivity in UPt_3.

The pressure effects on the superconducting properties were investigated through electronic transport measurements on two single crystalline whiskers of UPt_3.[46] Fig. 88

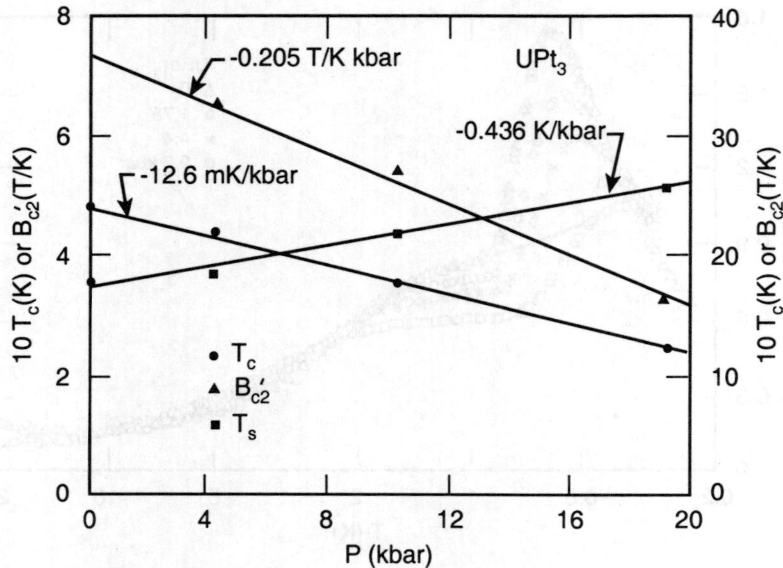

Fig. 87. The spin-fluctuation temperature T_s, the superconducting transition temperature T_c, and the initial slope of the upper critical field $B'_{c2} = dH_{c2}/dT$ vs. pressure for single crystal UPt$_3$ (from Willis *et al.*).[119]

presents T_c versus P at various field. The T_c depression rates (dT_c/dP) at P=0 was - 11.3 and - 13.2 mK/kbar for two samples, respectively.

4. URu$_2$Si$_2$

Maple *et al.*[66] measured the electrical resistivity in a polycrystalline sample of URu$_2$Si$_2$ at various hydrostatic pressure between 0 and 16 kbar. The measured data revealed that T_c decreased with pressure at the rate of $dT_c/dP = - 97$ mK/kbar. A second phase transition at $T_o = 17.5$ K was observed in the specific heat measurements at zero pressure. Maple *et al.* suggested that the transition at T_o was due to a charge-density-wave (CDW) or a spin-density-wave (SDW) that opened a BCS-like gap over part of Fermi surface. The onset of superconductivity at T_c caused the formation of another gap in the remainder of the Fermi surface.

McElfresh *et al.*[123] reported the pressure effect on a URu$_2$Si$_2$ polycrystal, which was cut from the same ingot as the one measured by Maple *et al.* The normalized

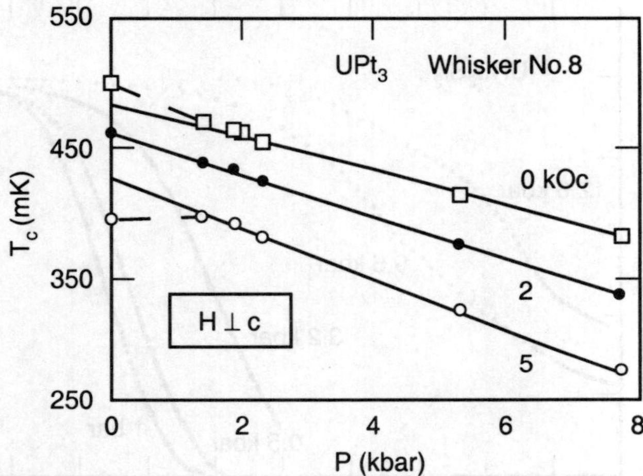

Fig. 88. Pressure dependence of the transition temperature (T_c) for three values of magnetic field in the basal plane of single crystalline whiskers of UPt_3. T_c is defined as the middle of the drop in resistivity between the normal state and the superconducting regime. Note the deviations from linearity at pressures below $P_c = 1.4$ kbar, upwards (by 20 mK) at H = 0 and downwards (by 27 mK) at H = 5 kOe. The disappearance of these opposite deviations at P = P_c suggests a symmetric collapse of the two transitions in zero field (from Behnia *et al.*).[46]

electrical resistivity versus temperature is plotted in Fig. 89 at various pressures. The values of T_c determined by the midpoints of the superconducting transition were reduced as increasing pressure. The transition width ΔT_c, defined as 10 - 90 % of the transition, increased approximately linearly with pressure from 0.3 K at 1 bar to 0.9 K at 12 kbar. The increase of ΔT_c with pressure might be associated with the different states of strain within the polycrystalline sample. The dependence of T_c on pressure is presented in Fig. 90. A rate of 95 mK/kbar was obtained from the linear decrease of T_c with pressure. A second transition at $T_o = 17.5$ K was observed in the specific heat measurements at zero pressure. This transition was suggested as the formation of a CDW or SDW. The T_o value was observed to increase with pressure at a rate of 130 mK/kbar (see Fig. 90). The nearly equal but opposite pressure dependences of T_o and T_c reflected a competition for electronic density of states at the Fermi level.

Fisher *et al.*[23] confirmed the experimental findings by Maple *et al.* and McElfresh *et al.*. With increasing pressure, as shown in Fig. 91, the superconducting transition observed

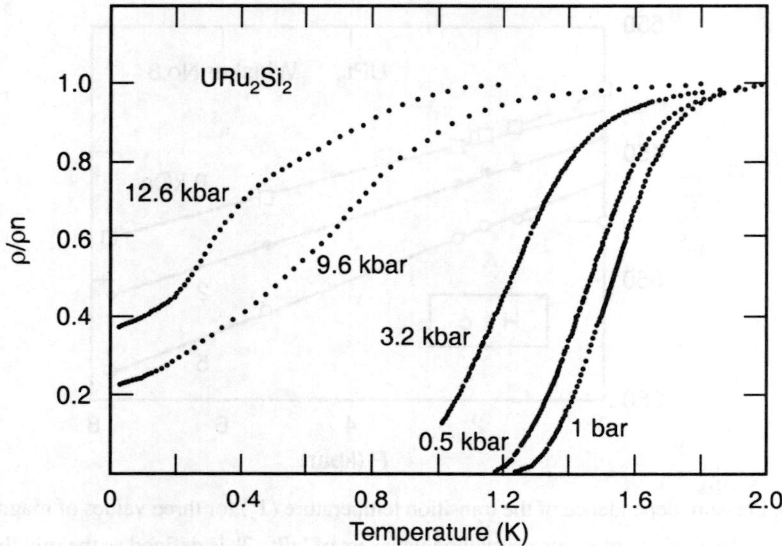

Fig. 89. Normalized electrical resistivity of URu_2Si_2 in the vicinity of the superconducting transition temperature T_c at various pressures. Note that the data shown for the two highest pressures were obtained from measurements performed with a three-lead configuration (from McElfresh *et al.*).[123]

in C/T versus T experiments was broadened, attenuated and shifted to lower temperatures. For the magnetic transition near $T_o = 18$ K, T_o increased with increasing pressure, accompanied by a broadening and attenuation of the specific heat anomaly (see Fig. 92). The values of dT_c/dP and dT_o/dP were - 56.2, 127 mK/kbar, respectively.

The effect of uniaxial stress on the superconducting (T_c) and magnetic (T_o) transition temperatures in the heavy fermion system URu_2Si_2 was investigated by means of resistivity measurements (Bakker *et al.*).[121] T_c was determined by the midpoint of the superconducting transition, while T_o was determined by the local minimum in the curve resistivity versus temperature. As shown in Figs. 93 and 94, T_c increased and T_o decreased under pressure when the pressure was applied along the c-axis of the crystal: $dT_c/dP = 25$ mK/kbar and $dT_o/dP = - 41$ mK/kbar. The opposite effects were found for the pressure applied along the a-axis, $dT_c/dP = - 35$ mK/kbar and $dT_o/dP = 126$ mK/kbar. the inverse correlations of T_c and T_o under pressure were observed. The inverse correlations of T_c and T_o for uniaxial pressure future supported the suggestion that superconductivity and antiferromagnetism competed for parts of the Fermi surface (McElfresh *et al.*).[123]

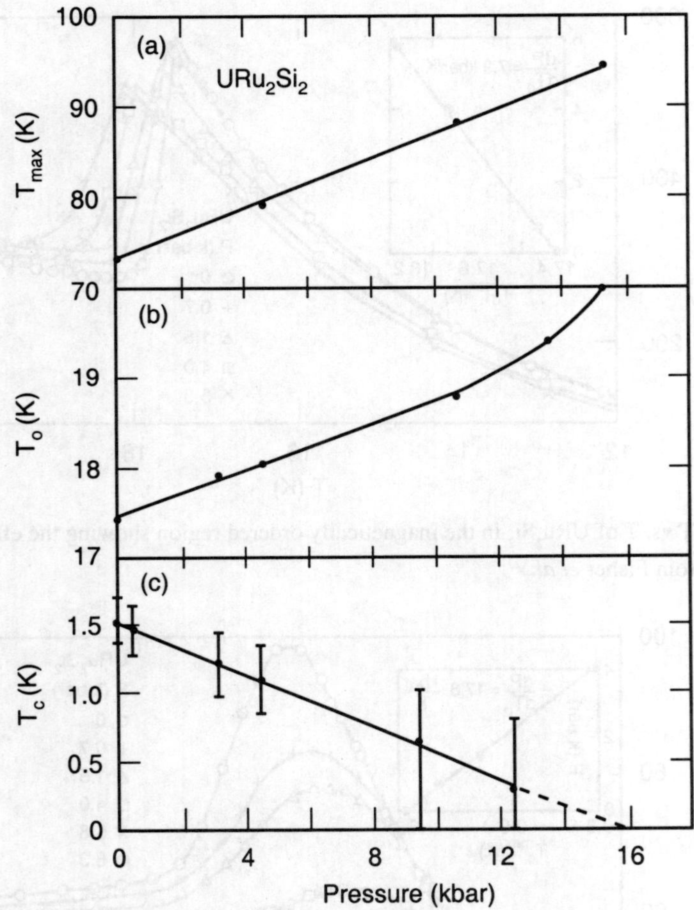

Fig. 90. Pressure dependence of T_{max}, T_o, and T_c in a URu_2Si_2 polycrystal. For (a) and (b), exemplary error bars are given at 1 bar (from McElfresh *et al.*).[123]

5. UPd_2Al_3

Caspary *et al.*[25] presented the measurements of the resistivity and the specific heat as a function of applied pressure on polycrystalline samples of UPd_2Al_3. Both resistivity and specific heat data showed weak depression of T_c and ΔC under pressure. The T_c depression shown in Fig. 16 yielded a dT_c/dP value of $- 7 \pm 1$ mK/ kbar, which was in good agreement with the value of $- 6.3$ derived from thermal expansion experiments (Modler *et al.*).[129-130]

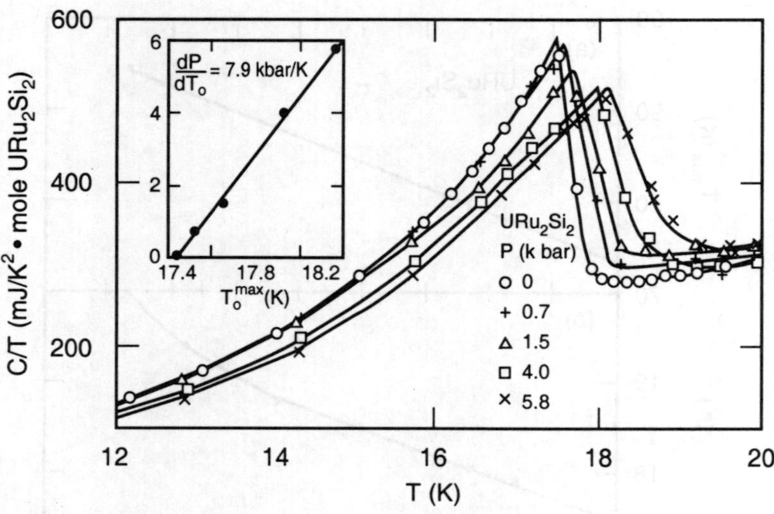

Fig. 91. C/T vs. T of URu_2Si_2 in the magnetically ordered region showing the effect of pressure (from Fisher *et al.*).[23]

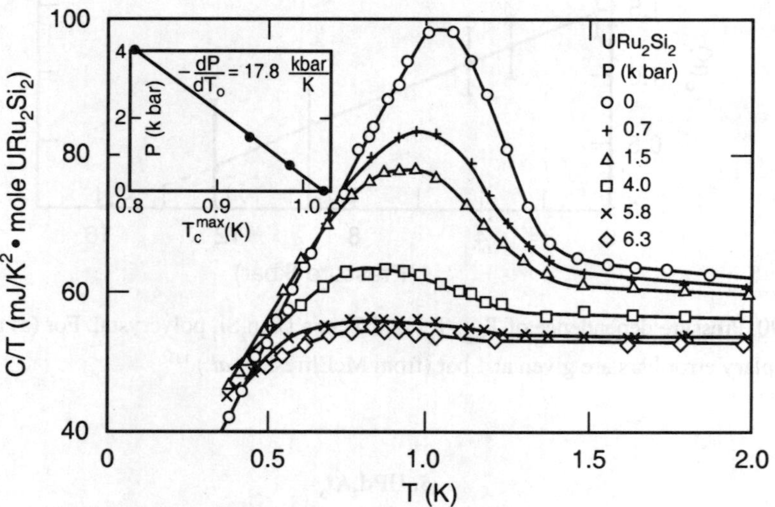

Fig. 92. C/T vs. T of URu_2Si_2 in the superconducting region showing the effect of pressure (from Fisher *et al.*).[23]

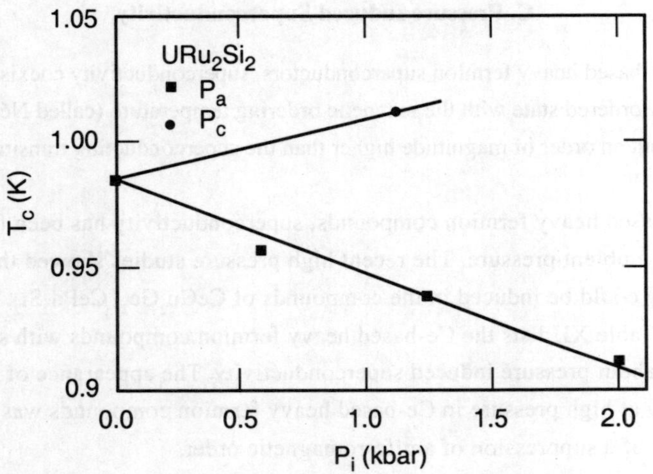

Fig. 93. T_c of single crystalline URu_2Si_2 as function of uniaxial pressure along the a-axis (■) and c-axis (●) (from Bakker *et al.*).[121]

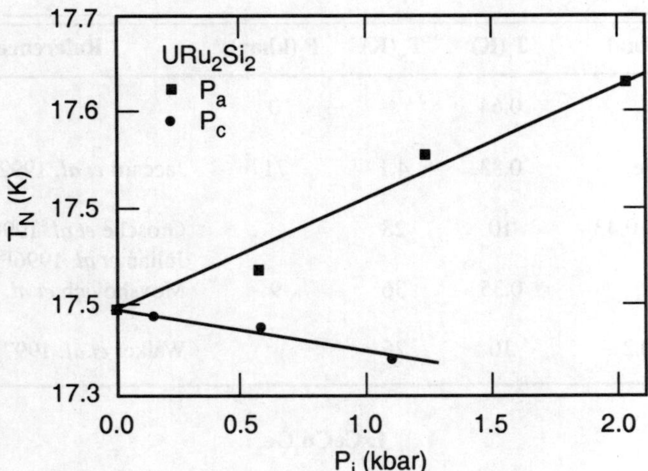

Fig. 94. T_N of single crystalline URu_2Si_2 as function of uniaxial pressure along the a-axis (■) and c-axis (●) (from Bakker *et al.*).[121]

C. Pressure Induced Superconductivity

In the U-based heavy fermion superconductors, superconductivity coexists with a magnetically ordered state with the magnetic ordering temperature (called Néel temperature, T_N) about an order of magnitude higher than the superconducting transition temperature, T_c.[49, 131, 132]

In Ce-based heavy fermion compounds, superconductivity has been found in $CeCu_2Si_2$ at ambient pressure. The recent high pressure studies showed that superconductivity could be induced in the compounds of $CeCu_2Ge_2$, $CePd_2Si_2$, $CeRh_2Si_2$ and $CeIn_3$. Table XII lists the Ce-based heavy fermion compounds with some parameters about pressure induced superconductivity. The appearance of the superconductivity at high pressure in Ce-based heavy fermion compounds was explained as the result of a suppression of antiferromagnetic order.

Table XII. Properties of Ce-based superconducting compounds where T_N is the Néel temperature at zero pressure and T_c is the onset superconducting transition temperature near the critical pressure P_c defined as the pressure to induce superconductivity.

Compound	T_c(K)	T_N(K)	P_c(kbar)	Reference
$CeCu_2Si_2$	0.64		0	
$CeCu_2Ge_2$	0.83	4.1	71	Jaccard *et al.* 1992, 1995[133, 41]
$CePd_2Si_2$	0.43	10	28	Grosche *et al.* 1996 and Julian *et al.* 1996[134, 135]
$CeRh_2Si_2$	0.35	36	9	Movshovich *et al.* 1996[132]
$CeIn_3$	0.2	10	25	Walker *et al.* 1997[136]

1. $CeCu_2Ge_2$

Jaccard *et al.*[133] reported the thermopower and transport properties of the $CeCu_2Ge_2$ polycrystals under high pressure. Both the resistivity and thermopower curves showed the antiferromagnetic order at T_N = 4.1 K at zero pressure. At higher pressure, T_N reduced to 4 K and 2.15 K at pressures of 3.5 kbar and 70 kbar, respectively. At 101 kbar, no sign of magnetic order was detected for temperature greater than 1.2 K. The resistivity data indicated that a possible superconducting transition appeared in $CeCu_2Ge_2$ at 101 kbar with the midpoint transition of 0.64 K and 10 -

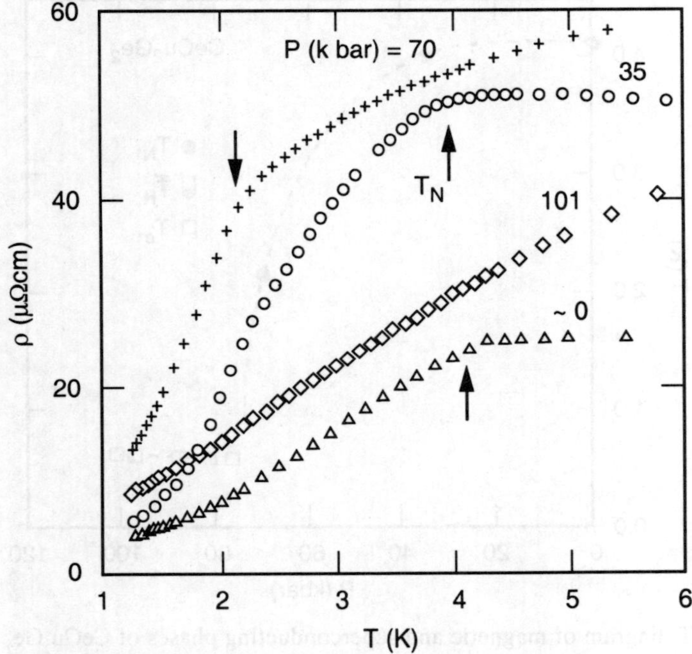

Fig. 95. Resistivity of CeCu$_2$Ge$_2$ at different pressures for 1.2 < T < 6 K. The arrows mark the temperature of magnetic ordering (from Jaccard *et al.*).[133]

90% transition width of 0.1 K. The resistivity became smaller than the threshold detection of 10^{-4} $\mu\Omega$cm at the temperatures below 0.55 K, which was four orders of magnitude smaller than that at normal state just above T_c (see Fig. 95). Jaccard *et al.* (1992) showed the pressure-temperature diagram of the magnetic and superconducting phases (Fig. 96) and concluded that the superconductivity appeared as the magnetic order vanished at a pressure above 70 kbar. The superconducting transition induced by high pressure in the CeCu$_2$Ge$_2$ compound was confirmed by the single crystal studies (Jaccard and Sierro 1995). The resistivity measurements showed that the superconducting transition at 71 kbar with midpoint transition of 0.68 K, onset transition of 0.83 K and 10 - 90% transition width of 0.25 K. The resistivity became smaller than the threshold detection of 10^{-3} mWcm at the temperatures below 0.23 K (see Fig. 97).

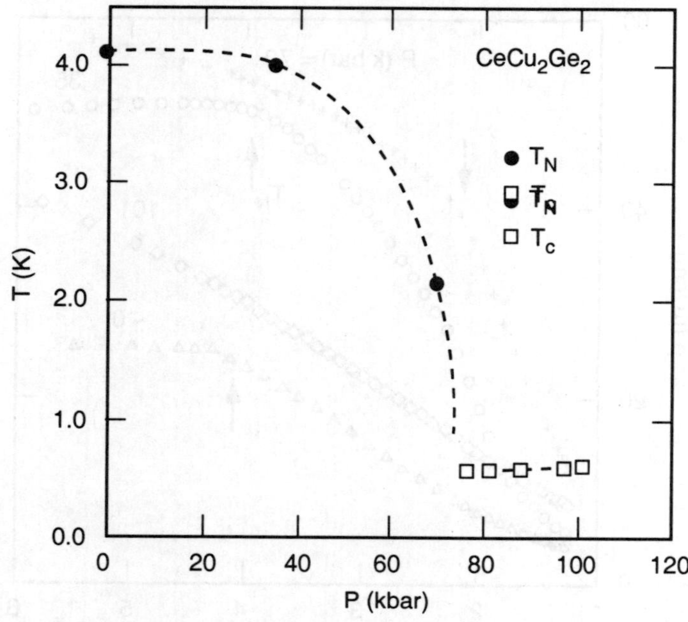

Fig. 96. P-T diagram of magnetic and superconducting phases of $CeCu_2Ge_2$ (from Jaccard *et al.*).[133]

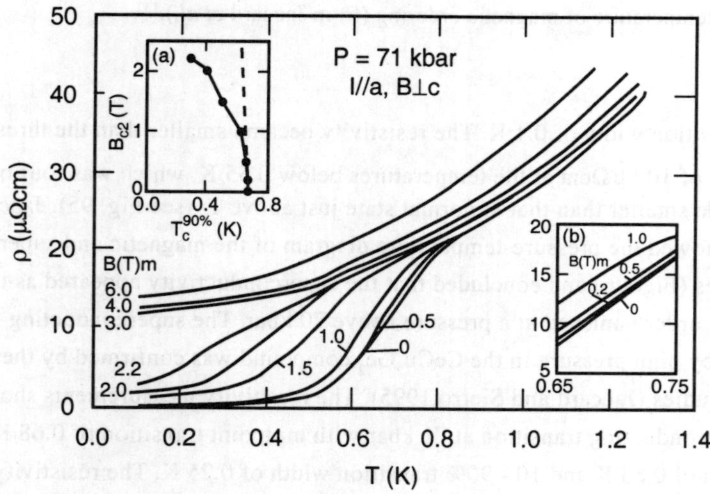

Fig. 97. Very low temperature resistivity of $CeCu_2Ge_2$ at 71 kbar and under different magnetic fields B. Inset (a): T-dependence of the upper critical field B-. Inset (b): low field data in the vicinity of the midpoint of the superconducting transition (from Jaccard *et al.*).[41]

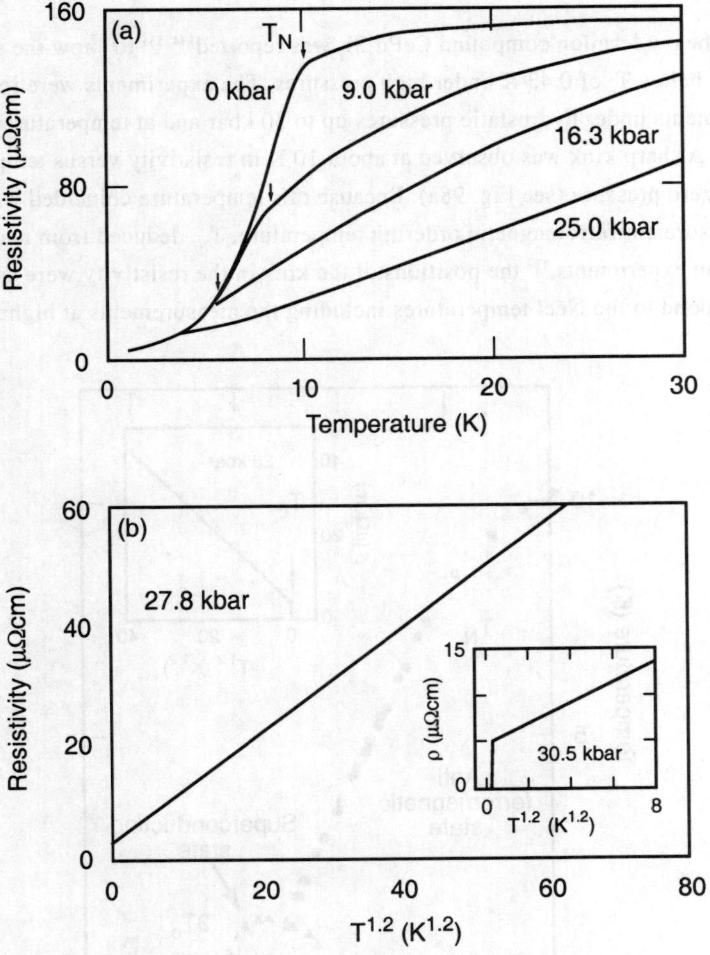

Fig. 98. (a) The temperature dependence of the resistivity, ρ, of $CePd_2Si_2$ along the a-axis at different pressures. The Néel temperature, T_N, marked by arrows, is visible as a sudden change in the slope of $\rho(T)$. (b) Near to and above the critical pressure, P_c, of ~ 28 kbar, a superconducting transition sets in at about 430 mK (inset of b) below a quasi-linear region at $\rho(T)$, which extends over two orders of magnitude in temperature (from Grosche et al.).[134]

2. CePd$_2$Si$_2$

The heavy fermion compound CePd$_2$Si$_2$ was reported[134, 135] to show the superconductivity below T$_c$ of 0.43 K under high pressures. The experiments were resistivity measurements under hydrostatic pressures up to 30 kbar and at temperatures down to 0.2 K. A sharp kink was observed at about 10 K in resistivity versus temperature curve at zero pressure (see Fig. 98a). Because this temperature coincided with the zero pressure antiferromagnetic ordering temperature, T$_N$, deduced from neutron-diffraction experiments,[138] the positions of the kink in the resistivity were assumed to correspond to the Néel temperatures including the measurements at higher

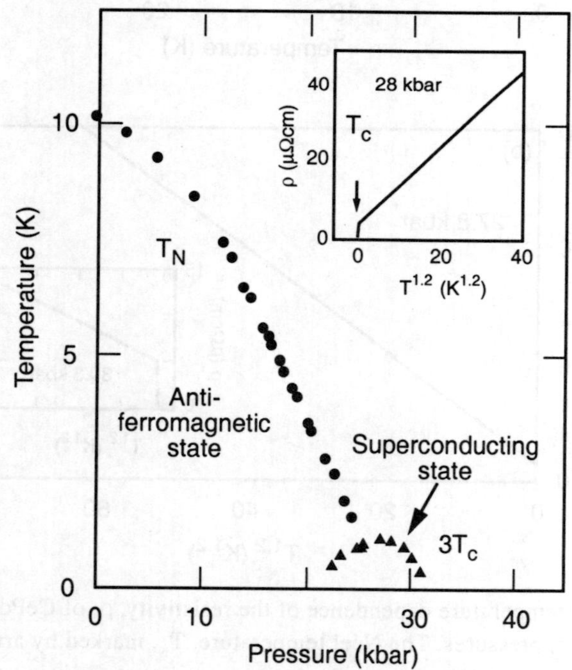

Fig. 99. Temperature-pressure phase diagram of high purity single crystal CePd$_2$Si$_2$. Superconductivity appears below T$_c$ in a narrow window where the Néel temperature T$_N$ tends to absolute zero. Inset: the normal state a-axis resistivity above the superconducting transition varies as T$^{1.2\pm0.1}$ over nearly two decades in temperature. The upper critical field B$_{c2}$ at the maximum value of T$_c$ varies near T$_c$ at a rate of approximately - 6 T/K. For clarity, the values of T$_c$ have been scaled by a factor of three, and the origin of the inset has been set at 5 K below absolute zero (from Mathur et al.).[139]

pressures. The Néel temperatures decreased with increasing pressure and fell below 1 K at around 26 kbar. Near to and above the critical pressure, P_c, of 28 kbar, a superconducting transition was determined from an abrupt drop of resistivity below the detection limit at temperatures below 0.4 K (see Fig. 98b). It was concluded that the formation of a superconducting phase was closed linked to the disappearance of antiferromagnetic order as shown in the phase diagram of Fig. 99.

Resistivity and thermopower measurements were reported for two samples of $CePd_2Si_2$ by Link et $al.$[140] The pressure range was from 21 to 34.5 kbar for the measurements at temperatures down to 0.03 K and higher pressure was performed up to 56 kbar at temperatures between 1.2 - 300 K. The Néel temperatures taken from the change of slope of the resistivity were determined to be 9, 8 and 6.7 K at 1, 9 and 14 kbar, respectively. In all the measurements, no trace of any superconducting transition was observed in $CePd_2Si_2$.

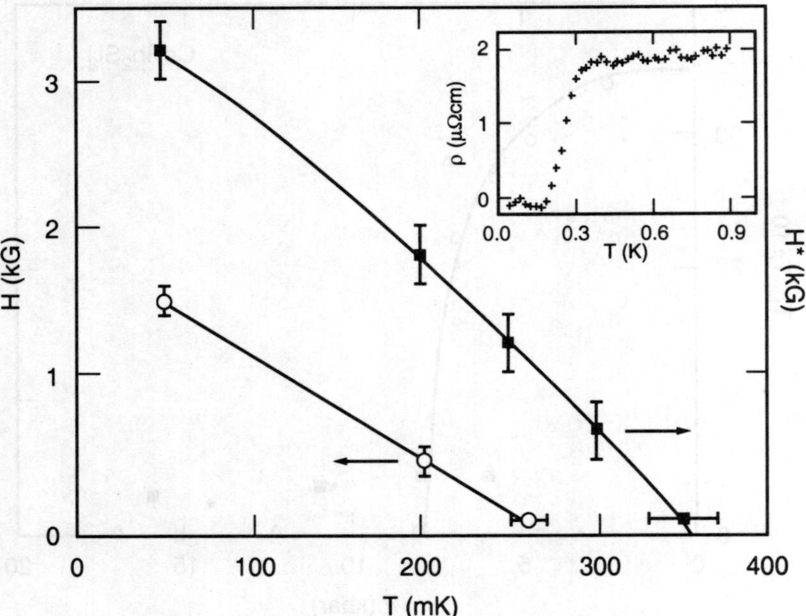

Fig. 100. Magnetic field at the onset of suppression of resistivity (H^*, solid squares; solid line is a guide to the eye) and the midpoint of the transition (H, O) as a function of temperature at a pressure of 11.0 ± 0.3 kbar in $CeRh_2Si_2$. Error bars represent systematic uncertainty in determining the onset and midpoints of the superconducting transitions. Inset: superconducting transition displayed in resistivity data at 11.0 ± 0.3 kbar (from Movshovich et $al.$).[132]

128

3. CeRh$_2$Si$_2$

A superconducting transition was observed in the heavy fermion compound CeCu$_2$Ge$_2$ under hydrostatic pressure of 9 kbar.[132] Fig.100 shows the resistivity data on a CeRh$_2$Si$_2$ sample at a pressure of 11.0 ± 0.3 kbar. The onset of superconductivity occurred at a temperature of 0.35 K and the resistivity dropped to zero within instrumental resolution about 0.2 K (see inset of Fig. 100). Superconducting transitions in CeCu$_2$Ge$_2$ under pressure were also confirmed by magnetic ac susceptibility measurements.[132] The Néel temperature at zero pressure of the CeCu$_2$Ge$_2$ compound was found to be 36 K (see phase diagram Fig. 101) and the corresponding pressure required to suppress antiferromagnetic order was about 8.5 kbar.[141] It was suggested by Movshovich *et al.* that the antiferromagnetic order in the superconducting

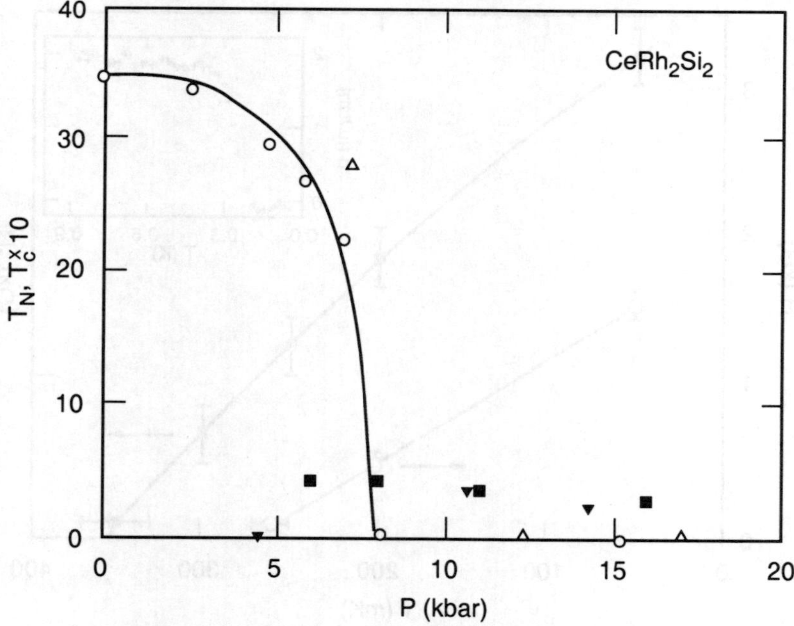

Fig. 101. P-T phase diagram of CeRh$_2$Si$_2$. Open symbols represent antiferromagnetic ordering temperature $T_N(P)$. Solid symbols are $T_c(P)$ points of the superconducting transitions obtained from the onset of suppression of resistivity for several samples that are represented by different symbols. The error bars are on the order of the size of the symbols and therefore omitted. Notice the different temperature scales for the two phase transitions (from Movshovich *et al.*).[132]

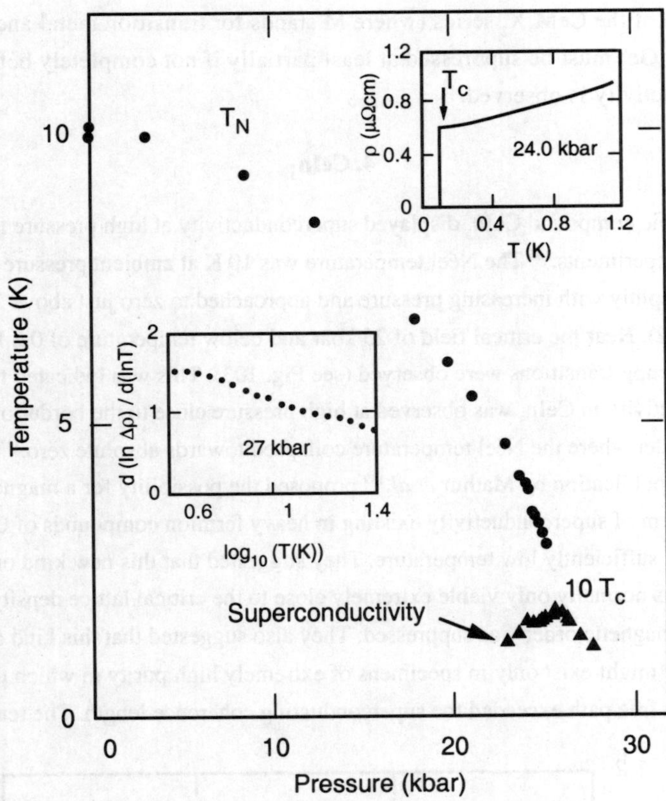

Fig. 102. Temperature-pressure phase diagram of high purity single crystal CeIn$_3$. A sharp drop in the resistivity consistent with the onset of superconductivity below T$_c$ is observed in a narrow window near P$_c$, the pressure at which the Néel temperature T$_N$ tends to absolute zero. Upper inset: this transition is complete even below P$_c$. Lower inset: just above P$_c$, where there is no Néel temperature, a plot of temperature dependence of d(lnΔρ)/d(lnT) is best able to demonstrate that the normal state resistivity varies as T$^{1.6\pm0.2}$ below several degrees K (Δρ is the difference between the normal state resistivity and its residual value, which is calculated by extrapolating the normal state resistivity to absolute zero. For clarity, the values of T$_c$ have been scaled by a factor of ten. The resistivity exponents of CeIn$_3$ and CePd$_2$Si$_2$ may be understood by taking into account the underlying symmetries of the antiferromagnetic states and using the magnetic interactions model. Superconductivity near the critical density n$_c$ in pure samples is expected to be a natural consequence of the same model (from Mathur et al.).[139]

compounds of the CeM_2X_2 series (where M stands for transition metal and X is either Si or Ge) must be suppressed at least partially if not completely before superconductivity is observed.

4. CeIn$_3$

The cubic compound CeIn$_3$ displayed superconductivity at high pressure from resistivity experiments.[136] The Néel temperature was 10 K at ambient pressure and decreased rapidly with increasing pressure and approached to zero just above 25 kbar (see Fig. 102). Near the critical field of 25 kbar and below temperature of 0.2 K, the superconducting transitions were observed (see Fig. 103). This was indicated that the superconductivity in CeIn$_3$ was observed at high pressure close to the border of antiferromagnetic order where the Néel temperature collapsed towards absolute zero.[136]

Recent publication by Mathur *et al.*[139] proposed the possibility for a magnetically mediated form of superconductivity existing in heavy fermion compounds of $CePd_2Si_2$ and CeIn$_3$ at sufficiently low temperature. They suggested that this new kind of superconductivity was normally only viable extremely close to the critical lattice density where long range magnetic order was suppressed. They also suggested that this kind of superconductivity might exist only in specimens of extremely high purity in which the charge carrier mean free path exceeded the superconducting coherence length. The temperature-

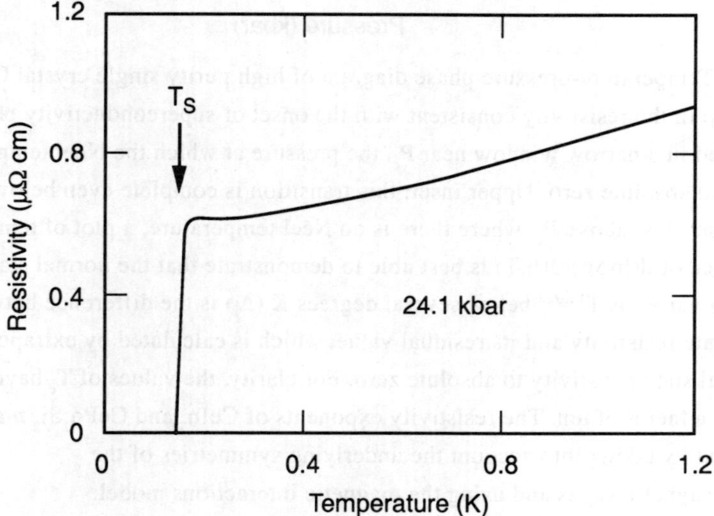

Fig. 103. The resistivity near the superconducting transition in the vicinity of the critical pressure for antiferromagnetic order in CeIn$_3$ (from Walker *et al.*).[136]

pressure phase diagrams of high purity single crystals of $CePd_2Si_2$ and $CeIn_3$ were displayed in Figs. 99 and 102. Superconductivity appeared below T_c in a narrow window near critical pressure P_c where Néel temperature tended to zero.

VIII. CONCLUSION

A. Experimental Findings

Most heavy fermion superconductors exhibit deviations of the specific heat discontinuity from the BCS weak coupling value of 1.43. The $\Delta C/\gamma T_c$ values for the compounds UPt_3, URu_2Si_2, UNi_2Al_3 and UPd_2Al_3 lie within 20 - 90% below the BCS result. The compound UBe_{13} shows the largest specific heat discontinuity of around 2.4. The specific heat in the superconducting state for the heavy fermion supercon-ductors cannot be described by an exponential form predicted by the BCS theory. A power-law temperature dependence can be obtained approximately in the com-pounds $CeCu_2Si_2$, UBe_{13} and UPd_2Al_3 with the exponents in the range of 2 to 3. All of the specific heat data for UPt_3 and URu_2Si_2 could be fitted to a quadratic form of $\gamma_s T + \beta_s T^2$. The entropy discrepancy between the superconducting and normal states is reported. The largest value of the entropy discrepancy in UPt_3 is around 6%. The compound URu_2Si_2 shows a large mismatch of entropy balance of 30% with broad superconducting transitions.

The temperature dependence of thermal conductivity studies of heavy fermion superconductors indicates that the thermal conductivity does not follow an exponen-tial temperature dependence at low temperatures. The thermal conductivity in the superconducting state can be described approximately by a temperature dependence in a power-law or a quadratic form. The single crystal study in the compound UPt_3 indicates that thermal conductivity has none or small anisotropic properties.

Ultrasonic attenuation studies in heavy fermion superconductors confirm a power-law temperature dependence. All the ultrasonic attenuation data show a maximum in the plot of $\alpha_s(T)$ versus T at temperatures at or below T_c. Two kinds of maxima were observed: one is a broad maximum and the other is a spike peak with a λ-shape. The ultrasound experiments also display anisotropic behavior in the ultrasonic attenuation of the compound UPt_3, depending on the polarization and direction of propagation of the sound wave relative to the crystal axes. At tempera-ture just below T_c, a pronounced λ-shaped peak was observed in the longitudinal ultrasonic attenuation, while a rather broad maximum appeared in the transverse sound. A power-law T^3-dependence was shown in the longitudinal sound propagated along the c-axis, while a linear T-dependence was found in the transverse sound propagated along the b-axis.

The heavy fermion superconductors are type II superconductors with κ as large as 50 or more. The large slopes of the upper critical field are observed in heavy fermion superconductors with the largest value of 100 T/K or more found in the UBe_{13} system. Anisotropic behavior is observed in some heavy fermion supercon-ductors. UPt_3 shows a large anisotropy of 30-40% in the initial slope of the upper critical field at T_c for the magnetic field applied parallel and perpendicular to the c-axis. URu_2Si_2 has no anisotropy in the slope of the upper critical field at T_c, but a marked difference of the upper critical field at temperatures lower than T_c.

Heavy fermion superconductivity is very sensitive to any kind of chemical impurities including non-magnetic impurities. The investigation of impurity effects in the compound $Ce_{1-x}M_xCu_{2.2}Si_2$ shows that the specific heat discontinuity at T_c was suppressed more strongly than T_c and the T_c depression rate depended upon the particular dopants. A peculiar feature in $U_{1-x}Th_xBe_{13}$ is a double-peak structure discovered in the specific heat measurements. This feature led to a complicated phase diagram of $U_{1-x}Th_xBe_{13}$ that shows at least two and possibly more phase transitions. In the off-stoichiometric studies of the series $CeCu_{2+x}Si_2$, the higher superconducting transition temperatures and sharper transitions are obtained in copper-rich samples. T_c reaches the highest value of 0.67 K with a 1 - 2 at.% Cu excess in $CeCu_{2+x}Si_2$.

B. Suggestions of Unconventional Superconductors

Experimental facts clearly show that heavy fermion superconductivity is unconventional in many aspects. The first evidence for an unconventional gap structure in heavy fermion superconductors comes from the observations of power-law temperature dependencies in the superconducting state of many physical quantities: specific heat, thermal conductivity, ultrasonic attenuation, NMR relax-ation rate and penetration depth. The studies of temperature dependence of the NMR relaxation rate and penetration depth in the heavy fermion superconductors can be found in MacLaughlin *et al.*,[142-143] Kitaoka *et al.*,[144] Einzel *et al.*,[145] Kohori *et al.*,[146-147] Asayama *et al.*,[148] and Tien and Jiang.[149] The three states of p-wave pairing are used to discuss the density of states of quasiparticles in unconventional supercon-ductors: gapless, line zeros and point zeros. Each of these states has its specific function in the density of states, which yields a specific power-law temperature dependence of various physical quantities at very low temperatures. Various power-law behaviors are reported in heavy fermion superconductors. The accumulated experimental data from the temperature dependencies indicate clearly that there are many low-lying excitations associated with nodes of gap functions in heavy fermion superconductors. However, in some cases there is no consistency about gap structures among the experimental results for different physical quantities. One possible explanation for this kind of inconsistency is that the temperature range that is accessible experimentally is not sufficiently low. Another explanation is this discrepancy can be resolved when the effect of impurity scattering is included. Impurities modify the low temperature properties, such as producing a quadratic temperature dependence rather than a simple power-law.

Another indication of unconventional s-wave pairing in heavy fermion systems occurs when dilute concentrations of nonmagnetic impurities are introduced. In conventional s-wave superconductors, magnetic impurities with stable local moments cause exchange scattering which breaks the Cooper pairing and yields a universal relationship between the superconducting transition temperature or the specific heat discontinuity and the impurity concentration. Therefore, dilute concentrations of nonmagnetic impurities in a s-wave superconductor have no or little effect on the superconducting properties. In heavy fermion superconductors, the nonmagnetic impurities are nearly as effective at breaking pairs as are the magnetic impurities.

Th-doped UBe_{13} has generated one of the most exciting results in heavy fermion superconductivity relatives to pair-breaking effects. Thorium, as a non-magnetic impurity, should theoretically suppress superconductivity with increasing concentration. Experimental results show decrease of the superconducting transition temperature with the increase of impurity concentration is not monotonic. In specific heat measurements an additional second-order transition below the onset of superconductivity has been discovered. Some experiments suggests an additional superconducting transition, while other observations point to the appearance of a magnetic transition within superconducting state.

The slope of the upper critical field at T_c is remarkably large. One of the key question concerning H_{c2} is the problem of Pauli limiting of H_{c2}. The large anisotropy observed in the H_{c2} slope at T_c in UPt_3 is argued as an evidence of a p-wave superconductor.

REFERENCES

1. F. Steglich, J. Aarts, C. D. Bredl, W. Lieke, D. Meschede, W. Franz and H. Schafer: "Superconductivity in the Presence of Strong Pauli Paramagnetism: $CeCu_2Si_2$", Phys. Rev. Lett. **43**, 1892-1895, (1979).

2. H. R. Ott, H. Rudigier, Z. Fisk, and J. L. Smith, "UBe_{13}: An Unconventional Actinide Superconductor," Phys. Rev. Lett. **50**, 1595-1598, (1983).

3. G. R. Stewart, Z. Fisk, J. O. Willis and J. L. Smith, "Possibility of Coexistence of Bulk Superconductivity and Spin Fluctuations in UPt_3", Phys. Rev. Lett. **52**, 679-682, (1984).

4. W. Schlabitz, J. Baumann, J. Diesing, W. Krause, G. Neumann, C. D. Bredl, U. Ahlheim, H. M. Mayer and U. Rauchwchwalbe,: Poster presented at the Fourth International Conference on Valence Flucturations ICVF, Cologne (1984).

5. C. Geibel, S. Thies, D. Kaczorowski, A. Mehner, A. Grauel, B. Shidel, U. Adlheim, R. Helfrich, K. Petersen, C. D. Bredl and F. Steglich, "A New Heavy-Fermion Superconductor: UNi_2Al_3", Z. Phys. B - Cond. Matt. **83**, 305-306, (1991).

6. C. Geibel, C. Schank, S. Thies, H. Kitazawa, C. D. Bredl, A. Böhm, M. Rau, A. Grauel, R. Caspary, R. Helfrich, U. Adlheim, G. Weber and F. Steglich, "Heavy-Fermion Superconductivity at $T_c = 2$ K in the Antiferromagnet UPd_2Al_3", Z. Phys. B - Cond. Matt. **84**, 1-2, (1991).

7. D. Rainer, "Remarks on the Theory of Heavy-Fermion Superconductivity", Phys. Scr. **T23**, 106-122, (1988).

8. J. R. Schrieffer, "Theory of Superconductivity", (Addison-Wesley, Redwood City, CA 1964).

9. R. D. Parks, Superconductivity, 2 volumes, (Marcel Dekker, New York, 1969).

10. W. Lieke, U. Rauchschwable, C. B. Bredl and F. Steglich, "Superconductivity in $CeCu_2Si_2$", J. of Appl. Phys. **53**, 2111-2116, (1982).

11. C. D. Bredl, H. Spille, U. Rauchschwalbe, W. Lieke, F. Steglich, G. Cordier, W. Assmus, M. Herrmann and J. Aarts, "Gapless Superconductivity and Variation of T_c in the Heavy-Fermion System $CeCu_2Si_2$." J. Magn Magn Mat, **31-34**, 373-376, (1983).

12. W. Assmus, M. Herrmann, U. Rauchschwalbe, S. Riegel, W. Lieke, H. Spille, S. Horn, G. Weber, F. Steglich and C. Cordier, "Superconductivity in $CeCu_2Si_2$ Single Crystals", Phys. Rev. Lett. **52**, 469, (1984).

13. H. R. Ott, H. Rudigier, T. M. Rice, K. Ueda, Z. Fisk and J. L. Smith, "p-Wave Superconductivity in UBe_{13}", Phys. Rev. Lett. **52**, 1915-1918, (1984).

14. H. M. Mayer, U. Rauchschwalbe, C. D. Bredl, F. Steglich, H. Rietschel, H. Schmidt, H. Wuhl and J. Beuers, "Normal-State and Superconducting Properties of the Heavy-Fermion Compound UBe_{13} in Magnetic Fields", Phys. Rev. B **33**, 3168-3171, (1986).

15. A. Ravex, J. Flouquet, J. L. Tholence, D. Jaccard and A. Mayer, "Thermal Conductivity and Specific Heat Measurements on UBe_{13}", J. Magn. & Magn Mater. **63&64**, 400-402, (1987).

16. J. J. M. Franse, A. Menovsky, A. de Visser, C. D. Bredl, U. Gottwich, W. Lieke, H. M. Mayer, U. Rauchschwalbe, G. Sparn and F. Steglich, "Specific Heat and Transport Properties of UPt_3", Z. Phys. B **59**, 15-19, (1985).

17. A. Sulpice, P. Gandit, J. Chaussy, J. Flouquet, D. Jaccard, P. Lejay and J. L. Tholence, "Thermodynamic and Transport Properties of UPt_3," J. of Low Temperature Physics **62**, 39-54, (1986).

18. A. de Visser, A. Menovsky and J.J.M. Franse, "UPt_3, Heavy Fermions and Superconductivity," Physica **147B**: 81-160, 1987.

19. R. A. Fisher, S. Kim, B. F. Woodfield, N. E. Phillips, L. Taillefer, K. Hasselbach, J. Flouquet, A. L.Giorgi and J. L. Smith, "Specific Hear of UPt_3: Evidence for Unconventional Superconductivity", Phys. Rev. Lett. **62**, 1411-1414, (1989).

20. E. Schuberth, G. Hofmann, F. Gross, K. Andres and J. Fufnagl, "UPt_3 at Very Low Temperatures: Specific Heat in High Magnetic Fields", Europhys. Lett. **11**, 249-254, (1990).

21. T. T. M. Palstra, A. Menovsky, J. van den Berg, A. J. Dirkmaat, P. H. Kes, G. J. Nieuwenhuys and J. A. Mydosh, "Superconducting and Magnetic Transitions in the Heavy-Fermion System URu_2Si_2, Phys. Rev. Lett **55**, 2727-2730, (1985).

22. W. Schlabitz, J. Baumann, B. Pollot, U. Rauchschwalbe, H. M. Mayer, U. Ahlheim and C.D. Bredl, "Superconductivity and Magnetic Order in a Strongly Interacting Fermi-System: URu_2Si_2," Z. Phys. B - Condensed Matter **62**, 171-177, (1986).

23. R. A. Fisher, S. Kim, Y. Wu, N. E. Phillips, M. W. McElfresh, M. S. Torikachvile and M. B. Maple: "Specific Heat of URu_2Si_2: Effect of Pressure and Magnetic Field on the Magnetic and Superconducting Transitions", Physica B **163**, 419-423, (1990).

24. K. Hasselbach, P. Lejay and J. Flouquet, "A Single Superconducting Transition in URu_2Si_2 Comparison with UPt_3", Phys. Lett. A **156**, 313-316, (1991).

25. R. Caspary, P. Hellmann, M. Keller, G. Sparn, C. Wassilew, R. Kohler, C. Geibel, C. Schank, F. Steglich, N. E. Phillips, "Unusual Ground-State Properties of UPd_2Al_3: Implications for the Coexistence of Heavy-Fermion Superconductivity and Local-Moment Antiferromagnetism", Phys. Rev. Lett. **71**, 2146-2149, (1993).

26. G. R. Stewart, "Heavy-Fermion Systems", Reviews of Modern Physics **56**, 755-87, (1984).

27. D. Scalapino, Superconductivity, Vol. 1, ed. by R.D. Parks, p.449 (Marcel Dekker, New York 1969).

28. F. Steglich, C. D. Bredl, W. Lieke, U. Rauchschwalbe and F. Aparn, "Heavy Fermion Superconductivity," Physica B **126**, 82-91, (1984).

29. R. Oscheroff, C. Richardson and D. M. Lee, Phys. Rev. Lett. **28**, 885, (1972).

30. A. J. Legget, Rev. Mod. Phys. **47**, 331, (1975).

31. J. C. Wheatley, Rev. Mod. Phys. **47**, 415 (1975).

32. P. W. Anderson and W. F. Brinkman, "The Physics of Liquid and Solid Helium", edited by K.H. Bennemann and J.B. Ketterson, Pt. II, p.177, (Wiley, New York, 1978).

33. G. E. Volovik, and L. P. Gor'kov, Zh. Eksp. Teor. Fiz. 39, 550, (1984 [translation: Sov. Phys.-JETP **39**, 674, 1984]).

34. H. R. Ott, E. Felder, A. Bernasconi, Z. Fisk, J. L. Smith, L. Taillefer and G. G. Lonzarich, "Specific Heat and Thermal Conductivity of Superconducting UBe_{13} and UPt_3 at Very Low Temperatures", Jpn. J. Appl. Phys. 26, Suppl. **26-3**, 1217-1218, (1987).

35. F. Steglich, U. Ahlheim, C. D. Berdl, C. Geibel, M. Lang, A. Loidl and G. Sparn, "Superconductivity and Magnetism in Heavy-Fermion Compounds", Frontiers in Solid State Sciences, Vol 1, L. C. Cupta and M. S. Multani, eds. (World Scientific, Singapore), page **527-583**, (1992).

36. F. Steglich, C. Geibel, A. Loidl, G. Sparn, C. D. Bredl and R. Caspary, "Heavy Fermions - Their Magnetism and Superconductivity", International J. of Modern Physics B **7**, 2-8, (1993).

37. Sato, N., T. Sakon, N. Takeda, T. Komatsubara, C. Geibel, and F. Steglich: "Anisotropy in a Heavy Fermion Superconductor: UPd_2Al_3", J. of the Physical Society of Japan **61**, 32-34, (1992).

38. Varma, C. M..: "Valence Fluctuations, Heavy Fermions and Their Superconductivity", Comments on Solid State Physics **11**, 221-243, (1985).

39. Steglich, F., U. Ahlheim, J.J.M. Franse, N. Grewe, D. Rainer and U. Rauchschwalbe: "$CeCu_2Si_2$ and UPt_3: Two Different Cases of Heavy Fermion Superconductivity," J. of Magn. & Magn Mat **52**, 54-60, (1985).

40. Jaccard, D., and J. Flouquet: "The Normal Phase of Heavy Fermion Compounds", J. Magn. & Magn. Mat. **47&48**, 45-50, (1985).

41. Jaccard, D., J. Flouquet, Z. Fisk, J.L. Smith, and H.R. Ott: "Thermal Conductivity and Thermoelectric Power of UBe_{13}", J. Physique Lett. **46**, L811-L817, (1985).

42. Jaccard, D., and J. Flouquet: "Transport Properties of Heavy Fermion Systems", Helv. Phys. Acta **60**, 108-121, (1987).

43. Behnia, K., L. Taillefer, J. Flouquet, D. Jaccard and Z. Fisk: "Anisotropic Field Dependence of Thermal Conductivity in Superconducting UPt_3", Physica B **165 &166**, 355-356, (1990).

44. Behnia, K., L. Taillefer, J. Flouquet, D. Jaccard, K. Maki, and Z. Fisk: "Thermal Conductivity of Superconducting UPt_3," ,J. of Low Temperature Physics **84**, 261-278, (1991).

45. Behnia, K., D. Jaccard, L. Taillefer, J. Flouquet, and K. Maki: "Thermal Conductivity of Superconducting UPt_3 Versus Magnetic Field: probing the gap structure", J. of Mahn. & Magn. Mat. **108**, 133-134, (1992).

46. Behnia, K., L. Taillefer and J. Fouquet: "The Superconducting Phases of UPt_3 Under Pressure", J. App. Phys. **67**, 5200-5202, (1990).

47. Varma, C.M.: "Phenomenological Aspects of Heavy Fermions", Phys. Rev. Lett. **55**, 2723-2726, (1985).

48. Sigrist, M.and Kaxuo Ueda: "Phenomenological Theory of Unconventional Superconductivity," Rev. Mod. Phys. **63**, 239-311, (1991).

49. R. H. Heffner, M. R. Norman, "Heavy Fermion Superconductivity," Comments on Condensed Matter Physics **17**, 361-408, (1996).

50. F. Steglich, U. Rauchschwalbe, U. Gottwick, H. M. Mayer, G. Sparn, N. Grewe, U. Poppe and J. J. M. Franse, "Heavy Fermions in Kondo Lattice Compounds", J. Appl. Phys. **57**, 3054-3059, (1985).

51. D. Jaccard, J. Flouquet, P. Lejay and J. L. Tholence, "Transport and Magnetic Measurements on UPt_3", J. Appli. Phys. **57**, 3082-3083, (1985).

52. N. Grewe and F. Steglich, "Heavy Fermions", Handbook on the Physics and Chemistry of Rare Earths, edited by Gschneidner K. A. Jr. and L. Eyring (Elsevier Science), Vol **14**, p343, (1991).

53. K. Behnia, D. Jaccard, J. Sierro, P. Lejay and J. Flouquet, "Thermal Conductivity of Heavy-Fermion Superconductor URu_2Si_2", Physica C **196**, 57-62 (1992).

54. C. M. Varma, "Heavy Fermion Superconductors and Some Associated Problems", J. Appl. Phys., **57**, 3064-3066, (1985).

55. B. Lussier, B. Ellman, and L. Taillefer, "Ansiotropy of Heat Conduction in the Heavy Fermion Superconductor UPt_3", Phys. Rev. Lett., **73**, 3294-3297, (1994).

56. D. J. Bishop, C. M. Varma, B. Batlogg, E. Bucher, Z. Fisk and J. L Smith: "Ultrasound-Attenuation in UPt_3", Phys. Rev. Lett. **53**, 1009-1011, (1984).

57. B. Golding, D. J. Bishop, B. Batlogg, W. H. Haemmerle, Z. Fisk, J. L. Smith and H. R. Ott, "Observation of a Collective Mode in Superconducting UBe_{13}", Phys. Rev. Lett. **55**, 2479, (1985).

58. V. Müller, D. Maurer, E. W. Scheidt, Ch. Roth, K. Lüders, E. Bucher and H. E. Mömmel: "Observation of a Lambda-Shaped Ultrasonic Peak in Superconducting UPt_3", Solid State Comm. **57**, 319-321, (1986).

59. B. S. Shivaram, Y. H. Jeong and T. F. Rosenbaum, "Anisotropy of Transverse Sound in the Heavy-Fermion Superconductor UPt_3", Phys. Rev. Lett. **56**, 1078-1081, (1986).

60. E. Bucher, B. Batlogg, D. J. Bishop, C. M. Varna, Z. Fisk and J. L. Smith, "Anisotropic (Triplet) Superconductivity in UPt_3", J. Appl. Phys., **57**, 3060-3063, (1985).

61. P. Thalmeier, B. Wolf, D. Weber, R. Blick, G. Bruls and B. Lüthi: "Sound Propagation in Heavy Fermion Compounds, J. of Magn. & Magn. Mat. **108**, 109-110, (1992).

62. D. W. Hess, P. S. Riseborough and J. L. Smith, Heavy-Fermion Phenomena, Encyclopedia of Applied Physics **7**, 435-463, (1993).

63. K. J. Sun, A. Schenstrom, Bimal K. Sarma, M. Levy and D. G. Hinks, "Ultrasonic Behavior of the Heavy-Fermion Superconductor URu_2Si_2," Phys. Rev. **B 40**, 11284-11286, (1989).

64. Z. Fisk, H. Borges, M. McElfresh, J. L. Smith, J. D. Thompson, H. R. Ott, G. Aeppli, E. Bucher, S. E. Lambert, M. B. Maple, C. Broholm and J. K. Kjems, "Superconductivity of Heavy-Electron Uranium Compounds", Physica C **153-155**, 1728-1733, (1988).

65. A. L. Fetter and P. C. Hohenberg, Superconductivity, Vol 2, ed. R. D. Parks , p. 817 (Marcel Rekker, New York 1969).

66. M. B. Maple, J. W. Chen, Y. Dalichaouch, T. Kohara, C. Rossel, M. S. Torikachvili, M. W. McElfresh and J. D. Thompson, "Partially Gapped Fermi Surface in the Heavy-Electron Superconductor URu_2Si_2", Phys. Rev. Lett. **56**, 185-188, (1986).

67. U. Rauchschwalbe, "Magnetic Properties of Heavy-Fermion Superconductors", Physica B&C **147**, 1-80., (1987).

68. S. E. Chen, J. W. Lambert, M. B. Maple, Z. Fisk, J. L. Smith, G. R. Stewart and J. O. Willis, "Upper Critical Magnetic Field of the Heavy Fermion Superconductor UPt_3", Phys. Rev. B **30**, 1583-1585, (1984).

69. U. Rauchschwalbe, W. Lieke, C. D. Bredl, F. Steglich, J. Aarts, K. M. Martini and A. C. Mota, "Critical Fields of the 'Heavy-Fermion' Superconductor $CeCu_2Si_2$", Phys. Rev. Lett. **49**, 1448-1451, (1982).

70. U. Rauchschwalbe, U. Ahlheim, F. Steglich, D. Rainer and J. J. M. Franse, "Upper Critical Fields of the Heavy Fermion Superconductors $CeCu_2Si_2$, UPt_3, and UBe_{13}: Comparison Between Experiment and Theory", Z. Phys. B - Condensed Matter **60**, 379-386, (1985).

71. U. Rauchschwalbe, U. Ahleim, C. D. Bredl, H. M. Mayer and F. Steglich, "Critical Magnetic Fields and Specific Heats of Heavy Fermion Superconductors", J. of Magnetism and Magnetic Materials **63&64**, 447-454, (1987).

72. J. W. Chen, S. E. Lambert, M. B. Maple, M. J. Naughton, J. S. Brooks, Z. Fisk, J. L. Smith and H. R. Ott, "Upper Critical Magnetic Field of the Superconducting Heavy Fermion System $(U_{1-x}Th_x)Be_{13}$", J. Appl. Phys., **57**, 3076-3078, (1985).

73. M. B. Maple, J. W. Chen, S. E. Lambert, Z. Fisk, J. L Smith, H. R. Ott, J. S. Brooks and M. J. Naughton, "Upper Critical Field of the Heavy-Fermion Superconductor UBe_{13}", Phys. Rev. Lett. **54**, 477-480, (1985).

74. J. O. Willis, Z. Fisk, J. L. Smith, J. W. Chen, S. E. Lambert and M. R. Maple, "Initial Slope of the Upper Critical Field of the Heavy Fermion Superconductor UPt_3", proceedings of the Seventeenth International Conference on Low Temperature Physics, edited by U. Eckern, A. Schmid and W. Weber, page 245-246, (Elesvier Science 1984).

75. B. S. Shivaram, T. F. Rosenbaum and D. G. Hinks, "Unusual Angular and Temperature Dependence of the Upper Critical Field in UPt_3", Phys. Rev. Lett. **57**, 1259-1262, (1986).

76. A. de Visser, J. J. M. Franse, A. Menovsky and T. M. Palstra, "Spin Fluctuations and Superconductivity in UPt_3", J. Phys. F: Met. Phys. **14**, L191-L196, (1984).

77. T. T. M. Palstra, P. H. Kes, J. A. Mydosh, A. de Visser, J. J. M. Franse and A. Menovsky, "Bulk Superconductivity in the Heavy-Fermion Superconductor UPt_3, Phys. Rev. **B 30**, 2986, (1984).

78. K. Kadowaki, A. Umezawa and S. B. Woods, "Magnetoresistance of the Heavy Fermion Superconductor UPt_3 near T_c", J. of Magn. & Magn. Materials **54-57**, 385-386, (1986).

79. F. Steglich, U. Ahlheim, A. Böhm, C. D. Bredl, R. Caspary, C. Geibel, A. Grauel, R. Helfrich, R. Köhler, M. Lang, A. Mehner, R. Modler. C. Schank, C. Wassilew, G. Weber, W. Assmus, N. Sato and T. Komatsubara, "Phase Transitions in the Heavy Fermion-Liquid State: $CeCu_2Si_2$, UNi_2Al_3 and UPd_2Al_3", Physica C **185-189**, 379-384, (1991).

80. N. Keller, J. L. Tholence, A. Huxley and J. Flouquet, "Angular Dependence of the Upper Critical Field of the Heavy Fermion Superconductor UPt_3", Physi. Rev. Lett. **73**, 2364-2367, (1994).

81. N. R. Werthamer, E. Helfand and P. C. Hohenberg, Phys. Rev. **147**, 295, (1966).

82. A. M. Clogston, Phys. Rev. Lett. **9**, 266, (1962).

83. T. P. Orlando, E. J. McNiff, S. Foner and M. B. Beasley, "Critical Fields. Pauli Paramagnetic Limiting, and Material Parameters of Nb_3Sn and V_3Si", Phys. Rev. B **19**, 4545, (1979).

84. R. R. Hake, "Upper-Critical-Field Limits for Bulk Type-II Superconductors", Appl. Phys. Lett. **10**, 189-192, (1967).

85. A. Amato, R. Feyerherm, F. N. Gygax, A. Schenck, M. Weber, R. Caspary, P. Hellmann, C. Schank, C. Geibel, F. Steglich, D. C. MacLaughlin, E. A. Knetsch and R. H. Heffner, "Magnetic and Superconducting Properties of the Heavy-Fermion Superconductor UPd_2Al_3", Europhys. Lett. **19**, 127-133, (1992).

86. R. Feyerherm, A. Amato, F. N. Gygax, A. Schenck, C. Geibel, F. Steglich, N. Sato and T. Komatsubara, "Coexistence of Local Moment Magnetism and Heavy-Fermion Superconductivity in UPd_2Al_3," Phys.Revi.Letters **73**, 1849-1852, (1994).

87. P. W. Anderson, J. Phys. Chem. Solids 11, **26**, (1959).

88. R. Balian and N. R. Werthamer. Phys. Rev. **131**, 1553, (1963).

89. H. R. Ott, E. Felder, Z. Fisk, R. H. Heffner and J. L. Smith, "Influence of Boron Impurities on the Superconducting Phase Transition of $U_{1-x}Th_xBe_{13}$", Phys. Rev. B **44**, 7081-7084, (1991).

90. H. Spille, U. Rauchschwalbe and F. Steglich, "Superconductivity in $CeCu_2Si_2$: Dependence of T_c on Alloying and Stoichmetry", Helv. Phys. Acta, **56**, 165-177, (1983).

91. F. Steglich, U. Ahlheim, U. Rauchschwalbe and H. Spille, "Heavy Fermions and Superconducting Spectroscopyof Non-Magnetic Impurities in $CeCu_2Si_2$", Physica **148B**, 6-13, (1987).

92. U. Ahlheim, M. Winkelmann, C. Schank, C. Geibel, F. Steglich and A. L. Giorgi, "Influence of Th Substitution in the Heavy Fermion Superconductor $CeCu_2Si_2$", Physica B **163**, 391-394, (1990).

93. U. Ahlheim, M. Winkelmann, P. van Aken, C. D. Bredl, F. Steglich and G.R. Stewart, "Pair Breaking in the Heavy Fermion Superconductors $Ce_{1-x}M_xCu_{2.2}Si_2$ and $U_{1-x}M_xBe_{13}$ (M:Th, La, Y and Gd)", J. Magn. & Magn. Mat. **76&77**, 520-522, (1988).

94. J. L. Smith, Z. Fisk, J. O. Willis, A. L. Giorgi, R. B. Roof, H. R. Ott, H. Rudigier and E. Felder, "Impurity Effects in UBe_{13}", Physica **135B**, 3-8, (1985).

95. J. L. Smith, Z. Fisk, J. O. Willis, B. Batlogg and H.R. Ott, "Impurities in the Heavy Fermion Superconductor UBe_{13}", J. Appl. Phys. **55**, 1996-2000, (1984).

96. H. R. Ott, "Characteristic Features of Heavy-Electron Materials", Progress in Low Temp. Phys., edited by D.F. Brewer (Elsevier Science), p **216**, (1987).

97. J. M. Lawrence, J. D. Thompson and Y. Y. Chen, Phys. Rev. Lett. **54**, 2537, (1985).

98. C. J. Pethick and D. Pines, "Transport Processes in Heavy-Fermion Superconductors", Phys. Rev. Lett. **57**, 118-121, (1986).

99. P. Hirschfeld, D. Vallhardt and P. Wölfle, Solid State Commun. **59**, 111, (1986).

100. K. Miyake, "Anisotropic Pairing in Heavy Fermion Superconductors - its Origin and Effects on Transport Properties", J. Magn. & Magn. Mat. **63&64**, 411-419, (1987).

101. S. Schmitt-Rink, K. Miyake and C.M. Varma, "Transport and Thermal Properties of Heavy-Fermion Superconductors: A Unified Picture" Phy. Rev. Lett. **57**, 2575-2578, (1986).

102. D. L. Cox and N. Crewe, Z. Phys. B **71**, 321, (1988).

103. F. Steglich, C. D. Bredl, F. R. de Boer, M. Lang, U. Rauchwchwalbe, H. Rietschel, R. Schefzyk, G. Sparn and G. R. Stewart, "Phenomenology of Heavy Fermion Systems", Phys. Scr. T19, 253-259, (1987).

104. A. A. Abrikosov and Gorkov and L.P. Gorkov, Sov. Phys. - JETP **12**, 1243, (1961).

105. A. A. Abrikosov and Gorkov and L.P. Gorkov, Zh. Eksp. Teor. Fiz. **39**, 1781, (1960).

106. S. Skalski, O. Betbeder-Matibet and P.R. Weiss, Phys. Rev. **136**, 1500, (1964).

107. E. Müller-Hartmann and J. Zittartz, Phys. Rev. Lett. **26**, 428, (1971).

108. I. F. Foulkes and B. L. Gyorffy, Phys. Rev. B **15**, 1395, (1977).

109. H. R. Ott, H. Rudigier, Z. Fisk and J. L. Smith, "Phase Transition in the Superconducting State of $U_{1-x}Th_xBe_{13}$ (x = 0 - 0.16)", Phys. Rev. B **31**, 1651-1653, (1985).

110. H. R. Ott, "Superconductivity in Heavy-Electron Metals", Journal of Low Temperature Physics, Vol. **95**, 95-108, (1994).

111. S. E. Lambert, Y. Dalichaouch, M. B. Maple, J. L. Smith and Z. Fisk, "Superconductivity Under Pressure in $(U_{1-x}Th_x)Be_{13}$: Evidence for Two Superconducting States", Phys. Rev. Lett., **57**, 1619-1622, (1986).

112. B. Batlogg, D. Bishop, B. Golding, C. M. Varma, Z. Fisk, J. L Smith and H.R. Ott: "λ-Shaped Ultrasound-Attenuation Peak in Superconducting (U, Th)Be$_{13}$", Phys. Rev. Lett. **55**, 1319. (1985).

113. A. L. Giorgi Z. Fisk, J. O. Willis, G. R. Stewart and J. L. Smith, "Substitution of Ga, Cu and B in Be Sublattice by Heavy Fermion Superconductor UBe$_{13}$", 17th International Conference on Low-Temperature Physics (LT-17), edited by U. Eckern, A. Schmid, W. Weber and H. Wuhl (Elesevier Science), p229, (1984).

114. K. Ueda and T. M. Rice, "p-Wave Superconductivity in Cubic Metals", Phys. Rev. B **31**, 7114-7119, (1985).

115. M. Ishikawa, H. F. Braun and J. L. Jorda, "Effect of Composition on the Superconductivity of CeCu$_2$Si$_2$", Phys. Rev. B **27**, 3092-3095, (1983).

116. F. G. Aliev, N. B. Brandt, R. V. Lutsiv, V. V. Moshchalkov and S. M. Chudinov, "Superconductivity of CeCu$_2$Si$_2$", Pis'ma Zh. Eksp. Teor. Fiz. **35**, 435-438, (1982). translation. (JETP 35, 539, 1982).

117. G. R. Stewart, Z. Fisk and J. O. Willis, "Characterization of Single Crystals of CeCu$_2$Si$_2$. A Source of New Perspectives", Physical Review B **28**, 172-176, (1983).

118. B. Bellarbi, A. Benoit, D. Jaccard, J. M. Mignot and H. F. Braun, "High Pressure Valence Instability and T$_c$ Maximum in Superconducting CeCu$_2$Si$_2$", Phys. Rev. B **30**, 1182-1187, (1984).

119. J. O. Willis, J. D. Thompson, Z. Fisk, A. de Visser, J. J. M. Franse and A. Menovsky, "Effect of Pressure on Spin Fluctuations and Superconductivity in Heavy Fermion UPt$_3$", Phys. Rev. B **31**, 1654-1657, (1985).

120. F. G. Aliev, N. B. Brendt, V. V. Moshchalkov and S. M. Chudinov, "Superconductivity in CeCu$_2$Si$_2$", Solid State Commun. **45**, 215-218, (1983).

121. K. Bakker, A. de Visser, E. Brück, A. A. Menovsky and J. J. M. Franse, "Anisotropic Variation of T$_c$ and T$_N$ in URu$_2$Si$_2$ by Iniaxial Pressure", J. Magn. & Magn. Matt. **108**, 63-64, (1992).

122. G. E. Brodale, R. A. Fisher, Norman E. Phillips, G. R. Stewart and A. L. Giorgi, "Pressure Dependence of Spin-Fluctuation Effects in the Specific Heat of the Heavy-Fermion Superconductor UPt$_3$", Phys. Rev. Lett. **57**, 234 -237, (1986).

123. M. W. McElfresh, J. D. Thompson, J. O. Willis, M. P. Maple, T. Kohana and M. S. Torikachvili, "Effect of Pressure on Competing Electronic Correlations in the Heavy-Electron System URu$_2$Si$_2$", Phys. Rev. B **35**, 43-47, (1987).

124. F. Thomas, J. Thomasson, C. Ayache, C. Geibel and F. Steglich, "Precise Determination of the Pressure Dependence of Tc in the Heavy Fermion Superconductor $CeCu_2Si_2$", Physica B **186-188**, (1993).

125. F. G. Aliev, N. B. Brandt, V. V. Moshchalkov and S. M. Chudinov, "Electric and Magnetic Properties of the Kondo-Lattice Compound $CeCu_2Si_2$", J. of Low Temperature Physics **57**, 61-93, (1984).

126. A. Bleckwedel and A. Eichler, "Calorimetric Investigation Under High Pressure of the Heavy Fermion Superconductor $CeCu_2Si_2$", Solid State Commun. **56**, 693-696, (1985).

127. N. E., Phillis, R. A. Fisher, J. Flouquet, A. L. Giorgi, A. Olsen and G. R. Stewart, "Pressure Dependences of the Specific Heats of UPt_3, UBe_{13} and $CeAl_3$", J. Magn. & Magn Matt. **63&64**, 332-334, (1987).

128. C. J. Pethick, D. Pines, K. F. Quader, K. S. Bedell, G. E. Brown, "One-Component Fermi-Liquid Theory and the Properties of UPt_3", Phys. Rev. Lett. **57**, 1955-1958, (1986).

129. R. Modler, K. Gloos, C. Geibel, T. Komatsubara, N. Sato, C. Schank and F. Steglich, "Thermal Expansion and Magnetostriction of the Heavy-Fermion Compound UPd_2Al_3", Int. J. Mod. Phys. B **7**, 42-45, (1993).

130. R. Modler, K. Gloos, H. Schimanski, C. Geibel, M. Günther, G. Bruls, B. Lüthi, T. Komatsubara, N. Sato, C. Schank and F. Steglich, "Thermal Expansion, Magnetostriction and Elestic Constants of UPd_2Al_3", Physica B **186-188**, 294-296, (1993).

131. F. Steglich, P. Gegenwart, C. Geibel, R. Helfrich, P. Hellmann, M. Lang, A. Link, R. Modler, G. Sparn, N. Buettgen and A. Loidl, "New Observations Concerning Magnetism and Superconductivity in Heavy-Fermion Metals", Physica B **223-224**, 1, (1996).

132. R. Movshovich, T. Graf, D. Mandrus, J. D. Thompson, J. L. Smith and Z. Fisk, "Superconductivity in Heavy-Fermion $CeRh_2Si_2$", Phys. Rev. **B8241**, (1996).

133. D. Jaccard, K. Behnia and J. Sierro: "Pressure Induced Heavy Fermion Superconductivity of $CeCu_2Ge_2$", Physics Letters A **163**, 475, (1992).

134. F. M. Grosche, S. R. Julian, N. D. Mathur and G. G. Lonzarich, "Magnetic and Superconducting Phases of $CePd_2Si_2$", Physica B **223-224**, 50, (1996).

135. S. R. Julian, C. Pfleiderer, F. M. Grosche, N. D. Mathur, G. J. McMullan, A. J. Diver, I. R.Walker and G. G. Lonzarich, "The Normal States of Magnetic d and f Transition Metals", Journal of Physics: Condensed Matter **8**, 9675, (1996).

136. I. R. Walker, F. M. Grosche, D. M. Freye and G.G. Lonzarich, "The Normal and Superconducting States of $CeIn_3$ Near the Border of Antiferromagnetic Order", Physica C **282-287**, 303, (1997).

137. D. Jaccard and J. Sierro, " Magnetic Instability and Superconductivity in Cerium Based HF Compounds", Physica B **206-207** 625, (1995).

138. B. H. Grier, J. M. Lawrence, V. Murgai and R. D. Parks, "Magnetic Ordering in CeM_2Si_2 (M=Ag,Au,Pd,Rh) Compounds as Studied by Neutron Diffraction", Phys. Rev. B **29**, 2664, (1984).

139. N. D. Mathur, F. M. Grosche, S. R. Julian, I. R. Walker, D. M. Freye, R. K. W. Haselwimmer and G. G. Lonzarich, "Magnetically Mediated Superconductivity in Heavy Fermion Compounds", Nature **394**, 39, (1998).

140. P. Link, D. Jaccard and P. Lejay, "High-Pressure Transport Properties of $CePd_2Si$.", Physica B **223-224**, 303, (1996).

141. J. D. Thompson, R. D. Parks and H. Borges, "Effect of Pressure on the Néel Temperature of Kondo-Lattice Systems", J. of Magn. and Magn. Mat. **54-57**, 377, (1986).

142. D. E. MacLaughlin, C. Tien, W. G. Clark, M. D. Lan, Z. Fisk, J. L. Smith and H. R. Ott, "Nuclear Magnetic Resonance and Heavy-Fermion Superconductivity in $(U,Th)Be_{13}$", Phys. Rev. Lett. **53**, 1833, (1984).

143. D. E. MacLaughlin, "Nuclear Magentic Resonance and the Coherent State of Unstable-Moment Rare-Earth Compounds at Low Temperatures", J. Magn. Magn. Mat. **47&48**, 121-126, (1985).

144. Y. Kitaoka, K. Ueda, T. Kohara, K. Asayama, Y. Onuki and T. Komatsubara, "Nuclear Magnetic Relaxation in the Heavy Electron Superconductor $CeCu_2Si_2$", J. Magn. Magn. Mat. **52**, 341-343, (1985).

145. D. Einzel, P. J. Hirschfeld, F. Gross, B. S. Chandrasekhar, K. Andres, H. R. Ott, J. Beuers, Z. Fisk and J. L. Smith:, "Magnetic Field Penetration Depth in the Heavy-Electron Superconductor UBe_{13}", Phys. Rev. Letters, **56**, 2513-2516, (1986).

146. Y. Kohori, T. Kohora, H. Shibai, Y. Oda, Y. Kitaoka and K. Asayama, "Nuclear Magnetic Relaxation in the Heavy Fermion Superconductor UPt_3", J. Phys. Soc. Japan **57**, 395-397, (1988).

147. Y. Kohori, H. Shibai, T. Kohora, Y. Oda, Y. Kitaoka and K. Asayama, "195Pt NMR Study of the Heavy Fermion Superconductor UPt_3", J. Phys. Soc. Japan **57**, 395-397, (1988).

148. K. Asayama, Y. Kitaoka and T. Kohori, "NMR Study of Heavy Fermion Materials", J. Magn Magn, Mat. **76&77** 449-454, (1988).

149. Cheng Tien and I. M. Jiang, "Magnetic Resonance of Heavy- Fermion Superconductors and High-T_c Superconductors", Phys. Rev. B **40**, 229-238, (1989).

3

MUON SPIN RELAXATION STUDIES OF SMALL-MOMENT HEAVY FERMION SYSTEMS

Robert H. Heffner

Los Alamos National Laboratory
Los Alamos, N.M. 87545, USA.

I. INTRODUCTION

As discussed in previous chapters, heavy fermion (HF) compounds can display paramagnetism, magnetic ordering (usually in an antiferromagnetic structure), semiconducting behavior and superconductivity. Antiferromagnetism (AFM) with either full-sized ($\mu \geq 1\mu_B$) or small ($\mu \cong 10^{-3}\mu_B$) magnetic moments has been observed. There are six known ambient-pressure HF superconductors (discussed in Chapter 2), and several of these exhibit weak magnetic signatures. In addition, although most HFs show low-temperature Fermi-liquid behavior in their transport and thermodynamic properties, a few disordered materials exhibit non-Fermi liquid behavior at low temperatures instead.

Although muon spin rotation or relaxation (μSR) is a relatively new technique, it has already made important contributions to the study of heavy fermion (HF) systems, particularly because of the technique's extreme sensitivity to small ($\cong 1$ Gauss) magnetic fields. Therefore, we focus here on those HF materials which exhibit very small ($\mu \leq 0.1\mu_B$) or undetectably small ($\mu << 0.001\mu_B$) moments, where μSR experiments are often able to make unique contributions. The materials covered include paramagnetic Fermi-liquid and non-Fermi-liquid systems, small-moment magnets and four of the ambient-pressure HF superconductors. For more complete coverage of μSR studies of magnetic materials in general, the reader is referred to recent reviews by Schenck and Gygax[1] and Amato.[2]

As a magnetic probe μSR is similar to NMR, the positive muon (μ^+) acting as a light interstitial proton (spin 1/2) with mass equal to 207 electron masses. (Strictly speaking one can do μSR with either μ^+ or μ^- particles, though almost all present-day experiments use positive muons.) The technique is also complementary to neutron scattering (discussed in Chapter 4), and is similar to the Mössbauer effect, also a local nuclear probe. There are several principal advantages of μSR:

(1) The muon can be deposited in any material of interest, limited only by the sample environment, i.e., the necessity to penetrate the cryostat or pressure cell walls, for example. Thus, unlike NMR, where a material with sizable nuclear moments is required, no such limitation for μSR exists.

(2) The technique has high sensitivity, advantageous for weak-moment magnetism studies. This is due to the large moment of the muon itself (3.2 × proton moment) and the fact that the muons are highly polarized (80-100%).

(3) Because of the high flux of muons produced by modern medium energy accelerators, up to 10^7 μSR events can be accumulated in less than one hour, resulting in high statistical precision.

(4) The high intrinsic polarization means that experiments are easily carried out in zero applied field; i.e., no applied field is necessary to polarize the probe spin.

(5) The muon has spin 1/2, and therefore there is no quadrupole splitting, as in NMR for nuclei with spin > 1/2. Thus the spectrum is usually a single line, making analysis easy compared to NMR. This can also be a relative disadvantage, because quadrupolar relaxation from lattice vibrations cannot be observed.

(6) Because a stationary muon is a local probe, it can only measure momentum-averaged quantities. While this is a disadvantage for some measurements, it is a distinct advantage in searching for very weak magnetism, because one does not have to locate a signal which is visible only in certain **k**-regions, as in neutron scattering experiments, for example.

Perhaps the biggest disadvantages of μSR are:

(1) The muon stopping site is not known a priori. Although the site(s) can often be determined, as described below this requires single crystals in order to measure the local symmetry of the muon position. Furthermore, the results are sometimes ambiguous.

(2) Because the muon lifetime is only 2.2 μs, slow spin-lattice relaxation is not observable. This is an important consideration for HF materials, where relaxation from the conduction electrons alone (Korringa mechanism), or from the weak electronic moments, is usually not observable. This is a major limitation compared to NMR.

(3) Like NMR, one cannot obtain the momentum or energy dependence of the dynamic susceptibility, as is available with neutron scattering

II. THE μSR TECHNIQUE

A. Experimental Setup

μSR experiments are carried out at the world's meson facilities and begin with the production of copious quantities of pi mesons. This occurs from the nuclear interaction of medium energy (500 - 800 MeV) protons with the nuclei in a production target, typically graphite. The pions decay with a lifetime of 27 ns into a muon and a neutrino:

$$\mu^+ \rightarrow \mu^+ + \nu_\mu \tag{1}$$

Most μSR experiments today utilize muons produced from pions at rest, that is, those that stop in the outer skin of the production target. These are referred to as 'surface' muons. Because of conservation of angular momentum and the near zero mass of the neutrino - which guarantees that it is 100% polarized along its own momentum - the muon is born essentially 100% polarized in the rest frame of the pion. Thus, a 'surface' muon beam is 100% polarized, with the polarization antiparallel to the momentum.

Muons are collected in an evacuated magnetic channel (consisting of bending and focusing magnets) and implanted into the material of interest where they thermalize and, at low enough temperatures, are self-trapped in the lattice. The typical range for 'surface' muon beams is about 0.15 g/cm^2 of material. The trapping site is one of low electrostatic energy and often of high structural symmetry, such as an octahedral or tetrahedral site in a cubic lattice. In metals (and in all heavy fermion materials of interest here) the muons do not lose their polarization during thermalization.

The muon itself decays into a positron and two neutrinos with a lifetime τ_μ of 2.2 μs:

$$\mu^+ \rightarrow e^+ + \bar{\nu}_\mu + \nu_e. \tag{2}$$

Because this is a weak decay which violates parity, the angular distribution of the decay positron with respect to the spin of the muon is asymmetric:

$$dN(t) \, / \, dt d \, \Omega = N_o (1/\tau_\mu) exp \, (-1/\tau_\mu) \{1 + <a> \, Gcos(\theta_e)\}. \tag{3}$$

Here θ_e is the angle between the muon spin and the positron momentum vector, $<a>$ is the average asymmetry of the decay process, typically $\leq 1/3$, and G is the muon spin polarization.

For convenience, we now review some of the salient features of muon spin relaxation phenomenology. The reader is referred to Schenck and Gygax[1] for additional details. Equation (3) is valid in free space. In a material, we are interested in monitoring the time-decay of the polarization G(t), due to the interaction of the muon's spin with the local magnetic field. This is accomplished by detecting the time of emission of the positron relative to the implantation time of the muon, for a particular positron direction.[1] This direction determines either a 'longitudinal-field' or 'transverse-field' geometry, depending on whether the applied field is directed along the axis of the initial muon spin direction or perpendicular to it. In a longitudinal- or zero-field geometry, the positrons are always detected along the initial direction of the muon spin, as will be made clear below.

There are two primary types of accelerator sources for muons, those which produce a quasi-continuous beam (in time) and those which produce a pulsed muon beam. A continuous beam requires the rejection of all multiple muon stopping events which occur within a typical time window set by the muon lifetime, usually 10 - 15 μs. Furthermore, the time spectrum always contains a background from random positron events. The highest precession frequencies which can be measured in this setup are limited by the time resolution of the electronics and scintillation counters, typically a few ns. (In specially designed spectrometers, a factor of ten better is achieved, allowing GHz frequencies to be measured.) In a pulsed-beam experiment, the muon beam is literally turned off between pulses and so there is no random background. This type of experimental arrangement is therefore best for measuring very small relaxation rates. The highest frequency one can measure with a pulsed beam is determined by the muon pulse width, 70 ns at ISIS (Rutherford Appleton Lab), limiting the measurable frequencies to the MHz range and the applied or internal fields to less than about 500 Gauss.

B. Muon Spin Relaxation

To appreciate how a μSR experiment works, we consider a muon in free space with its spin $\mathbf{I}$ along the +z-axis at t = 0 in the presence of a static field $\mathbf{B}_0$, which makes an angle θ with $\mathbf{I}(t = 0)$. If we choose to detect the positrons along the +z direction, then we are measuring the time evolution of $G_z(t)$, which is given by

$$G_z(t) = \cos^2 \theta + \sin^2 \theta \cos (\omega_\mu t). \tag{4}$$

Here $\omega_\mu = \gamma_\mu |\mathbf{B}_0|$ and $\gamma_\mu = 2\pi(13.55 \times 10^3 \text{ G}^{-1} \text{ s}^{-1})$ is the muon's gyromagnetic ratio.

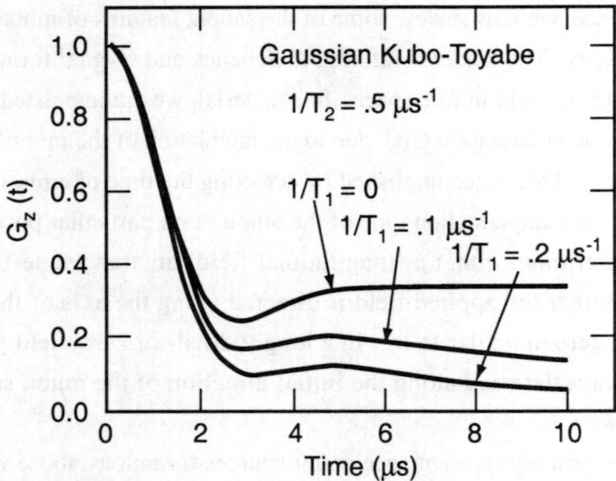

Fig. 1. Gaussian Kubo-Toyabe relaxation function $G_z(t)$ appropriate for zero-field μSR (Equation 8), with $\langle \cos^2\theta \rangle = 1/3$ appropriate for random field directions. Various spin-lattice relaxation rates $1/T_1$ are shown to demonstrate how the tail of $G_z(t)$ falls off for slow relaxation.

In a real material the internal field **B** experienced by the muon arises from the nuclear and/or electronic dipole fields, as well as a transferred hyperfine interaction between the electronic and muon moments. We take $\mathbf{B}_0$ as the static component of the internal field and $\delta\mathbf{B}(t)$ as its fluctuating component. By 'static' we mean that any time variation of $\mathbf{B}_0$ is much slower than the muon Larmor frequency, defined above as ω_μ.

Suppose $|\delta\mathbf{B}(t)| = 0$ and that $\mathbf{B}_0$ is randomly distributed in direction and magnitude according to $P(|\mathbf{B}_0|)$, but with zero mean ($\langle|\mathbf{B}_0|\rangle = 0$). This yields the well-known Kubo-Toyabe[3] zero-field-type relaxation, arising from the dephasing of the muon spin due to the distribution of x and y $|\mathbf{B}_0|$ components. In this case we have

$$G_z(t) = \int \{ \langle \cos^2\theta \rangle + \langle \sin^2\theta \rangle\, P(|\mathbf{B}_0|) \cos(\gamma_\mu |\mathbf{B}_0| t) \} d\,|\mathbf{B}_0| \qquad (5)$$

The random orientation yields $\langle \cos^2\theta \rangle = 1/3$ and $\langle \sin^2\theta \rangle = 2/3$. If Δ is the width of a Cartesian component of $\mathbf{B}_0$ and $\Delta_x = \Delta_y = \Delta_z$, then one has, for Gaussian[4] and Lorentzian[5] distributions $P(|\mathbf{B}_0|)$, respectively:

$$G_z(t) = \langle \cos^2\theta \rangle + \langle \sin^2\theta \rangle \{ 1 - (t/T_2)^2 \} \exp(-0.5\,(t/T_2)^2), \qquad (6)$$

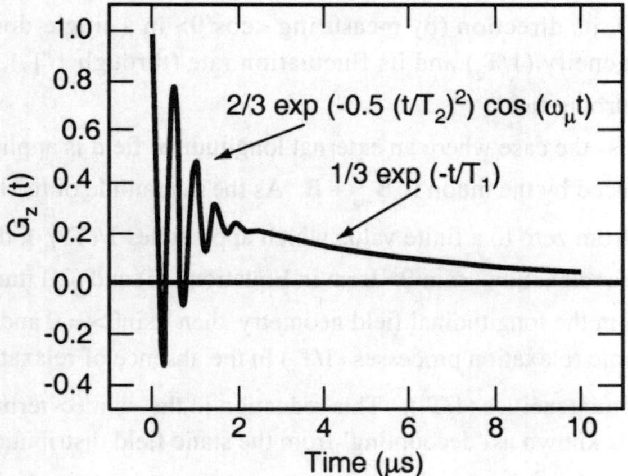

Fig. 2. Zero-field μSR function $G_z(t)$ typical of relaxation in a magnetically ordered powder (Equation 8), assuming a Gaussian distribution of local static fields. The early time oscillations reflect the magnitude of the internal static field at the muon site. The tail reflects the spin lattice relaxation rate $1/T_1$.

and

$$G_z(t) = \langle\cos^2\theta\rangle + \langle\sin^2\theta\rangle\{1 - t/T_2\}\exp(-t/T_2), \qquad (7)$$

where $1/T_2 = \gamma_\mu \Delta$ For slowly fluctuating fields ($|\delta\mathbf{B}(t)| > 0$ and $\omega_\mu\tau \gg 1$) with an autocorrelation time τ, the $\langle\cos^2\theta\rangle$ component of $G_z(t)$ decays like $\exp(-t/T_1)$, for example, with $1/T_1 = 1/\tau$ (as shown in Fig. 1). We are using the familiar notation developed for NMR in terms of the homogeneous and inhomogeneous lifetimes T_1 and T_2, respectively.

If the internal fields undergo spontaneous magnetic order so that $\langle|\mathbf{B}_0|\rangle$ is greater than zero, we obtain a finite precession frequency:

$$G_z(t) = \langle\cos^2\theta\rangle \exp(-t/T_1) + \langle\sin^2\theta\rangle \exp\{-t/T_2\}\cos(\omega_\mu t + \phi_\mu), \qquad (8)$$

Here we have assumed an exponential inhomogeneous relaxation function (see below). The function for Gaussian inhomogeneous relaxation is shown in Fig. 2. Therefore, in principle we can simultaneously measure the magnitude of the internal field (through ω_μ), its direction (by measuring $<\cos^2\theta>$ in a single-domain sample), its static inhomogeneity $(1/T_2)$ and its fluctuation rate (through $1/T_1$), all in the absence of a perturbing field.

We now discuss the case where an external longitudinal field is applied, so that the total field experienced by the muon is $\mathbf{B}_{app} + \mathbf{B}$. As the magnitude of the longitudinal field is increased from zero to a finite value which approaches $1/(T_2\gamma_\mu)$, the amplitude of the inhomogeneous relaxation ($<\sin^2\theta>$ term in Equations (6) and (7)) must decrease. When $|\mathbf{B}_{app}| >> |\mathbf{B}|$ in the longitudinal field geometry, then $<\sin\theta> = 0$ and we are able to measure the dynamic relaxation processes $(1/T_1)$ in the absence of relaxation from the intrinsic field inhomogeneities $(1/T_2)$. This reduction in the $<\sin^2\theta>$ term with applied longitudinal field is known as 'decoupling' from the static field distribution, and is an important signature of quasi-static fields.

The relaxation rate $1/T_1$ is proportional to the Fourier transform of the autocorrelation function for the fluctuating field at the muon site:[1]

$$1/T_1 \propto \sum_k \int dt \langle \delta B^+ (-k,0) \bullet \delta B^- (k,t)\rangle \cos (\omega_\mu t) \qquad (9)$$

where $\delta B\pm(t)$ is the amplitude of the transverse component of the fluctuating local field. The summation over the momentum k of the excitations is because of the local nature of the probe. In the fast fluctuation limit $\gamma_\mu(\delta B)\tau << 1$, the correlation time of the fluctuating field τ is related to $1/T_1$ as follows: $1/T_1 \propto \gamma_\mu^2 (\delta B)^2 \tau/((\omega_\mu\tau)^2 + 1)$. For slow fluctuations, $T_1 \propto \tau$.

In the transverse-field geometry, $\mathbf{B}_{app} \parallel \mathbf{z}$-axis $\perp \mathbf{I}(0)$, the muon spin precesses in the static field $\mathbf{B}_{app} + \mathbf{B}_0$. The spin polarization is relaxed either by dephasing in the x-y plane due to inhomogeneities in the z-component of the local field, or by dynamic spin-flip processes from transverse components of $\delta\mathbf{B}(t)$. Usually, the former process dominates the latter $(1/T_2 >> 1/T_1)$. This gives rise to a transverse-field relaxation function

$$G_x(t) = \exp(-0.5(t/T_2)^2) \text{ and} \qquad (10a)$$

$$G_x(t) = \exp(-t/T_2), \qquad (10b)$$

for a Gaussian or Lorentzian distribution of quasi-static fields with Cartesian width $\Delta_z = 1/(T_2\gamma_\mu)$. Note that only Δ_z dephases the spin in transverse field (Equation 10), while both Δ_x and Δ_y are effective in zero applied field (Equation 6).

C. Form of the Relaxation Functions G(t)

For quasi-static relaxation, the form of the inhomogeneous part of the relaxation function is just the Fourier transform of the internal field distribution $P(|\mathbf{B}_0|)$, as shown in Equation 5. Thus if the internal field has a Gaussian distribution, the muon spin will exhibit Gaussian relaxation; a Lorentzian field distribution yields exponential relaxation in time. Physically, a Gaussian distribution of local fields arises from a concentrated spin system, such as in the case of nuclear dipoles. By contrast, a dilute system of spins, such as might arise in a typical RKKY spin glass (a few per cent Mn in Ag), yields a Lorentzian local field distribution.

Dynamical relaxation of the muon spin by stochastically fluctuating fields yields an exponential relaxation function $\exp(-t/T_1)$.[1] In some physical systems there may be a distribution of coupling constants between the muon and the internal spins (denoted above as $\delta \mathbf{B}(t)$, the amplitude of the fluctuating field) or a distribution of fluctuation rates τ. This gives rise to a distribution of T_1 values which can be modeled by a stretched exponential relaxation function $\exp(-(t/T_1)^\beta)$ (Ref. 1). Often a stretched exponential can also be fit as a sum of two exponentials, which is a special case consisting of two and only two assumed T_1 values.

III. μ^+ POSITION IN THE LATTICE

For some μSR experiments it is not necessary to have a precise knowledge of the muon position in order to achieve a meaningful result. This is true for measurements of the local field distribution in a type II superconductor, which yields the temperature dependence and magnitude of the magnetic field penetration depth. The muon location is not essential because the flux lattice is incommensurate with the crystal lattice and so the muon randomly samples the local field distribution surrounding the flux vortices. Also the vortex lattice constant is many atomic spacings. (It is necessary to know that the muon is stationary, however.) In other kinds of experiments, knowledge of the muon position is very important to interpret the data properly, however. For example, in the search for small-moment magnets one must be sure that the *absence* of a local field is not because the muon sits at a site where the dipole fields from the ordered moments accidentally cancel. Although this is rare, it does happen, as illustrated below in the case of UPt$_3$. Also, in certain non-Fermi liquid compounds, one must calculate the dipolar coupling in order to properly interpret measurements of the linewidth broadening. This requires a knowledge of the muon site.

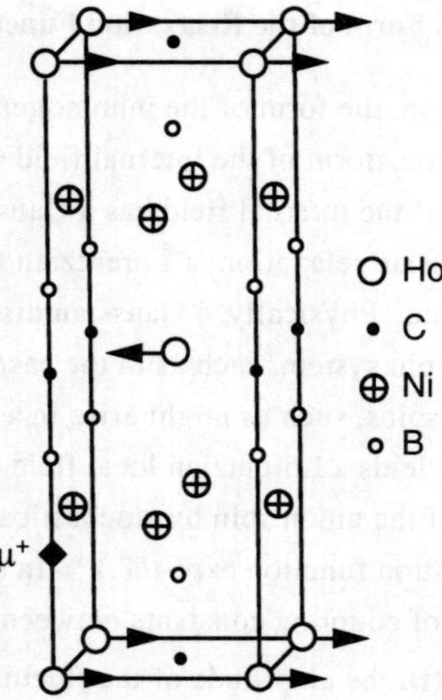

Fig. 3.   Crystal and magnetic structure in the commensurate state of HoNi$_2$B$_2$C. The muon site is shown as a black diamond (from L. P. Le *et al.*).[6]

The most reliable method to determine the μ^+ position is a measurement of the average muon precession frequency or frequencies in the presence of an ordered magnetic structure, followed by a comparison of the measured field(s) to the calculated dipole field(s) for the candidate muon site(s). This method works in principle because the dipole field is exactly calculable for a specific muon site in the presence of a known moment aligned in a known direction. The ordered magnetic structure can be produced by a magnetic phase transition or by an applied field inducing a ferromagnetic alignment of ionic spins. Either way, it is important to make measurements in single crystals with the muon spin or the applied field having a known orientation with respect to the principal axes of the crystal. As nice as this method sounds, it often yields ambiguous results, primarily because either the magnetic phase is complicated or not well enough determined (with neutron scattering) or because of multiple muon stopping sites.[1]

As an example, we consider the case of HoNi$_2$B$_2$C, a borocarbide superconductor (T_c = 5 K) with two magnetic phase transitions, an incommensurate AFM phase below 6 K and a commensurate phase below 5 K. The crystal structure is shown in Fig. 3, with the Ho atoms occupying a body-centered tetragonal structure. The low-temperature AFM phase is also illustrated. Zero-field μSR experiments[6] have been carried out by Le *et al.,*[6] with the initial muon spin both parallel and perpendicular to the c-axis. The observed relaxation function is given in Equation (8). In the $I(0)$ $\parallel$ c-axis configuration the full amplitude precession was observed below 6 K, meaning that $<\cos^2\theta> = 0$. This allowed the authors to conclude that the dipole fields from the Ho ions lay in the basal plane. For $I(0) \perp$ c-axis it was found that $<\cos^2\theta> = <\sin^2\theta> = 1/2$, which is also consistent with this field direction. Considering that only one muon frequency was observed, only two possible muon sites exist, both with axial symmetry: $(0, 0, z)$ and $(x, x, 0)$. Because the calculated dipole field for the $(x, x, 0)$ site is an order of magnitude larger than derived from the observed precession frequency, this site can be eliminated.

The next step in the site determination for $HoNi_2B_2C$ was a measurement of the temperature dependence of the Knight shift.[6] As mentioned above, the μ^+ - ion coupling consists of two components: a dipole field and a transferred hyperfine field. In the presence of an external field, the Ho electronic spins develop a moment μ given by, $\vec{\mu} = \overset{\leftrightarrow}{\chi} \bullet \vec{B}_{app}$, where $\overset{\leftrightarrow}{\chi}$ is the atomic susceptibility tensor. Thus the fields at the muon site are given by: $\vec{B}_{dip}(r_\mu) = \vec{\mu} \bullet \overset{\leftrightarrow}{A}_{dip}(r_\mu)$ and $\vec{B}_{hyp}(r_\mu) = \vec{\mu}\,\overset{\leftrightarrow}{A}_{hyp}(r_\mu)$, where the dipolar and hyperfine coupling constants are denoted by A. From the axial crystal symmetry one has $\chi(\theta) = \chi_\parallel \cos^2\theta + \chi_\perp \sin^2\theta$. The Knight shift K is defined as $K = (B - B_{app}) / B_{app}$, where here $B = B_{app} + B_{dip} + B_{hyp}$. We are neglecting possible contributions from the conduction electrons, and assume that all corrections for the Lorentz and demagnetization fields have been made.[1] This yields

$$K(\theta) = A_{hyp}(\chi_\parallel \cos^2\theta - \tfrac{1}{2}\chi_\perp \cos^2\theta). \tag{11}$$

We note that the dipole field depends on the muon site in the lattice, as well as the direction of the applied magnetic field relative to the crystalline axes.

Measuring the temperature dependence of $K(\theta)$ for $\theta = 0$ and 45 degrees in the paramagnetic state, as well as $\chi_\parallel(T)$ and $\chi_\perp(T)$, enabled Le $et\ al.$ to extract both $A_{dip} = 2.03(7)$ kG/μ_B and $A_{hyp} = 0.15(2)$ kG/μ_B. From the dipolar coupling a muon position of $(0, 0, 0.20)$ was determined. Note that in order for this technique to work, one must have a situation where both $\chi_\parallel(T)$ and $\chi_\perp(T)$ depend significantly on temperature, so that one can extract the slopes in a K versus χ plot. Otherwise one cannot independently extract the two different hyperfine couplings.

A second way that the muon position is sometimes inferred is by measuring the nuclear dipolar linewidth and comparing that to the calculated second moment of the nuclear dipolar field distribution for various assumed muon sites. A similar technique involves extracting the transverse-field lineshape from paramagnetic ions with an anisotropic susceptibility in a powder sample and comparing that to the dipolar lineshape. Although these methods can be employed, they are typically more useful in confirming a muon site determined by other means, than in providing a unique determination by themselves.

It should be mentioned that hopping of the muon from its low-temperature equilibrium position will generally occur via thermal activation at high enough temperatures. This leads to motional narrowing of the nuclear dipolar linewidth (and the linewidth from the ions, if the diffusion is fast enough) when the hopping rate $1/\tau_h$ exceeds the nuclear or ionic linewidth $1/T_2$, i.e., $\tau_h \ll T_2$. This is easily detected in almost all cases. For the investigation of HF compounds, where the temperature range of interest is generally quite low, muon diffusion almost never presents a problem for the interpretation of the data.

IV. SMALL-MOMENT MAGNETIC COMPOUNDS

As discussed in other chapters, HF behavior arises from the interaction between localized electronic moments and itinerant conduction electrons. In the materials of interest here, this interaction involves local f-moments and takes two predominant forms: the RKKY interaction, which gives rise to magnetic coupling between the f-electrons, and the Kondo interaction, which quenches the local f-moments through hybridization with the conduction electrons, forming a many-body spin-singlet state. These interactions both arise from the same exchange coupling J. This coupling results in a characteristic temperature, which for the RKKY and Kondo interactions, is given by: $T_{RKKY} \propto J^2 N(0)$ and $T_K \propto N(0)^{-1} \exp(-1/N(0)J)$, respectively. Here $N(0)$ is the conduction electron density-of-states. As the examples below suggest, it is the competition between the RKKY and the Kondo interaction which is thought to give rise to HF behavior.

This competition was first addressed quantitatively by Doniach[7] in the so-called 'Kondo-necklace' model, schematically illustrated in the T-J phase diagram of Fig. 4. For smaller values of J the RKKY interaction dominates, yielding AFM order. When the Kondo interaction dominates (larger J), the magnetic moments are quenched and Fermi-liquid (FL) properties are observed. HF behavior occurs in the region where T_{RKKY} and T_K are comparable ($J \cong J_c$), so that the Kondo interaction is just strong enough to quench the magnetic order. This can leave the system in the proximity of a zero-temperature magnetic phase transition, which is one of the scenarios believed to give rise to so-called non-Fermi liquid behavior, as discussed in the introductory chapter by MacLaughlin. It is instructive to begin our discussion with a review of μSR experiments in the paramagnetic HF system $CeCu_6$.

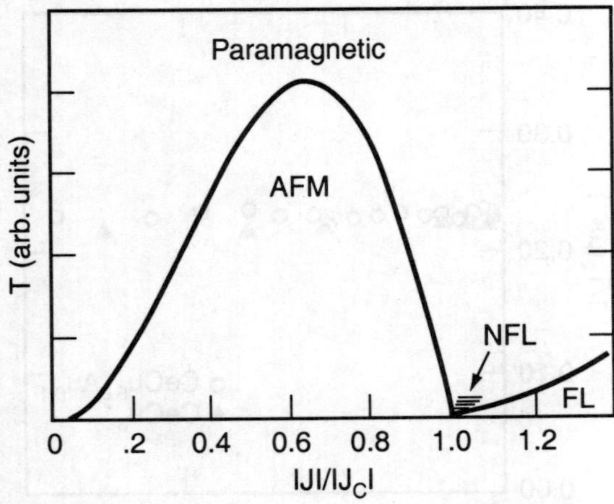

Fig. 4. Modified Doniach model for a Kondo lattice. The vertical axis is temperature and the horizontal axis is the s-f exchange J, normalized to the critical exchange J_c where the Néel temperature vanishes. (From J. D. Thompson, Physica B **223&224**, 643 (1996).) Regions of expected antiferromagnetism (AFM), Fermi-liquid (FL) and non-Fermi liquid (NFL) behavior are shown.

A. CeCu$_6$

Although one of the hallmarks of heavy fermion systems is the disappearance of their high-temperature local moments at low temperatures, there are very few systems that actually exhibit this behavior. Most HFs which show Curie-like susceptibilities at high temperatures exhibit reduced, but not zero, moments below their coherence temperature. One exception to this is the compound CeCu$_6$. This compound[8] forms in the *Pnma* orthorhombic structure, has a low-temperature linear coefficient of specific heat γ of 1600 J/mol K^2 and exhibits Fermi-liquid properties at low temperatures. Although both the magnetoresistivity[9] and thermal power[10] indicate possible low-temperature magnetic anomalies, as we shall see the μSR data give no indication of either magnetic ordering or appreciable slowing down of the Ce moments above 40 mK.

The muon stopping site for this compound has been determined in what has now become the classical way for paramagnets, by measuring the temperature dependence of the muon Knight shift in a transverse field applied along each of the three principal axes of the crystal. The data yielded the dipolar and contact hyperfine coupling constants in a

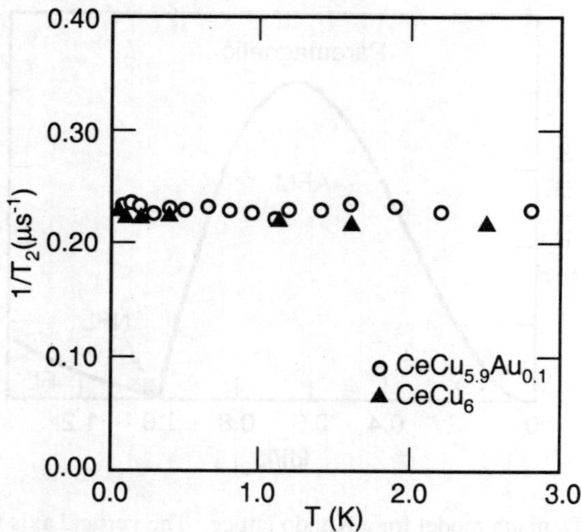

Fig. 5. Temperature dependence of zero-field μSR rate in $CeCu_6$ and $CeCu_{5.9}Au_{0.1}$. (from Amato *et al.*[21]

manner similar to that described above for $HoNi_2B_2C$. The results[11] are $A_{hyp} = 0.17$ kG/μ_B, $A^{xx}_{dip} = 0.89$ kG/μ_B, $A^{yy}_{dip} = -0.10$ kG/μ_B and $A^{zz}_{dip} = -0.79$ kG/μ_B. The muon site was determined to be the (0, 0, 1/2) site on the basis of a calculated dipole field tensor at that site of $\overleftrightarrow{A}_{dip,\,calc,} = (0.97, -0.11, -0.86)$ kG/μ_B. Only this site gives dipolar field components close to the observed ones.

The zero-field relaxation function at T = 2.2 K shows[12] a typical Gaussian Kubo-Toyabe form (Equation 6) with $1/T_2 \cong 0.23$ μs⁻¹. This rate is typical of nuclear dipolar broadening, and in fact is consistent with the calculated dipolar linewidth at the (0, 0, 1/2) site from randomly oriented ^{63}Cu and ^{65}Cu nuclear moments. Furthermore, the application of 1 kOe longitudinal field produces a zero (statistically non-measurable) relaxation rate, indicating that the zero-field relaxation is from quasi-static fields. No temperature dependence for the relaxation rate $1/T_2$ is observed between 40 mK and 3.0 K, as shown in Fig. 5. A tendency toward magnetic ordering (either FM, AFM or spin-glass like) would at least produce an appreciable slowing down of the ionic spins (if not actual muon precession in a static field), leading to an additional exponential relaxation $(\exp(-1/T_1))$ multiplying the observed Gaussian function. Neither a change in functional form of the relaxation nor an increasing linewidth was observed, however, yielding an upper limit of $(1 - 3) \times 10^{-3}$ μ_B/Ce ion for a static moment in $CeCu_6$ above 40 mK. Thus $CeCu_6$ is one of

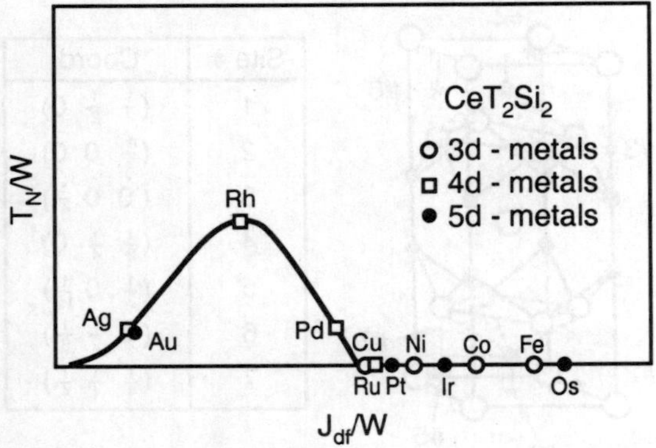

Fig. 6. Schematic modified Doniach phase diagram (see Fig. 4) for the CeT$_2$Si$_2$ series discussed in the text. (from Endstra *et al.*[21] The "T-elements" are labeled along the curve of the Neel temperature vs. d-f exchange. Each axis is normalized to the band width (W).

the few HF systems displaying a paramagnetic ground state; the high-temperature moments have been completely quenched.

B. (U,Ce)X$_2$Si$_2$ Systems

The systematics of the (U,Ce)X$_2$Si$_2$ compounds (where X is a 3d, 4d or 5d element) illustrate the development of HF behavior and the role that μSR can play in elucidating HF properties. (U,Ce)X$_2$Si$_2$ forms mostly in either the body-centered tetragonal I4/mmm or the primitive tetragonal P4/nmm structures. The structural and other magnetic properties have been systematized by Endstra *et al*,[13] using a phenomenological f-d electron hybridization scheme within the context of the Doniach Kondo-necklace model. (The authors also cover the series (U,Ce)X$_2$Ge$_2$.)

According to the Doniach model[7] there is a critical coupling J$_c$ at which a T = 0, second-order AFM transition occurs (see Fig. 4). For J $\leq$ J$_c$, AFM with partially compensated moments occurs, while for J $<<$ J$_c$ the system exhibits full-moment AFM. When J $>>$ J$_c$ the model yields a paramagnetic state with highly reduced moments. The coupling constant J is given by $J = V_{fd}^2 / (E_F - E_f)$, where E$_F$ is the Fermi energy, E$_f$ is the f-electron energy and V$_{fd}$ is the f-d hybridization. The latter depends on the d-band filling and the d-f ion distance. Using calculated values of V$_{fd}$, together with the experimentally measured lattice parameters and ordering temperatures, Endstra *et al.*

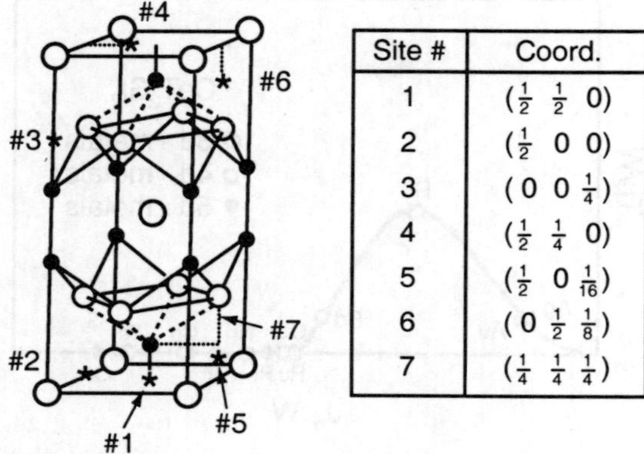

Site #	Coord.
1	$(\frac{1}{2}\ \frac{1}{2}\ 0)$
2	$(\frac{1}{2}\ 0\ 0)$
3	$(0\ 0\ \frac{1}{4})$
4	$(\frac{1}{2}\ \frac{1}{4}\ 0)$
5	$(\frac{1}{2}\ 0\ \frac{1}{16})$
6	$(0\ \frac{1}{2}\ \frac{1}{8})$
7	$(\frac{1}{4}\ \frac{1}{4}\ \frac{1}{4})$

Fig. 7. Schematic picture of the ThCr$_2$Si$_2$ crystal structure with possible μ^+ sites denoted by stars. The coordinates of the sites are shown in the accompanying table. The large open circles denote Th, the smaller open circles Cr and the filled circles Si. (From Schenck and Gygax)[1]

placed the individual members of the (U,Ce)X$_2$Si$_2$ compounds on a T-J Doniach-type phase diagram. This is illustrated in Fig. 6 for the Ce based materials. One expects to find full-moment AFM materials on the left of the diagram, high γ HF materials near the magnetic/non-magnetic boundary, and small-moment materials to the right side. μSR experiments have been carried out[1] on the following compounds: UX$_2$Si$_2$ and CeX$_2$Si$_2$, with X = Rh, Ru and Pt. In addition, the HF superconductor CeCu$_2$Si$_2$ has been studied extensively.

Because the Rh compounds possess large-moment AFM they should be good systems in which to identify candidate muon sites. Experiments carried out by Dalmas de Réotier *et al*[14] have produced uncertain results, however, principally because of the limited time resolution at the ISIS pulsed source where the work was performed. Thus the precession frequencies in the AFM state were not always observable.

For example, in URh$_2$Si$_2$ the AFM phase transition ($T_N = 137$ K, $\mu = 1.95\ \mu_B$) was observed only as a decrease in the overall asymmetry; no muon precession was observed below T_N, presumably because of the limited time resolution of the pulsed beam.[14] For CeRh$_2$Si$_2$ ($T_N = 35$ K) the μSR spectrum at 22 K suggests two possible muon sites, one with no dipolar field (and hence no precession) and one with a net field of about 110 Gauss. Dalmas de Réotier *et al.* suggest[14] that these signals in CeRh$_2$Si$_2$ correspond to the (1/2 , 0, 0) and (1/2, 1/4, 0) positions shown in Fig. 7. One might expect, therefore, that

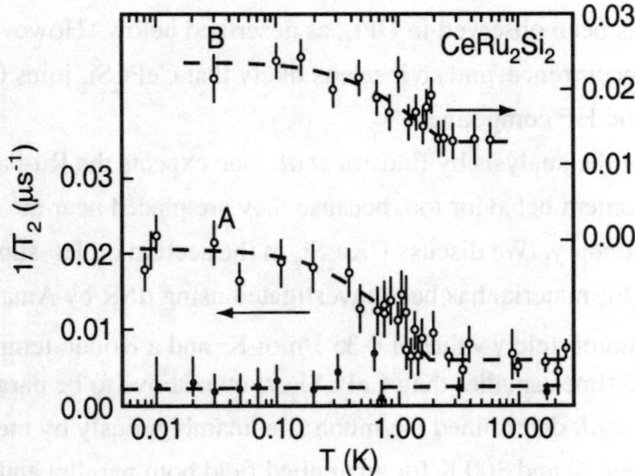

Fig. 8. Temperature dependence of the zero-field inhomogeneous relaxation rate $1/T_2$ for $CeRu_2Si_2$ in two different samples (open symbols) with the initial muon spin $I(0) \parallel$ c-axis.. Data taken in a longitudinal field of 6 kG are shown as closed circles. (Amato et al.)[21]

these sites (one, or the other, or both) would be the muon stopping sites for all of the isomorphic $(U,Ce)X_2Si_2$ series discussed here. This is not the case, however, because small differences in the electrostatic energies from one chemical composition to another may cause different muon equilibrium positions. This becomes more apparent in some of the following examples.

$CePt_2Si_2$ crystallizes[15] in the P4/nmm $CaBe_2Ge_2$ structure and possesses a Kondo temperature of about 70 K on the basis of specific heat measurements.[16] The susceptibility is large at low temperatures (3.3×10^{-3} emu/mol), the specific-heat γ value reaches 0.12 J/mol-K^2 near 2 K, and the resistivity appears to show non-Fermi liquid behavior below 4 K. Zero-field μSR measurements by Dalmas de Réotier[17] were carried out in $CePt_2Si_2$ at the ISIS facility, an accelerator well suited to searches for low-moment systems, because the pulsed nature of the source guarantees zero background signal. The authors observed a Gaussian relaxation function with a temperature-independent relaxation rate $1/T_2 = 0.05 - 0.06$ μs^{-1} over the entire temperature range between 0.06 - 50 K. In a longitudinal field of 100 Gauss, no relaxation was observed. This zero-field relaxation is entirely consistent with that from nuclear dipole fields, and an upper limit of 2×10^{-4} μ_B/Ce for any quasi-static electronic moment can be set. However, because the muon site for this system was not determined, it is possible that the muon sits at a site of high symmetry, where the dipole fields from an undiscovered AFM structure exactly cancel.

Such a situation has been observed in UPt_3, as described below. However, this cancellation is a very rare occurrence, and so it seems likely that $CePt_2Si_2$ joins $CeCu_6$ as the only known paramagnetic HF compounds.

On the basis of the analysis by Endstra *et al.*, one expects the Ru-based compounds to exhibit small-moment behavior too, because they are placed near the magnetic/nonmagnetic phase boundary. We discuss URu_2Si_2 in the section on HF superconductors below. The $CeRu_2Si_2$ material has been investigated using μSR by Amato *et al.*[18] This material[19] has a Sommerfeld γ value of 0.35 J/mol-K^2 and a Kondo temperature near 25 K, which is about 3 times smaller than $CePt_2Si_2$, shown above to be paramagnetic down to 60 mK. Amato *et al.* determined the muon site unambiguously by measuring the Knight shift between 10 and 300 K for an applied field both parallel and perpendicular to the tetragonal axis. This yielded the muon stopping site (1/2, 1/2, 0), determined from the calculated dipole fields for a ferromagnetic alignment of the Ce spins.

The temperature dependence of the zero-field relaxation rate in $CeRu_2Si_2$ for $\mathbf{I}(0) \parallel$ c-axis is given in Fig. 8, showing a small, but clear increase below about 2 K. This relaxation is decoupled in a longitudinal field, proving its static origin. Furthermore, the initial muon spin orientation implies relaxation from a distribution of local fields in the basal plane. A similar enhancement in the μSR rate below 2 K was found in a transverse-field experiment with $\mathbf{B}_{app} \parallel$ c-axis, showing that the field distribution also exists along the c-axis. The rms increase in linewidth is about 0. 2 Gauss, corresponding to small-moment ordering with a moment $\mu \cong 10^{-3}$ μ_B/Ce, one of the smallest ever found in HF materials. No evidence for this transition has been seen in specific heat, presumably because of the extremely small moment. There is an indication of a possible magnon drag effect in thermoelectric power below 2 K, however.[20]

The zero-field relaxation function in $CeRu_2Si_2$ was found to be exponential both above and below the 2 K transition; i.e., the Gaussian function typical of antiferromagnetic order from concentrated moments was not observed below the transition. This may be an indication that the magnetic order is incommensurate or possesses a limited coherence length.

Furthermore, a small but finite dynamic relaxation rate was observed in longitudinal field at temperatures above and below the transition (Fig. 8). This longitudinal μSR rate is also consistent with that observed in transverse field at higher temperatures. From these data a Ce-ion correlation time τ was derived and was found to scale with the quasi-elastic neutron scattering linewidth,[18] yielding an effective fluctuating moment of 0.6 μ_B. One interpretation of the μSR data above and below

2 K is that there may be two coexisting f-electron components, one fluctuating and one static (below 2K).[18] These two components could possibly derive from multiple f-electron bands, though at present no such explanation has been convincingly advanced. Certainly the coexistence of magnetism and superconductivity found in some U-based HF materials suggests two different f-electron components, however the case for Ce-based materials has yet to be made. This issue is discussed further below (c.f., $CeCu_2Si_2$). The reader is also referred to the chapter by Barry Cooper in this book for a theoretical discussion of related topics. Finally, it is important to note that the data in $CeRu_2Si_2$ may reflect just a single f-ion moment fluctuating around a non-zero static value.

C. CeX_2Sn_2 Compounds

This series of materials crystallizes in the tetragonal $CaBe_2Si_2$, space group P4/nmm, which is similar to the I4/mmm structure of the CeX_2Si_2 materials discussed above, except one of the X layers has changed places with the Si layer. Beyermann *et al.* have carried out resistivity, susceptibility and heat capacity measurements[21] and deduced that these materials have comparable RKKY and Kondo energy scales, although they do not possess particularly small moments. One of these materials ($CePt_2Sn_2$) has a relatively high Sommerfeld $\gamma > 3$ J/mol-K^2 and gives rise to very interesting μSR relaxation phenomena. This material also provides a good introduction to another small-moment HF compound YbBiPt.

Beyermann *et al.* found that polycrystalline $CePt_2Sn_2$ undergoes AFM order below about 0.88 K and possesses a small monoclinic distortion. Above this transition they found a large specific heat $\gamma \cong 3.5$ J/mol-K^2, which they attributed to HF behavior. μSR experiments by Luke *et al.* in a single crystal with 100 Gauss longitudinal-field parallel to the a-axis revealed[22] an exponential relaxation function above about 0.70 K, which became a stretched exponential $\exp(-(\lambda t)^\beta)$ and then a Gaussian function ($\beta \cong 2$) at the lowest temperatures. The relaxation rate reached a plateau below about 0.2 K with a value $\cong$ 4-5 μs^{-1}. Usually a temperature-independent Gaussian rate implies a quasi-static field distribution, and a value of 5 μs^{-1} corresponds to a field of about 60 Gauss. Most remarkably, however, at 0.02 K the Gaussian functional form, its amplitude and the relaxation rate remained largely unchanged in 1 kG longitudinal field. Except for the most unusual circumstances this lack of decoupling implies that the observed relaxation rate is not due to quasi-static fields, which would be decoupled in a 1 kG longitudinal field. Luke *et al.* thus concluded that the spins in single-crystalline $CePt_2Sn_2$ fluctuate very slowly down to 20 mK without ordering.

Subsequent zero-field µSR experiments by Lidström *et al.* reached a different conclusion, however.[23] Lidström's data, obtained on polycrystals, are almost identical to those of Luke, but Lidström *et al.* fit the Gaussian short-time relaxation below about 1.8 K to a rapidly-damped oscillating term, corresponding to spontaneous precession in an inhomogeneous, but magnetically ordered state. (That these fitting procedures are comparable can be easily seen by comparing $\exp(-(\Delta t)^2)$ to $\cos(\omega\tau)$ for small Δt and small ωt.) Lidström *et al.* thus concluded that, like its neighbors with X = Cu, Ni and Pd, $CePt_2Sn_2$ undergoes AFM order. They attributed the lack of decoupling observed in longitudinal field to a canting of the spin alignment in the applied field. Lidström *et al.* also concluded that the ordered moment is about 1 μ_B, assuming a likely (though undetermined) μ^+ site at the (1/2, 1/4, 0) position. This assignment of AFM order is also in line with the specific heat data of Beyermann.

Finally, Luke *et al.* carried out a later series of measurements on a monoclinically distorted polycrystalline sample of $CePt_2Sn_2$.[24] These new data were essentially identical to the single-crystal results reported earlier above 1 K, but at lower temperatures they observed static moments corresponding to an ordering temperature of 0.85 K, in excellent agreement with Beyermann's specific heat results. The relaxation function below T_N was typical of quasi-static ordering, with its "1/3 tail" showing slow fluctuations (see Equation 6). Furthermore, the early-time Gaussian relaxation was decoupled in fields smaller than for the case of the single-crystal material. To explain these different results in single- and poly-crystalline $CePt_2Sn_2$ Luke *et al.* surmised that the ground state of the (tetragonal) single-crystal material has a number of degenerate spin configurations and is therefore geometrically frustrated. This frustration prevents magnetic ordering in the single crystals, but the monoclinic distortion in the polycrystals removes the degeneracy and allows magnetic ordering to occur.

D. YbBiPt and YbSbPd

YbBiPt crystallizes in the cubic half-Heusler structure $F\bar{4}3$ m, with all three ionic sites possessing tetrahedral symmetry. Interest in YbBiPt arose because of the discovery by Fisk *et al.*[25] of an extremely large and temperature independent specific heat $\gamma \cong 8J/mol\text{-}K^2$ below about 4 K. Fisk *et al.* also concluded that this compound exhibits nearly equal characteristic energy scales for the Kondo, RKKY and crystalline-electric-field (CEF) effects. In part, these low energy scales ($\cong 1$ K) occur because of a low carrier density of about 0.05 carrier/unit cell. The specific heat data also showed a small, but

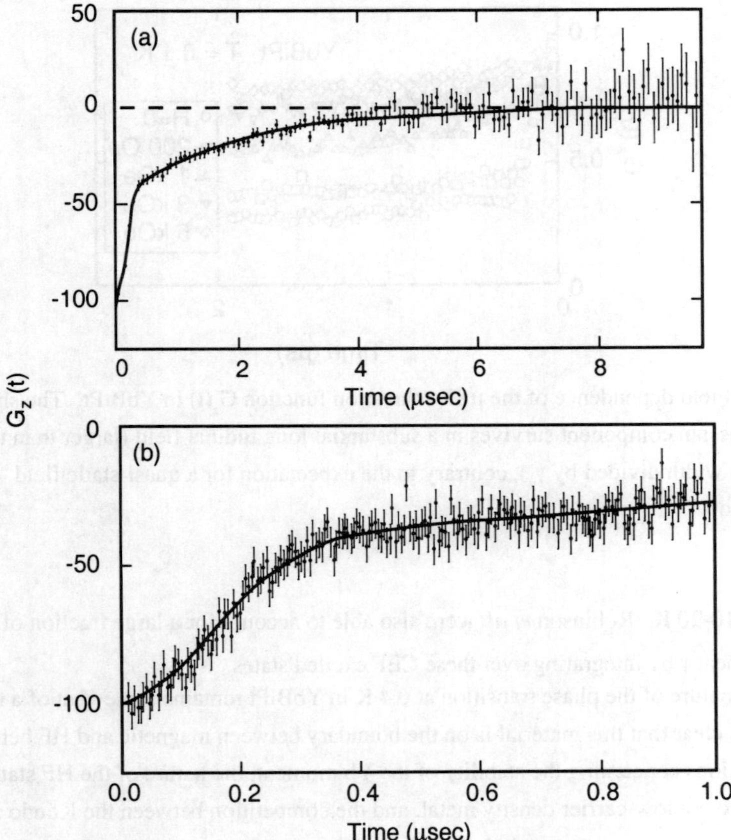

Fig. 9. Zero-field μSR function $G_z(t)$ for YbBiPt at T = 0.06 K (from Amato *et al.*)[21] A two-component structure, the sum of a long-time exponential and a short-time Gaussian is evident.

sharp anomaly at 0.4 K. Subsequently, Movshovich *et al.*'s low-temperature resistivity[26] data displayed an anisotropically-gapped Fermi surface from a possible spin-density-wave, which is suppressed by $\geq$ 1 kbar pressure or $\geq$ 3 kG applied field. Because there was no observation of a low-temperature Yb nuclear Schottky anomaly in the specific heat, it was concluded that the phase transition at 4 K must involve small ordered moments ($\leq$ 0.1 μ_B/Yb ion). Consistent with this, neutron scattering has failed to observe magnetic ordering, setting an upper bound of 0.25 μ_B/Yb.[27] Neutron scattering studies[28] by Robinson *et al.* also have established that the lowest CEF splittings are small: the ground state is six-fold degenerate, consisting of a doublet (G_7) and a quartet (G_8), split at

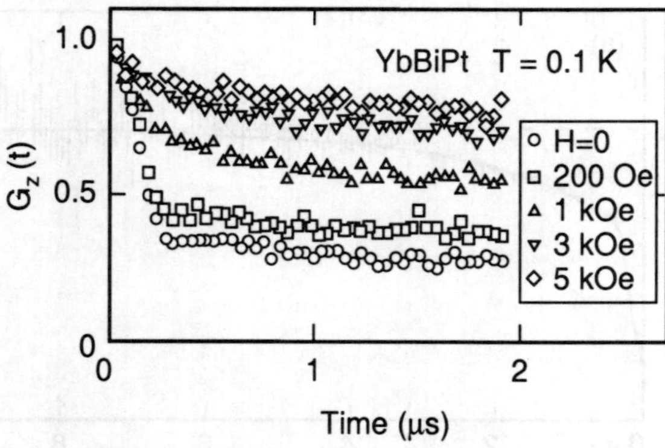

Fig. 10. Field dependence of the µSR relaxation function G_z(t) in YbBiPt. The short-time Gaussian component survives in a substantial longitudinal field (larger than the Gaussian width divided by γ_μ), contrary to the expectation for a quasi-static field distribution.

most by 10-20 K. Robinson *et al.* were also able to account for a large fraction of the specific heat γ by integrating over these CEF excited states.

The nature of the phase transition at 0.4 K in YbBiPt remains somewhat of a mystery. It is clear that this material is on the boundary between magnetic and HF behavior, and questions concerning the stability of the Yb moment, the nature of the HF state which can arise out a low-carrier density metal, and the competition between the Kondo and RKKY interactions come to mind. The first µSR experiments were carried out by Amato *et al.*[29] on crushed powders; later studies by Heffner *et al.* and Amato *et al.*[30] were performed on crystalline materials with qualitatively similar results. Above about 30 K Amato *et al.* observed[29] a single exponential relaxation function in zero-applied field. Below 20-30 K, however, a two-component relaxation function was found, which could be fit with a sum of two exponentials, one fast-relaxing and the other more slowly-relaxing. The relative amplitudes of these signals were found to be temperature dependent, with the more rapidly relaxing component gaining strength as the temperature was decreased below 20-30 K. Below about 0.4 -0.5 K the fast relaxing component was observed to become Gaussian in time with a relative amplitude of about 1/2.

These data were taken to signify the development of an inhomogeneous sample environment, where a fraction of the sample developed magnetic order (and showed rapid µSR relaxation above the ordering temperature) and a fraction of the sample remained paramagnetic down to the lowest temperatures measured. The relaxation function at 0.06

K is shown in Fig. 9. This is not the typical Kubo-Toyabe function shown in Fig. 1 for quasi-static relaxation from randomly oriented moments. Despite the initial Gaussian relaxation and exponential tail, the YbBiPt data do not show the characteristic dip following the short-time Gaussian that the Kubo-Toyabe function displays. Instead of the Gaussian fit presented by Amato *et al.*, the early time data below 0.4 K can alternatively be well described using a strongly-damped oscillating component, as was done by Lidström *et al.*[23] in the case of $CePt_2Sn_2$ described above. Either way, the μSR data are consistent with a magnetic phase transition in YbBiPt and, to date, provide the only indication from a microscopic magnetic probe that such a phase transition exits.

There are several interesting points which need to be addressed with further research. The first concerns the size of the ordered moment and the unusual longitudinal field dependence of the μSR rate observed in the ordered state. The second has to do with the apparent extreme inhomogeneity observed in YbBiPt. Fig. 10 shows the effect of an applied field between 100 G and 10 kG on the μSR relaxation function at $T = 0.10$ K. One observes that the amplitude of the initial Gaussian rate remains finite in fields of several kG. In the standard picture described above, the absence of decoupling implies that the muon experiences a local (dipolar) field $\cong 1$ kG, which generally implies an ordered moment $\cong 1$ μ_B/Yb. This would contradict the neutron scattering experiments (unless the spin freezing is disordered or the ordered state possesses an incommensurate **k**-vector not found by neutron scattering), as well as the absence of a Yb nuclear Schottky anomaly and the small jump in specific heat ΔC found at 0.4 K. The magnitude of the lowest-temperature μSR relaxation rate (or the precession frequency in a fit to an oscillating function) indicates a small local dipolar field of the order of only 50 Gauss, assuming that the muon does not sit at a site where the dipole fields from the ordered state nearly cancel. Amato *et al.* estimated that the muon site is close to the (1/2, 1/2, 1/2) position by modeling the transverse-field powder lineshape for an anisotropic susceptibility. No such cancellation occurs at this site. This leads one to conclude that decoupling should occur in longitudinal fields ≥ 100 Gauss, which was not observed. Because the ordered magnetic structure is not known, however, it is difficult to definitively derive the magnitude of the ordered moment from the μSR data alone. Nevertheless, Amato *et al.* did assign a magnitude of $\cong 0.1$ μ_B/Yb by calculating the second-moment of the dipolar field distribution, assuming randomly oriented Yb moments and a site near (1/2, 1/2, 1/2).

Because the totality of evidence indicates that the ordered moment in YbBiPt is small ($\cong 0.1$ μ_B), the absence of decoupling observed in moderate fields by μSR might be explained if either the muon itself or the applied field perturbed the ground state configuration in some undetermined way. For example, the positive muon exerts a force on the electronic quadrupolar moments in its surroundings through its electric field gradient. In a least one case, $PrNi_5$, this effect has been observed to alter the rare-earth ground state.[31] The possible strength of such an effect in YbBiPt has not been quantified, however. Although the application of a field in YbBiPt may indeed perturb the ground state because of the small and competing energy scales in YbBiPt described above, the fact that the μSR Gaussian rate itself is relatively independent of field up to 5 kG at 0.10 K indicates that the magnitude of the local field is not strongly effected by the applied field.

It is worth noting that there have been other cases of non-decouplable Gaussian relaxation functions in μSR experiments besides YbBiPt and $CePt_2Sn_2$, most notably in the Kagomè system $SrCr_8Ga_4O_{19}$ investigated by Uemura *et al.*[32] Here the authors attributed this phenomenon to a large and sporadic local field which exists only for a small fraction of the time ($\cong 5$ %); thus the relaxation rate is reduced by this fraction, but the decoupling field remains of the order of the local field. In $SrCr_8Ga_4O_{19}$, Uemura *et al.* surmised that the zero-field times were produced by the formation of local spin singlets, interrupted occasionally by the passage of unpaired spins. Because this system may possess multiple (magnetically) inequivalent muon sites near its oxygen atoms, comparing the muon relaxation in $SrCr_8Ga_4O_{19}$ to that in the more structurally simple metals, such as YbBiPt, may be speculative, however.

Recently, Bonville *et al.* have performed μSR experiments on a material related to YbBiPt, namely YbSbPd.[33] This material has a large Sommerfeld coefficient $\gamma \cong 3$J/mol-K^2 and undergoes a first-order magnetic transition at $T \cong 1$ K with an ordered moment of 1.3 μ_B/Yb.[34] The μSR experiments found a Kubo-Toyabe type relaxation at $T > 100$ K, with a value of $1/T_2 \cong 0.16$ μs^{-1}, which corresponds nicely to the calculated nuclear dipole linewidth, assuming a (1/2, 1/2, 1/2) stopping site. At the lowest temperatures ($T \cong 0.05$ K) Bonville also reported an undecouplable Gaussian relaxation rate corresponding to a local field of only about 50 Gauss, just like in YbBiPt. The authors attributed this to a mechanism similar to that invoked by Uemura *et al.* for $SrCr_8Ga_4O_{19}$. In YbSbPd Bonville surmised that the zero-field times arise from a magnetic ground state whose field cancels at the muon site, with occasional fluctuations out of the ground state producing a large, finite muon local field. They further suggested that the various configurations arise from a spin liquid state. More definitive modeling of the candidate magnetic state(s) needs to be done, however.

Bonville *et al.* fit their data to the dynamic Kubo-Toyabe function (given for the static case in Equation (6) and extracted a dynamic relaxation rate $1/T_1$. Note that this implies that the Gaussian and exponential components arise from the static and dynamic

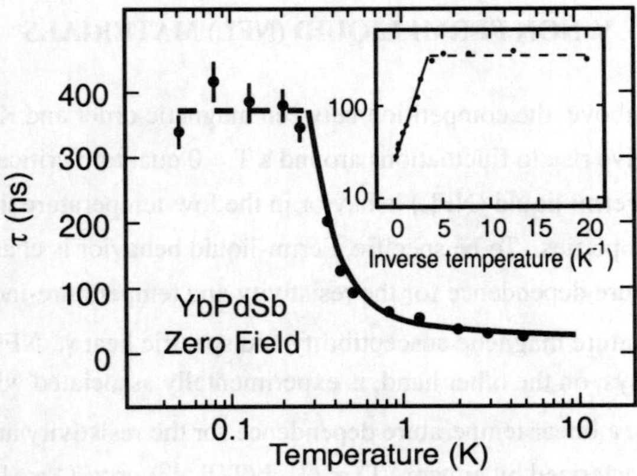

Fig. 11. Temperature dependence of Yb ion correlation time τ in YbSbPd. The insert shows τ plotted vs. inverse temperature (from Bonville *et al.*)[21]

behavior of a single magnetic environment. This is contrary to the interpretation put forth by Amato *et al.* for YbBiPt, where separate magnetic domains was assumed. The temperature dependence of $1/T_1$ for YbSbPd is shown in Fig. 11. The steep temperature dependence above the low temperature plateau was fit to an activated behavior, with an activation energy of 0.75 K. If the sporadic-field model is invoked, then the correlation time shown in Fig. 11 should be reduced by a factor of about 20 to account for the intermittency.

Clearly, the experimental situation in the YbBiPt-type materials is still open to interpretation. One important question is the degree of magnetic inhomogeneity present in the material. If there are both magnetically ordered and paramagnetic domains at low temperatures in YbBiPt, they begin to form only below about 20-30 K. One would expect such an effect to be driven by structural effects, such as Jahn-Teller distortions. Unfortunately, the best neutron diffraction experiments have only been carried out near room temperature. Likewise, NMR experiments by Reyes *et al.*[35] showed no ^{209}Bi quadrupolar splitting above about 35 K, which is consistent with the neutron experiments indicating no disorder. Below this temperature, the ^{209}Bi NMR lines were too broad to measure. Perhaps the best indication of two possible domains comes from the synthesis of the YbSbPd material using two different processing techniques: one synthesis route[34] leads to a magnetic phase transition and a large γ coefficient, while the other produces a large Sommerfeld γ but no magnetic order.[36] This is strong supporting evidence for the existence of two types of magnetic domains in YbBiPt also.

V. NON FERMI LIQUID (NFL) MATERIALS

As discussed above the competition between magnetic order and Kondo screening can theoretically give rise to fluctuations around a $T = 0$ quantum critical point, producing so-called non Fermi liquid (NFL) behavior in the low-temperature transport and thermodynamic properties. To be specific, Fermi-liquid behavior is characterized by a quadratic temperature dependence for the resistivity and temperature-independent values for the low-temperature magnetic susceptibility and specific heat γ. NFL behavior in heavy fermion alloys, on the other hand, is experimentally associated with a logarithmic divergence of $\gamma(T)$, a linear temperature dependence for the resistivity and a magnetic susceptibility characterized by either $\chi(T) \propto (1 - b(T/T_K)^{1/2})$ or $\chi(T) \propto -\ln(bT/T_K)$. To date almost all materials exhibiting NFL behavior are disordered alloys and are found near the phase boundary for magnetic order in a temperature-concentration phase diagram. HF compounds are found to exhibit both ordinary and NFL low-temperature properties.

Recent μSR experiments have played an important role in the study of NFL materials in two fundamental ways: (1) First, it is always important to establish the *absence* of magnetic order in a material before ascribing anomalous low-temperature properties to NFL behavior. Here, the sensitivity to small-moment magnetism which μSR provides is essential. (2) Second, recent discoveries by Bernal *et al.*[37] have suggested that sometimes NFL behavior can be attributed to a simple microscopic distribution of Kondo temperatures arising from disorder, and not, for example, due to proximity to a $T = 0$ quantum critical point. (See the introductory chapter by MacLaughlin, for a brief discussion of other possible sources of NFL behavior.)

A. $Y_{1-x}U_xPd_3$

The earliest μSR experiments to investigate NFL systems were carried out by Wu *et al.* in the alloy series $Y_{1-x}U_xPd_3$, where $x = 0.1, 0.2$ and 0.4.[38] Investigations by Seaman *et al.* revealed NFL behavior in the resistivity, electronic specific heat and entropy for $x \leq 0.2$.[39] The source of this behavior was attributed to the possible existence of a two-channel quadrupolar Kondo effect, requiring a non-magnetic ground state (such as a Γ_3 crystal-field-split level).[40] The results of Wu *et al.* conclusively demonstrated magnetic ordering in the $x = 0.4$ compound; the shape of the relaxation function was found to be consistent with spin-glass type freezing below 11 K (see Equation 7). Unfortunately, because the muon site in these materials is not known definitely, the size of the ordered moment is uncertain. Nevertheless, on the basis of a likely candidate site, Wu *et al.*

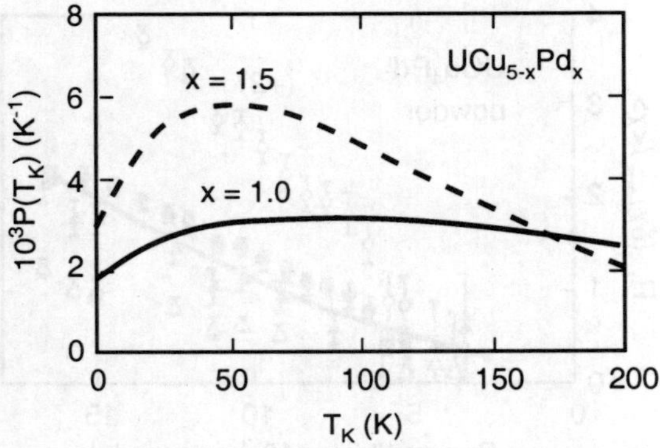

Fig. 12. Distribution of Kondo temperatures $P(T_K)$ used to fit the susceptibility in $UCu_{5-x}Pd_x$ using the Kondo disorder model. (from Bernal *et al.*)[21]

estimated a moment $\cong 1\ \mu_B/U$. As the U concentration was reduced to 0.2, the freezing temperature decreased to about 1 K, but the magnetic volume fraction dropped to about 40 %, with the remaining volume showing no evidence of magnetic behavior. In the *x* = *0.1* sample, fully 90 % or more of the sample was found to be non-magnetic.   The magnetic fraction showed a decreased relaxation rate compared to the higher concentrations of U, indicating the disappearance of magnetic order with decreasing *x*. Subsequent metallurgical studies by Süllow[41] have shown that the $0.02 \le x \le 0.20$ systems contain macroscopically inhomogeneous distributions of Y and U ions, however. These results are qualitatively consistent with the µSR experiments of Wu *et al.*, and also indicate that these compounds are not the most ideal for investigating exotic sources of NFL behavior.

B. $UCu_{5-x}Pd_x$

The first evidence for the Kondo-disorder model was found to occur in the materials $UCu_{5-x}Pd_x$, x = 1.0 and 1.5. Experiments by Bernal *et al.*[37] found an additional contribution to the low-temperature Cu NMR linewidth, exceeding that from simple powder broadening. This additional linewidth was shown to be consistent with a local distribution of Kondo temperatures $P(T_K)$, assuming a susceptibility given by $\chi = C\ /\ (T + \alpha T_K)$ and $P\ (T_K \ne 0) \ne 0$. Physically, this means that at all temperatures $T \ge 0$, there exist Kondo spins with $T_K < T$. These spins are paramagnetic, giving rise to a finite χ even at the lowest temperatures, therefore contributing

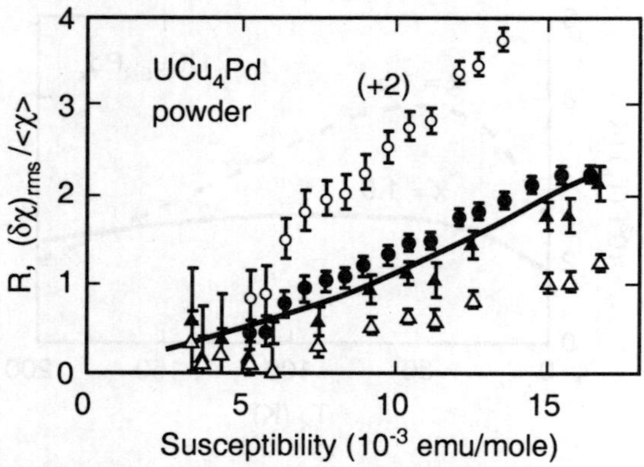

Fig. 13. Curve: Relative rms width of the susceptibility distribution in UCu$_4$Pd. Points: normalized relative linewidths R in the long-range-correlation (open symbols) and short-range-correlation (filled symbols) models from μSR (circles) and NMR (triangles). From Bernal *et al.*[37]

to the linewidth.

We briefly sketch the analysis procedure[37] required to reveal the presence of Kondo disorder. The Kondo temperature is denoted as $T_K = E_F \exp(-1/\lambda)$, with $\lambda = N(0)J$. Taking a Gaussian distribution $P(\lambda)$ for λ, one can then fit the bulk χ data, obtaining both a mean and width for $P(\lambda)$. The resultant $P(T_K)$ is shown in Fig. 12 for the x = 1.0 and 1.5 materials in UCu$_{5-x}$Pd$_x$. These parameters are then fixed for the analysis of the NMR (and μSR) data.

The Knight shift K for the ith spin probe is given by $K_i = \sum_j a^*_{ij} \chi_j$. The

inhomogeneous linewidth is $(\delta K)_{rms} = a^*(\delta\chi)_{rms}$, where $(\delta\chi)_{rms} = \sqrt{\langle\chi^2\rangle - \langle\chi\rangle^2}$. This assumes that only the susceptibility, and not the coupling constant a^* from the hyperfine and/or dipole fields is disordered. If χ is ordered but a^* is disordered, then $(\delta K)_{rms} \propto \langle\chi\rangle$. In other words, if the quantity $R \equiv (\delta K)_{rms} / (a^*\langle\chi\rangle)$ depends on $\langle\chi\rangle$, then there exists a distribution in $\langle\chi\rangle$.

The effective coupling a^* is obtained from the calculated dipole fields and a separate measurement of $\langle K\rangle$, the powder-averaged Knight shift, which yields the isotropic hyperfine coupling constant. A very important point is that the calculated value of a^* depends on the assumed correlation length ξ associated with the disorder. For short range correlations (SRC, $\xi \leq$ lattice constant) $a^* = a^*_{SRC} = (\sum_j a^2_{ij})^{1/2}$, while for long-range correlations (LRC, $\xi \gg$ lattice constant) $a^* = a^*_{LRC} = |\sum_j a_{ij}|$.

The NMR data of Bernal *et al.* for $UCu_{5-x}Pd_x$, $x = 1$, is shown in Fig. 13, assuming both the SRC and LRC limits. One sees that the ratio R defined above is larger than 1 and varies with $\langle\chi\rangle$, implying that there is significant disorder in the susceptibility. Furthermore, as shown by the solid line in the figure (which is the relative width $(\delta\chi)_{rms} / \langle\chi\rangle$, the NMR data are more compatible with the SRC limit.

This limit is not definitively determined by the NMR data alone, however. Only by combining the data from *two* spin probes, with different values of a^*, can a definite conclusion can be drawn. Consequently, transverse-field μSR data were accumulated on the same powdered samples and the inhomogeneous Gaussian linewidth $1/T_2$ deter-mined.[37] In this case one has $(\delta K)_{rms} = 1/(T_2\omega_\mu)$ and the analysis proceeds the same as in the case of NMR. (Additionally, zero-field μSR measurements placed an upper bound of 0.01 μ_B/U atom for an ordered magnetic moment down to about 2 K.) Possible muon stopping sites are known from experiments[42] on single-crystalline UCu_5, and so the dipolar linewidths are calculable. One obtains a spread in the values of a^*_{SRC} of only about 20 % from the uncertainty in the stopping site for the muon. Furthermore, an upper bound can be set on a^*_{LRC} from the value of the isotropic Knight shift. When both the NMR and μSR data are plotted together in a graph of R versus $\langle\chi\rangle$ (Fig. 13), one sees that the two probes agree *only* in the SRC limit. This then provides very strong evidence for Kondo disorder as the source of NFL behavior in the *x = 1.0* $UCu_{5-x}Pd_x$ material. In the *x = 1.5* material, a similar result is obtained, but it is necessary to include disorder in the μSR coupling a^* to obtain good agreement.

Further evidence for disorder-induced NFL behavior in UCu_4Pd comes from x-ray absorption fine structure (XAFS) measurements by Booth *et al.*[43] The XAFS experiments examined the U L_{III} and Pd and Cu K edges to probe the near-neighbor bond-length distributions. The experiments were carried out on the same sample used for the μSR experiments discussed above. UCu_4Pd can be described in terms of interlocking U and Pd fcc lattices, with an interstitial vertex-sharing network of Cu octahedra. The XAFS measurements established that about 1/4 of the Pd atoms sit at the nominal Cu sites. Booth *et al.* furthermore showed that the distribution of Kondo temperatures $P(T_K)$ derived from the μSR and NMR measurements could be described in terms of a simple model of f-d hybridization V_{fd}, coupled to the site interchange found in the XAFS studies. Here $T_K = E_F \exp(-1/(\rho J))$, where E_F is the Fermi energy, ρ is the density of states at the Fermi level, $J = V_{fd}^2/\varepsilon_f$ and ε_f is the f-electron energy. The magnitude of V_{fd} depends ladts strongly on the U-Cu and U-Pd bond lengths. The dominant effect for $P(T_K)$ in this model is the bond-length

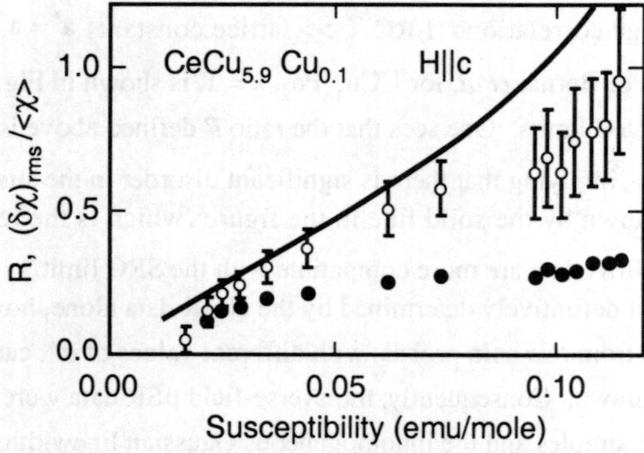

Fig. 14. Curve: Relative rms width of the susceptibility distribution in CeCu$_{5.9}$Au$_{0.1}$. Points: normalized relative linewidths R in the long-range-correlation (open symbols) and short-range-correlation (filled symbols) from μSR. (from Bernal et al.)[21]

disorder between the U-Cu and U-Pd atoms which results from the site interchange measured with XAFS

C. CeCu$_{6-x}$Au$_x$

As discussed above, CeCu$_6$ is a HF compound which displays Fermi-liquid behavior and remains paramagnetic down to at least 0.05 K. When alloyed with Au above a critical concentration x $\cong$ 0.1, CeCu$_{6-x}$Au$_x$ exhibits long-range AFM. For x = 0.10, however, NFL behavior has been well established.[44] Furthermore, when antiferromagnetic CeCu$_{6-x}$Au$_x$ (x =0.3) is subjected to applied pressures greater than 8 kbar, the AFM disappears and NFL behavior is recovered.[45] This is strong evidence that the NFL properties of CeCu$_{5.9}$Au$_{0.1}$ arise from proximity to a T = 0 quantum critical point, because pressure is known to suppress magnetic order in Ce HF materials. It is nonetheless necessary to investigate the possibility that Kondo disorder could be present in this material as well.

To date no NMR data have been reported on CeCu$_{5.9}$Au$_{0.1}$, primarily because of the low symmetry of the Cu sites, resulting in a potentially complicated NMR spectrum with up to 30 lines. μSR data[37] provide insight into this material, however. The muon site in the parent compound CeCu$_6$ has been well determined (see above), and possesses relatively low symmetry with non-vanishing dipole fields.

The value of R for an applied field of 5 kG for $\mathbf{B}_{app}$ || c-axis is plotted against $\langle\chi\rangle$ in Fig. 14, assuming no disorder in a*. The curvature in 1/T$_2$ vs. $\langle\chi\rangle$ was attributed to a number of possible sources, including a temperature-dependent a* or crystalline-electric-field splitting of the Ce energy levels. The downward curvature in R for small $<\chi>$ may be due to muon diffusion at high temperatures.   The value of a*$_{LRC}$ was determined from the Knight shift data: a*$_{LRC}$ = K/ $<\chi>$, while the value of a*$_{SRC}$ was obtained from the calculated dipolar couplings (and a small correction from the hyperfine interaction reported for $CeCu_6$). One sees that under the assumption of short-range correlations in a*, the value of R is quite small compared to $UCu_{5-x}Pd_x$; even for the case of long range correlations in a* the data always give $R \leq 1$. In the latter case, the calculated value of $<\delta\chi>/<\chi>$ is indeed closer to the data. There are, however, experiments which show that some properties of $CeCu_{5.9}Au_{0.1}$ depend on whether or not a Au atom is a near neighbor to a Ce atom.[46] This would imply a short-range correlation of the disorder in $<\chi>$, and hence, on the basis of the µSR linewidths, that Kondo disorder is not the dominate source of NFL behavior in this material. This is a conclusion which really needs to be checked with NMR data, however, if such data are attainable.

D. CeRhRuSi$_2$

In the previous section it was pointed out that $CeRh_2Si_2$ is an AFM with $T_N = 35$ K and $T_K \cong 33$ K, while $CeRu_2Si_2$ was found to possess very small moment magnetic ordering. The magnetic order in $CeRh_2Si_2$ can be suppressed to $T = 0$ either through the application of pressure ($P_c \cong 9$ kbar)[47] or by alloying[48] with Ru in $CeRh_{2-x}Ru_xSi_2$ ($x_c \cong$ 0.95). The compound $CeRu_2Si_2$ has recently been investigated[49] by Graf et $al.$, and it has been determined that the pressure-dependence of the specific heat γ ($d\gamma/dP$) changes sign near the critical value P_c, indicating that the balance between the Kondo and RKKY temperatures is being shifted just as the magnetic order is suppressed. The authors have found no evidence for NFL behavior near P_c, however. By contrast Graf et $al.$ found that $CeRhRuSi_2$, for which the (Rh, Ru) sublattice is disordered, exhibits NFL behavior in the specific heat and susceptibility in the temperature range 1-11 K, but Fermi liquid behavior below 1 K. They have analyzed their susceptibility data using the Kondo disorder model described above and found that the data are well described by a Gaussian distribution of Kondo temperatures with a width which is $only$ about 12 % of the mean Kondo temperature. This implies that the disorder in $CeRhRuSi_2$ is not sufficient to give NFL behavior at zero temperature. Preliminary µSR experiments to test the Kondo disorder model in this material have been carried out by MacLaughlin et $al.$[50] The results are more consistent with a distribution of muon coupling constants a^* than a distribution of susceptibilities arising from Kondo disorder.

VI. HEAVY FERMION SUPERCONDUCTORS

HF superconductivity has been a subject of very strong scientific interest since $CeCu_2Si_2$ was found to be superconducting by Steglich and co-workers 20 years ago.[51] A principal reason for this continuing interest has been the fascinating interplay between superconductivity and magnetism displayed in these materials. Initially, the question of how superconductivity could be supported in a system with large local Ce moments was addressed. This was eventually resolved when the hybridization between the conduction electrons and f-electrons was appreciated, so that it is now well established that the f-electrons themselves take part in the superconducting transition. More recently, the interplay between magnetism and superconductivity has been seen in terms of the development of the pairing state itself (where spin fluctuations give rise to the pairing force), and in terms of the competition, coupling or coexistence of the magnetic and superconducting ground states. All three of the latter have been observed in HF super-conductors.

An additional feature of HF superconductivity is the observation from a variety of experiments that the superconducting order parameters are unconventional. That is, they possess a symmetry which is lower than that of the lattice, or they break time-reversal symmetry and hence possess either spin or orbital moments, or they violate parity conservation.[52] Because the conventional s-wave superconductors (like Nb) cannot possess multiple superconducting states, two or more superconducting transitions in a single material constitute irrefutable evidence for an unconventional order parameter. Two of the HF superconductors, UPt_3 and $(U,Th)Be_{13}$, are now known to possess multiple superconducting transitions, and the boundaries between their various phases have been mapped. In the case of UPt_3 these boundaries occur in the field-temperature plane, while in $U_{1-x}Th_xBe_{13}$ they occur in the x -T plane.

There are six known ambient-pressure HF superconductors and three materials which become superconducting under applied pressure. Of the six ambient pressure superconductors, four have been found to possess small moments: UPt_3, URu_2Si_2, $CeCu_2Si_2$ and UBe_{13} doped with Th. In this chapter we discuss the four superconductors which possess small moments. A recent detailed review of HF superconductivity has been given by Heffner and Norman.[52]

A. UPt_3

This material is a HF superconductor[53] ($T_c = 0.55$ K) with a γ value of 0.45 J/mol-K^2. Until 1986 UPt_3 was also considered to be paramagnetic at all temperatures above the

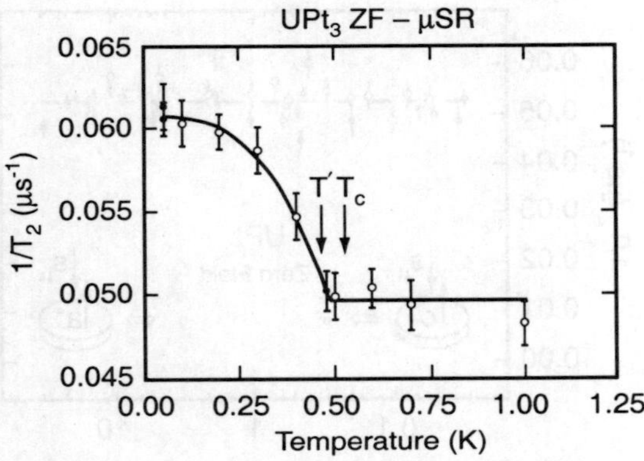

Fig. 15. Temperature dependence of the zero-field μSR linewidth $1/T_2$ in UPt$_3$ for the muon spin $\mathbf{I}(0) \parallel$ a-axis. The lower (T ') and upper (T$_c$) superconducting transition temperatures are marked. A spontaneous magnetic field develops below T ' (from Luke et al.)[21]

lowest temperatures measured $\cong 0.05$ K. This assumption was based on specific heat measurements, for example, which showed a jump only at the superconducting transition near 0.5 K. Cooke et al. then reported[54] an increased zero-field μSR linewidth below about 5 K in polycrystalline UPt$_3$, which corresponded to a very small quasi-static moment of the order of 0.01 μ_B. Subsequently, Aeppli et al. identified[55] the magnetic structure associated with this phase using neutron diffraction, and showed that the order parameter had a mean-field temperature dependence. In 1991 Fisher et al. discovered[56] that UPt$_3$ actually had *two* superconducting transitions in zero field, split by $\cong 0.05$ K (T$_{c1}$ = 0.55 K and T$_{c2}$ = 0.50 K). Later Bruls et al.[57] and Adenwalla et al.[21] using sound velocity and sound attenuation measurements, mapped out the superconducting phase diagram in the field-temperature plane. This was extended to the field-temperature-pressure regime by Boukhny et al.[59] Luke et al., using zero-field μSR, found that a spontaneous magnetic signal corresponding to a small-moment of around 10^{-3} μ_B/U atom arose *below* the lower superconducting transition temperature at around 0.50 K.[60] These data are shown in Fig. 15. One interpretation of Luke's data is that the superconducting order parameter below T$_{c2}$ breaks time-reversal symmetry, giving rise to small orbital or spin moments. This dictates a superconducting state with a finite angular momentum between pairs and/or a possible spin-triplet pairing state. (Thus, S = 0 and L = 2,4, ... or S =1 and L = 1,3, ..., for the case of weak spin-orbit coupling. If one has strong spin-orbit coupling then parity and total spin are the good quantum numbers.)

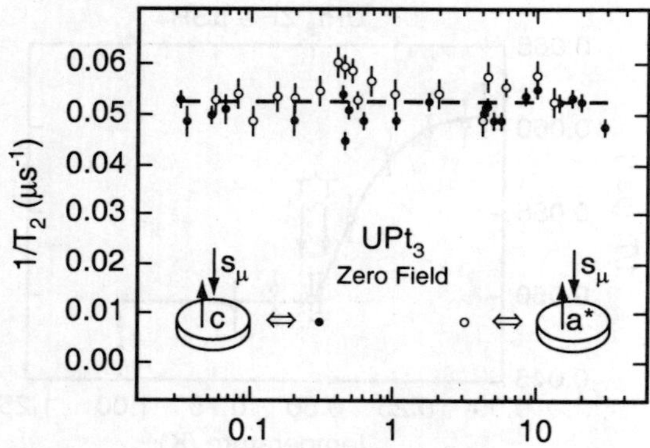

Fig. 16. Temperature dependence of the zero-field μSR linewidth $1/T_2$ in UPt$_3$ for the muon spin $\mathbf{S}_\mu \parallel$ and $\perp$ c-axis (from P. Dalmas de Réotier *et al.*)[21] No spontaneous field is seen at either 5 K or in the superconducting state below about 0.55 K. (Note that the muon spin is denoted $\mathbf{I}(0)$ in the text, whereas it is denoted $\mathbf{S}_\mu$ in the figure.)

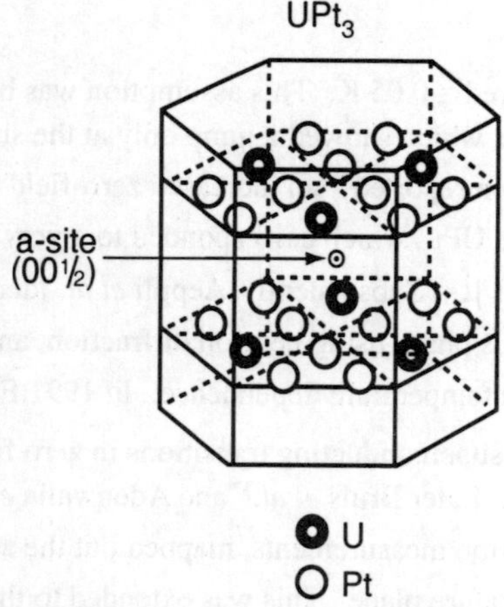

Fig. 17. Crystal structure of UPt$_3$. The likely muon site is shown as the (0, 0, 1/2) site. (from Schenck and Gygax)[1]

The 5 K magnetic transition plays a key role in the interpretation of the superconductivity in this system. One model[61] for multiple superconducting transitions involves a two-component order parameter, each component having a different T_c. The splitting of these transition temperatures must be caused by a symmetry-breaking field which couples to the superconducting order parameter. It has been suggested that the AFM order below

5 K, which breaks the hexagonal symmetry in the basal plane of UPt_3, is the field which splits the two components of the superconducting order parameter.[61] The most decisive evidence for this hypothesis is the observation[62] by Hayden *et al.* that both the 5 K AFM transition and the splitting of the two superconducting transitions disappear at an applied pressure of about 4 kbars.

This particular story does not end here, however, at least for the μSR measurements. Recently, Dalmas de Réotier *et al.*[63] performed μSR experiments in UPt_3 with single crystal samples which were of much greater purity than had been measured previously. Their surprising result is shown in Fig.16; namely, the zero-field relaxation rate $1/T_2$ is independent of temperature from 0.02 K to 30 K. These same samples were also measured with neutron diffraction and the 5 K transition characterized previously by Aeppli *et al.* was also found in this sample. Thus these most recent μSR measurements in UPt_3 now appear to be in contradiction with neutron diffraction, whereas the earliest μSR experiments by Cooke *et al.* actually led the way to the discovery and full characterization of the small-moment 5 K transition.

The explanation for this paradox appears to lie in differing sample qualities. Presumably, the muon stopping site is (0, 0, 1/2), shown in Fig. 17, where the dipole fields from the U moments exactly cancel for the AFM structure below 5 K. Most probably the μ^+ site in the samples used for the earliest experiments was slightly perturbed from this high symmetry position by impurities and/or strain, so that the dipole fields from the antiferromagnetically aligned moments did not cancel by symmetry. This example illustrates the importance of carrying out measurements on the highest-quality single crystals, and of the necessity of determining the muon site whenever possible.

The measurements by Dalmas de Réotier *et al.* also raise the question whether the small magnetic signature below $T_{c2} = 0.50$ K seen by Luke *et al.* is actually associated with the magnetism which sets in at 5 K, since *both* signals disappear in the μSR experiments on higher quality materials, implying a common symmetry for the magnetic state.

B. URu_2Si_2

This material is also a HF superconductor ($T_c = 1.2$ K) with a γ value of 0.13 J/mol-K^2 and an unusual AFM transition at 17.5 K.[52] Both neutron scattering[64] and x-ray diffraction[65] measurements have been carried out in URu_2Si_2, and the magnitude of the ordered moment at 3K ($<<T_N$) has been determined quite well, yielding a value of 0.03 μ_B. The AFM structure is quite simple, with alternating FM planes of U moments along the tetragonal axis. In contrast to UPt_3, which shows no anomaly in the specific heat ΔC at 5K, URu_2Si_2 exhibits a large jump in specific heat at 17.5 K, even though the moments

in the two materials are of comparable size. Physically, because ΔC is caused by the entropy released in the transition, one does not expect a large ΔC for a small-moment system, because most of the entropy has already been removed in the 'screening' process (see discussion below regarding UBe_{13}). Thus it is apparent that the magnetism at 17.5 K in URu_2Si_2 is not the primary order parameter driving the phase transition.

The earliest μSR experiments on URu_2Si_2 by MacLaughlin *et al.* found an increase in the transverse-field linewidth below T_N which was quite small, corresponding to a local dipole field spread of only about 1.2 Gauss.[66] No coherent muon precession in zero field was observed. These authors calculated the powder-pattern muon linewidths from the known AFM structure for 11 plausible muon sites in the tetragonal lattice. Except for the (1/4 , 1/4, 1/4) position, where the dipole fields vanish by symmetry, all of the candidate sites yield fields $\geq$ 50 Gauss, for $\mu = 0.03\ \mu_B/U$. Thus the μSR rate is an order of magnitude smaller than expected, unless the muon occupies a position near the (1/4 , 1/4, 1/4) site. It is important to note that asymmetric muon zero-point motion or local lattice distortion could yield a finite dipolar field even for this site.

Additional muon site information comes from experiments on single crystals by Knecht *et al.*[67] These authors carried out Knight shift measurements for $\mathbf{B}_{app} \parallel$ c-axis and $T > T_N$, obtaining the hyperfine coupling $(A_c + A_{dip})$, as given in Equation (11). Combined with the isotropic Knight shift obtained in a polycrystal by MacLaughlin *et al.* (which yields A_c), the dipolar coupling A_{dip} was obtained. This was compared to that calculated for various muon sites, and the best agreement was obtained for the (0, 1/2, 1/8) site. The calculated dipole field in the AFM state for this site is again an order of magnitude larger than measured. Similar results are obtained for other sites, such as that obtained for the isomorphic compound $CeRu_2Si_2$ discussed above.

There is no current resolution of this discrepancy between the moment measured by μSR and by neutron and x-ray scattering, but it may be that part of the muon signal below T_N is dynamical rather than quasi-static. This is possible because of the different time scales for the μSR and scattering experiments. Indeed, the temperature dependence of the μSR signal below T_N does not follow that of a typical static order parameter, as found in other small-moments magnetic transitions described above. The most recent μSR measurements in URu_2Si_2 by Yaouanc *et al.* even appear to show a change in the shape of the relaxation function above and below the 17.5 K transition.[68] Above T_N the typical Gaussian shape is found, characteristic of nuclear dipolar broadening, while below T_N the functional form is that of a stretched exponential $\exp(-(\lambda\tau)^\beta)$ with $\beta \cong 0.60$. This may indicate either a dynamical contribution to the linewidth or[68] a very short correlation length.

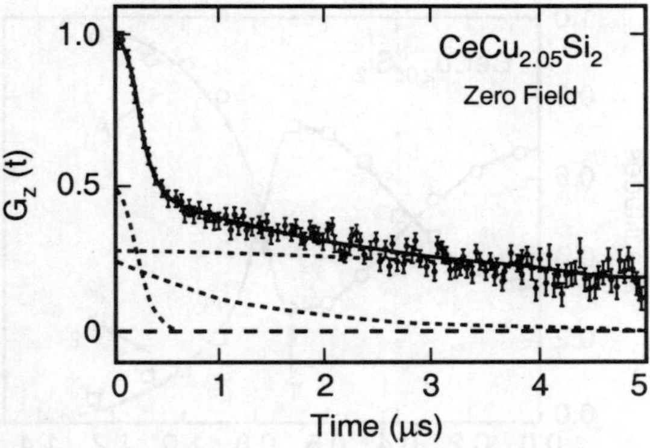

Fig. 18. Measured zero-field relaxation function $G_z(t)$ in $CeCu_{2.05}Si_2$, showing the decomposition into magnetic (rapidly decaying) and paramagnetic (slowly decaying) components. (From the Ph. D. thesis of R. Feyerherm, ETH Zurich, Switzerland, Nr. 11249, 1995).

Alternatively, the answer to the discrepancy between the μSR and other measurements of the ordered moment may involve a "primary," but hidden order parameter mentioned above. Following a definitive determination of the symmetry of the ordered magnetic phase below 17.5 K by Walker *et al.* using neutron scattering,[69] it has been suggested by Agterberg and Walker[70] that the dominant order parameter involves a triple-spin correlation which couples weakly to the AFM magnetic dipole order parameter. If this were true an estimate of the expected local field detected by the muon from the dipole ordering alone would not be likely to yield a correct result.

C. CeCu₂Si₂

Superconductivity in $CeCu_2Si_2$ was discovered in 1979.[20] Since that time only one superconducting phase ($T_c \cong 0.7K$) has been identified. The γ value for $CeCu_2Si_2$ is $\cong 0.73 - 1.1$ J/mol-K², qualifying it as a HF superconductor. The first μSR experiments in this material by Uemura *et al.* found evidence for some kind of quasi-static magnetism setting in just above T_c, with a small moment $\mu \cong 0.1\ \mu_B$.[71] Subsequent experiments using μSR, specific heat and microstructural analysis by Luke *et al.*[72] and Feyerherm *et al.*[73] have established that this magnetism persists below the superconducting transition in zero-applied field and competes with the

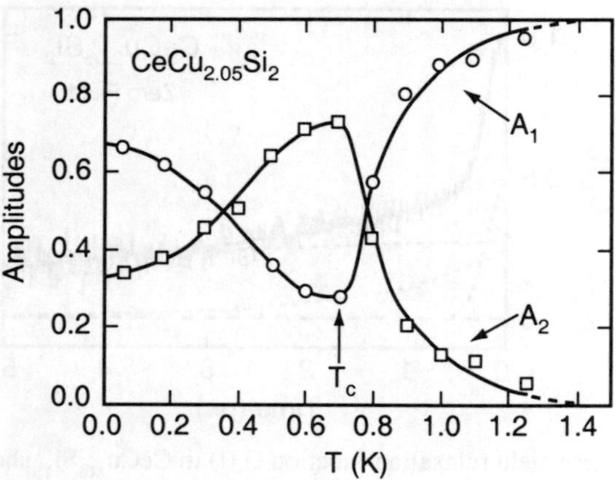

Fig. 19. Temperature dependence of the amplitudes of the magnetic (A_2) and paramagnetic (A_1) components of the zero-field relaxation function shown in Fig. 18. (From R. Feyerherm *et al.*)[21]

superconductivity. In non-zero fields the magnetism persists above H_{c2} for temperatures below T_c.

The zero-field μSR relaxation function in CeCu$_{2.05}$Si$_2$ (Fig. 18) exhibits a two-component structure below 1.3K, where the magnetic signature sets in. Thus,

$$G_z(t) = A_1(T)\, G_z(t) + A_2\,(T)\,[2/3\,\exp\,(-0.5(\Delta t)^2) + 1/3\,\exp\,(-\lambda t)]. \quad (12)$$

This segregation into two relaxation components occurs in an otherwise homogeneous sample. The Gaussian Kubo-Toyabe component $G_z(t)$ (Equation 6), proportional to $A_1(T)$ in Equation (12), was attributed to paramagnetic domains (with only nuclear dipolar relaxation) which become superconducting. The second component, proportional to $A_2(T)$ in Equation 12, was associated with magnetic domains which do not possess superconductivity. These two components are separable because their relaxation rates differ by an order of magnitude. The second form is possibly characteristic of an incommensurate spin-density wave.

It is therefore possible to track the temperature dependence of the volume fractions of the magnetic and non-magnetic portions of the sample, as shown in Fig. 19. The magnetic fraction increases below 1.3K, reaching a maximum of about 3/4 at T_c. Below T_c the magnetic fraction decreases to about 1/3 at $T = 0.05$ K, while the superconducting (paramagnetic) volume fraction increases to about 2/3. Corroborating evidence for this effect is found in the height of the specific heat jump at T_c, which correlates well with the fraction of the sample which is superconducting, as obtained from the μSR measurements. This effect has been studied in several different samples with varying Cu stoichiometries. In each case the magnetic relaxation rate is unaffected by the onset of superconductivity and is relatively independent of the sample, indicating a common origin from one sample to the next.

$CeCu_2Si_2$ therefore displays a competition between magnetism and superconductivity, each existing in its own macroscopic domain. This effect is unique among HF superconductors and may be related to the fact that the existence of magnetism and/or superconductivity in $CeCu_2Si_2$ is very sensitive to subtle changes in stoichiometry, which may be related to changes in unit-cell volume or internal strain. Extensive experiments inducing volume changes have been carried out by La doping and Cu deficit (where $\Delta V/V > 0$) or Cu excess and hydrostatic pressure (where $\Delta V/V < 0$).[74] Samples with large unit-cell volume tend to be magnetic and non-superconducting, reflecting a reduction of T_K. When the volume is reduced, the f-electrons delocalize resulting in increased hybridization so that T_K is raised and superconductivity is found.

D. UBe_{13}

Pure UBe_{13} exhibits[75] only a single superconducting phase transition, ($T_c \cong 0.86$ K) although there have been reports of structure in $H_{c2}(T)$ below T_c by Rauchschwalbe *et al.*[76] and of anomalous magnetic torque measurements by Schmiedeshoff *et al.*[21] possibly indicating a second superconducting phase. No clear evidence for two phases has been seen in the specific heat, however. When either magnetic (Gd) or non-magnetic (La) impurity atoms are substituted for U in UBe_{13}, T_c is depressed as expected for a superconductor with an anisotropic order parameter. However, when doped with Th, Ott *et al.* found[78] that the T_c for $U_{1-x}Th_xBe_{13}$ becomes a non-monotonic function of Th concentration x. For $0.019 < x < 0.043$ two specific heat transitions occur. Early ultrasonic attenuation measurements by Batlogg *et al.* were consistent with an AFM phase at the lower transition temperature T_{c2}.[79]

The existence of magnetic correlations in $U_{0.965}Th_{0.035}Be_{13}$ was later directly observed[80] in zero-field μSR experiments by Heffner *et al.*, revealing a mean-field like magnetic transition below $T_{c2} \cong 0.40$ K, with an approximate moment $\mu \cong 10^{-3} \mu_B$. No such signature was found at $T_{c1} \cong 0.60$, the onset of superconductivity, as seen in Fig. 20, where the temperature dependence of $1/T_2$ is shown. The muon position was not well

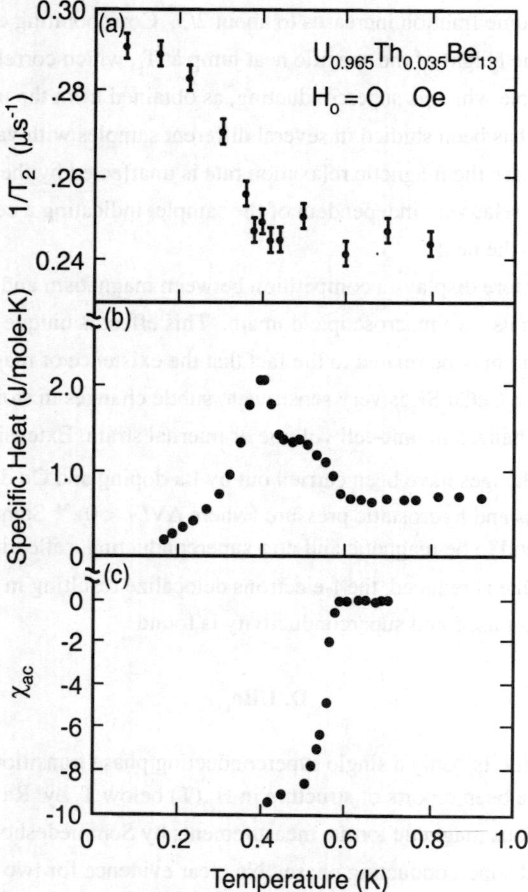

Fig. 20. Temperature dependence of the zero-field inhomogeneous relaxation rate $1/T_2$ (top), the specific heat (middle) and ac susceptibility in (U,Th)Be$_{13}$. (From R. H. Heffner et al.)[21] Two transitions are observed from the specific heat data. The μSR data show that a spontaneous internal magnetic field develops below the lower transition.

established in UBe$_{13}$ at the time (due to a lack of sizable single crystals), however, this ignorance does not invalidate the finding of small-moment magnetism.

Additional μSR experiments in the x = 0.0, 0.01, 0.024, 0.035, and 0.06 materials have established that the magnetic signature occurs only in the region where two specific heat jumps have been observed, resulting in a T-x phase diagram of Fig. 21.[81] The vertical dotted lines were not actually observed by μSR, but only postulated from μSR measurements for various Th concentrations.

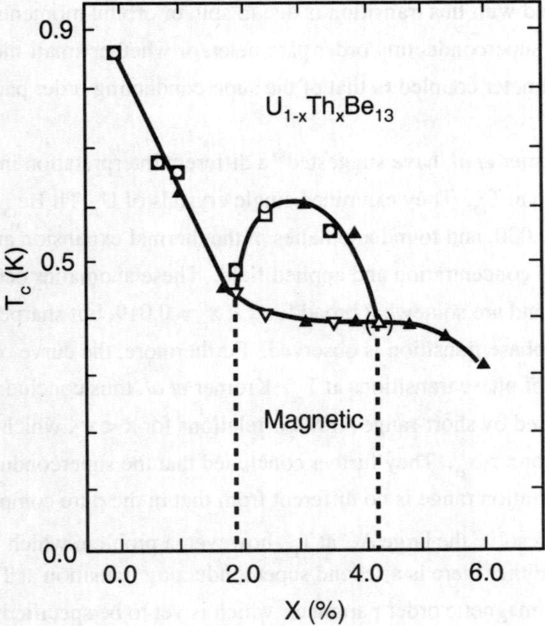

Fig. 21. Phase diagram for $U_{1-x}Th_xBe_{13}$. The magnetic region was mapped out from data similar to that shown in Fig. 20 (from R. H. Heffner *et al.*)[21]

It is important to know whether the lower transition in $(U, Th)Be_{13}$ is associated with a transition to a different superconducting state (possibly violating time-reversal symmetry because of the magnetic signature) or is simply associated with a small-moment AFM transition. Part of the answer lies in the large specific-heat jump at the lower transition: a small local-moment AFM transition is unlikely to produce the large observed jump in specific heat ΔC because most of the entropy associated with the exchange-coupled conduction-electron/local-moment system would already have been removed in reducing the value of the moments to 10^{-3} μ_B. If the magnetic order occurs through a spin-density-wave transition, the large ΔC would again be unlikely because the Fermi surface would already have been largely consumed by the superconducting transition at T_c. For a spin-density-wave transition, a large ΔC would require an exceptional enhancement of the density of states near the zeros of the superconducting energy gap. One can therefore conclude that the large ΔC at the lower transition is not caused primarily by the development of the small-moment magnetic order parameter. If instead, a second superconducting transition is involved, one must then answer the question whether the magnetic signature associated with this transition is due to spin or orbital moments from a time-reversal-violating superconducting order parameter, or whether small-moment AFM - with its order parameter coupled to that of the superconducting order parameter - is involved.

Recently, Kromer $et\ al.$ have suggested[82] a different interpretation involving only a magnetic transition at T_{c2}. They examined single crystals of $U_{1-x}Th_xBe_{13}$, $x = 0.00$, 0.01, 0.017, 0.022 and 0.030, and found anomalies in the thermal expansion and specific heat as a function of Th concentration and applied field. These anomalies occur at temperatures T_L below T_c and are somewhat broad for $x < x_{cr} = 0.019$, but sharper for $x > x_{cr}$ where the second phase transition is observed. Furthermore, the curve of T_L vs. x intersects the line of phase transitions at T_{c2}. Kromer $et\ al.$ thus concluded that these anomalies are caused by short-range AFM correlations for $x < x_{cr}$, which undergo true long-range order for $x > x_{cr}$. They further concluded that the superconducting state in the critical Th concentration range is no different from that in the pure compound. This scenario does not resolve the large ΔC at T_{c2}, however, a problem which is not addressed by Kromer $et\ al.$ Either there is a second superconducting transition at T_{c2}, or there is another "primary" magnetic order parameter which is yet to be specified, similar to the AFM transition at 17.5 K in URu_2Si_2 discussed above.

Thus, UPt_3 and possibly $(U,Th)Be_{13}$ exhibit multiple superconducting transitions, providing unambiguous evidence for unconventional superconductivity in HF systems. There is still no conclusive picture of the symmetry of the order parameter in either UBe_{13} or $(U,Th)Be_{13}$, however. Nevertheless, more recent μSR experiments appear to show a new phenomenon which highlights the continuing role that weak magnetic signatures play in our understanding of these unusual superconductors. Early experiments by Luke $et\ al.$[83] and more recent experiments by Heffner $et\ al.$[84] found evidence for a paramagnetic frequency shift below T_c in polycrystalline UBe_{13} and $UBe_{12.91}B_{0.09}$, respectively. As discussed below this is atypical behavior, and may give evidence for the field-induced alignment of spin or orbital moments associated with an unconventional superconducting order parameter.

To illustrate this point, we discuss the most recent transverse- and zero-field μSR data[84] in polycrystalline $UBe_{12.91}B_{0.09}$ ($T_c = 0.8$ K) of Heffner $et\ al.$ Experiments were carried out in the temperature range 0.05 -1.2 K in applied fields of 0 - 1000 Gauss. The zero-field data were fit with a standard Gaussian Kubo-Toyabe relaxation function. In transverse field the data were fit with a two-component relaxation function given by $G_x(t) = A_1 \exp(-\lambda_1 t) \cos(\omega_1 t + \phi_1) + A_2 \exp(-\lambda_2 t) \cos(\omega_2 t + \phi_2)$. These two terms account for the signals from the sample and from the Ag cryostat cold-finger, respectively.

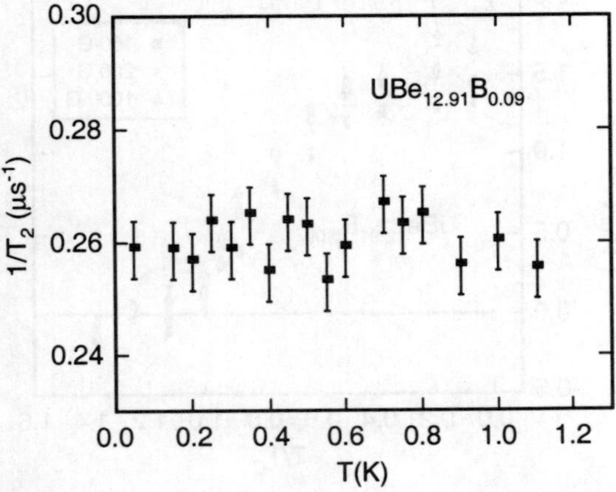

Fig. 22. Temperature dependence of the zero-field inhomogeneous relaxation rate $1/T_2$ in B-doped UBe_{13} (from R. H. Heffner *et al.*)[21] No evidence for quasi-static magnetic ordering is observed.

Fig. 22 shows the temperature dependence of the zero-field linewidth $1/T_2$, which measures the width of the local field distribution at the muon site in $UBe_{12.91}B_{0.09}$. The magnitude of $1/T_2$ is entirely consistent with that generated by nuclear dipole fields. There is no indication of any additional electronic fields induced by the onset of super-conductivity, contrary to the case of $(U,Th)Be_{13}$, where electronic fields were observed below the lower transition.

The magnitude of the internal field B is given by ω_1/γ_μ. Fig. 23 gives a plot of $\Delta B \equiv B(T) - B(T_c)$ as a function of the reduced temperature T/T_c for three applied fields: $B_{app} \cong$ 100, 200 and 1000 Gauss. The data were taken by cooling in the applied field from above T_c. One sees that the local magnetic field increases below T_c and that, despite a change in applied field by a factor of 10, ΔB is essentially field-independent. (More recent experiments on single-crystalline UBe_{13} at fields of several tesla do show a field dependence to the frequency shift, but the important paramagnetic phenomenon persists.)

Explanations of this effect are in the speculative stage. The fact that a change in frequency is observed very near T_c indicates that moments associated with the supercon-ducting state may be playing a role. Alternatively, the data may suggest AFM correla-tions of the type suggested by Kromer *et al.*[82] It is unlikely that inhomogeneously trapped flux is the cause of the shifts because of the insensitivity to applied field. A paramagnetic response in the dc field-cooled magnetization below T_c has been observed

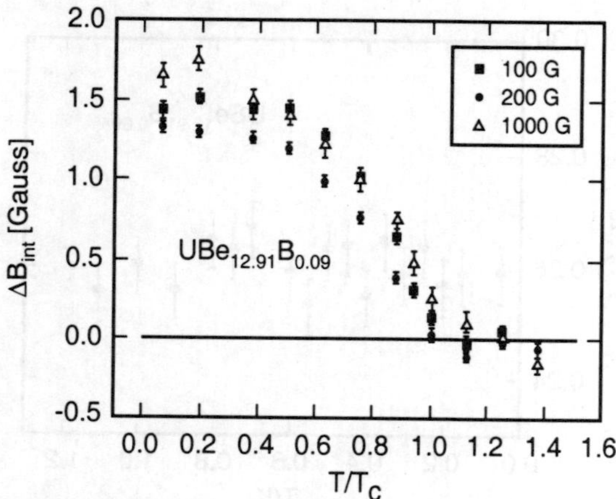

Fig. 23. Temperature dependence of the growth of an internal field below the superconducting transition in B-doped UBe_{13} for three different applied fields (from R. H. Heffner *et al.*)[21]

in high-temperature superconductors[85] and in superconducting Nb.[86] Typically, the paramagnetic response decreases with applied field, becoming diamagnetic at fields below H_{c1}. These paramagnetic effects have been variously attributed to inhomogeneously trapped flux,[87] paramagnetic supercurrents[88] in a system with Josephson junctions possessing spin-flip scattering in the tunneling barrier, or a Josephson network of d-wave superconductors,[89] wherein a π-junction arises from the possibility of an intrinsic phase difference in the d-wave order parameter on the two sides of the junction. In these latter two cases an applied field could align the orbital currents, producing a field-independent magnetization.

Orbital currents can also arise in the case of a time-reversal-violating order parameter when the electrons scatter from impurities in the system.[90] The theory of this effect yields currents proportional to the superfluid density, which could qualitatively explain the temperature dependence seen in Fig. 23. As pointed out above, however, a similar effect has been observed in undoped UBe_{13}, where impurities are expected to play a smaller role than in the B-doped samples. Perhaps the muon itself could provide a source of impurity scattering. The fact that no broadening is observed in zero field seems to be inconsistent with an array of orbital moments, which would give rise to an inhomogeneous internal field (in zero applied field), unless an applied field is required to produce the currents in the first place.

VII. SUMMARY

This chapter has been devoted to explaining how μSR can be used to elucidate the small-moment properties of selected HF systems. The value of the technique for such studies derives from its high sensitivity to small magnetic fields, being at least an order of magnitude more sensitive in this regard than NMR or neutron scattering. We thus began by describing how a μSR experiment is carried out, and then discussed the phenomenology of analyzing the experimental μSR data. The important issue of how to find the muon position in the lattice was also addressed, using specific illustrative examples.

Three principle categories of materials have been surveyed: small-moment magnetic compounds, non-Fermi liquid (NFL) materials and heavy fermion superconductors. The first category involves materials where the conduction electron/local moment exchange interaction is very near the critical value where the Kondo and RKKY interactions are of equal strength. This is the arena in which paramagnetic and small-moment HF compounds are found. In this regime μSR has demonstrated that $CeCu_6$ and $CePt_2Si_2$ remain paramagnetic down to 0.05 K, making these compounds the only two non-ordering HF materials known at the time of this writing. A close second in this regard, $CeRu_2Si_2$ has been found to possess a tiny static moment of only 10^{-3} μ_B below about 2K, the smallest measured ordered electronic moment to date. These Ce-based materials thus apparently have f-d exchange interactions very close to their critical values, giving rise to completely or nearly completely compensated f-electron moments. A phenomenological model for such effects was developed long ago by Doniach and others, and a phase diagram for the $(Ce,U)X_2Si_2$ system based on Doniach's f-d exchange model has been formulated by Endstra *et al.*, with reasonable success. Nevertheless, a complete understanding of these issues, including that of mixed-valence in general, is still under development. For a modern theoretical exposition of these and related concepts, the reader is referred to the chapter in this book by Barry Cooper.

The next class of materials addresses the role that disorder and distortion play in the character of the magnetic ground state. In single crystal $CePt_2Sn_2$ ($\gamma \cong 3.5$ J/mol-K^2) a temperature-independent, Gaussian relaxation rate was observed below 0.2 K. A Gaussian relaxation function typically signifies a static Gaussian local magnetic field distribution. The fact that the Gaussian rate persists in a sizeable longitudinal field indicates the absence of static internal fields, however. This was interpreted in terms of very slowly fluctuating spins which do not order down to 0.02 K. By contrast experiments in polycrystals of $CePt_2Sn_2$ showed a clear signature of magnetic order below 0.85 K, suggesting that this compound possesses geometrical frustration which prevents freezing in single crystals, but that this frustration is removed by strain in polycrystals.

Frustration effects may also be responsible for similar phenomena in YbBiPt, which is a HF ($\gamma \cong 8$ J/mol-K^2) exhibiting nearly equal energy scales for Kondo, RKKY and CEF effects. μSR experiments are consistent with small-moment magnetic ordering below 0.4 K, where transport and specific heat studies also show anomalies. In addition, μSR finds evidence for magnetic inhomogeneity from the observation of a two-component relaxation function. A rapid Gaussian relaxation observed below the ordering temperature in YbBiPt corresponds to a small moment ($\leq 10^{-1} \mu_B$), but, as in single crystal $CePt_2Sn_2$, the Gaussian function is not decoupled in a longitudinal field. Similarly, an undecouplable Gaussian relaxation is found below the magnetic ordering temperature $\cong 1$ K in YbSbPd ($\gamma \cong 3$ J/mol-K^2). In the latter material this behavior is ascribed to occasional fluctuations out of a magnetic ground state whose field cancels at the muon site. Such behavior could arise from a spin liquid state.

Paramagnetic non-Fermi-liquid (NFL) materials do not show the characteristic low-temperature transport (resistivity $\propto T^2$) and magnetic behavior (temperature independent magnetic susceptibility) found in simple metals. μSR is an ideal probe for one source of NFL behavior, namely a local distribution of Kondo couplings $P(T_K)$ due to atomic disorder. The $P(T_K)$ distribution gives rise to a broadened inhomogeneous μSR linewidth, which scales with the bulk susceptibility and is easily detectable with μSR. The best known example to date is $UCu_{5-x}Pd_x$, $x = 1.0$ and 1.5. By contrast, the system $CeCu_{5.9}Au_{0.1}$ displays classical NFL behavior but does not show the degree of magnetic inhomogeneity found in $UCu_{5-x}Pd_x$. Thus, the NFL behavior in $CeCu_{5.9}Au_{0.1}$ most likely arises from proximity to a quantum critical point. The study of disorder-induced NFL behavior has spawned theoretical effort to incorporate disorder into a picture including both the Kondo and RKKY interactions, giving rise to possible Griffith phases.

The small magnetic moments in HF superconductors can arise from two sources, magnetic ordering or from the superconducting state itself, if the superconducting order parameter breaks time-reversal symmetry. Only UBe_{13} seems to possess no small-moment magnetic ordering, either above or below its superconducting transition. By contrast, UPt_3 possesses small-moment AFM ordering at 5 K, and this ordering is thought to provide a symmetry breaking field which breaks the degeneracy of the superconducting order parameter, yielding multiple superconducting states. A small spontaneous magnetic field has also been found by μSR below the lower of the two zero-field superconducting transitions in UPt_3, indicating a time-reversal breaking superconducting order parameter. A similar effect was observed by μSR in Th-doped UBe_{13}, although its origin is less clear. These examples are discussed in the last portion of this chapter, which ends with a speculative discussion of unexpected paramagnetic field shifts below the superconducting state in boron-doped and undoped UBe_{13}.

REFERENCES

1. A. Schenck and F. N. Gygax, Handbook of Magnetic Materials, **9**, 57 (1995).
2. A. Amato, Rev. Mod. Phys. **69**, 1119, (1997).
3. R. Kubo and T. Toyabe, "Magnetic Resonance and Relaxation", ed. R. Blinc (North Holland, Amsterdam) (1967).
4. R. S. Hayano *et al.*, Phys. Rev. B**20**, 850, (1979).
5. Y. J. Uemura, Sol. St. Comm. **36**, 369, (1980).
6. L. P. Le *et al.*, Phys. Rev. B**53**, R510, (1996).
7. S. Doniach, "Valence Instabilities and Related Narrow-Band Phenomena", ed. R. D. Parks (Plenum, New York, 1977) p.169; M. A. Continento, Phys. Rep. **239**, 179, (1994).
8. Y. Onuki, J. Phys. Soc. Jpn. **53**, 1210 (1984); G. R. Stewart *et al.*, Phys. Rev. B**30**, 482, (1984).
9. Y. Onuki, *et al.*, J. Phys. Soc. Jpn. **54**, 304, (1985).
10. A. Amato *et al.*, J. Low Temp. Phys. **68**, 371, (1987).
11. A. Amato *et al.*, Hyperf. Interactions **104**, 115, (1997).
12. A. Amato *et al.*, Phys. Rev. B**52**, 54, (1995).
13. T. Endstra *et al.*, Phys. Rev. B**48**, 9595, (1993).
14. P. Dalmas de Réotier *et al.*, Hyperf. Int. **64**, 457, (1990).
15. K. Hiebl and P. Rogl, J. Magn. Magn. Mater. **50**, 39, (1985).
16. C. Ayache *et al.*, J. Magn. Magn. Mater. **63-64**, 329, (1987).
17. P. Dalmas de Réotier *et al.*, Phys. Rev. B**55**, 2737, (1997).
18. A. Amato, *et al.*, Phys. Rev. B**50**, 619, (1994).
19. M. J. Besnus *et al.*, Sol. St. Comm. **55**, 779, (1985).
20. F. Steglich *et al.*, J. Appl. Phys. **57**, 3054, (1985).
21. W. P. Beyermann *et al.*, Phys. Rev. B**43**, 13130, (1991); Phys. Rev. Lett. **66**, 3289, (1991).
22. Luke *et al.*, Physica B **206&207**, 222, (1995).

194

23. Erik Lidström *et al.*, Physica Scripta **54**, 210, (1996).

24. G. M. Luke *et al.*, Hyperf. Int. **104**, 199, (1997).

25. Z. Fisk *et al.*, Phys. Rev. Lett. **67**, 3310, (1991).

26. R. Movshovich *et al.*, Phys. Rev. Lett. **73**, 492, (1994).

27. R. A. Robinson *et al.*, Phys. Rev. B**50**, 9595, (1994).

28. R. A. Robinson *et al.*, Phys. Rev. Lett. **75**, 1194, (1995).

29. A. Amato *et al.*, Phys. Rev. B**46**, 3151, (1992).

30. R. H. Heffner *et al.*, Physica B**199&200**, 113, (1994); A. Amato *et al.*, B**186-188**, 615, (1993).

31. N. Kaplan *et al.*, Hyperf. Int. **85**, 271, (1994).

32. Y. J. Uemura *et al.*, Phys. Rev. Lett. **73**, 3306, (1994).

33. P. Bonville, *et al.*, Physica B**220-232**, 266, (1997).

34. G. Le Bras *et al.*, Physica B**199&200**, 542, (1994).

35. A. P. Reyes *et al.*, Physica B**206&207**, 332, (1995).

36. F. G. Aliev *et al.*, JETP Lett. **60**, 734, (1994).

37. O. O. Bernal *et al.*, Phys. Rev. Lett. **75**, 2023, (1995); O. O. Bernal *et al.*, Phys. Rev. B **54**, 13 000, (1996); D. E. MacLaughlin *et al.*, J. Phys.: Condens. Matter **8**, 9855, (1996).

38. W. D. Wu *et al.*, Phys. Rev. Lett. **72**, 3722, (1994).

39. C. L. Seaman *et al.*, Phys. Rev. Lett. **67**, 2882, (1991); C. L. Seaman *et al.* J. Alloys and Compounds **181**, 327, (1992).

40. D. L. Cox, Phys. Rev. Lett. **59**, 1240, (1987).

41. Stefan Süllow, Ph. D. thesis, University of Leiden, 1995.

42. A. Schenck *et al.*, Hyperf. Int. **64**, 511, (1990).

43. C. H. Booth *et al.*, Phys. Rev. Lett. **81**, 3960, (1998).

44. H. v. Löhneysen *et al.*, Phys. Rev. Lett. **72**, 3262, (1994).

45. H. v. Löhneysen *et al.*, Physica B**223&224**, 471, (1996); Phys. Rev. Lett. **72**, 3262, (1994); B. Bogenberger and H. v. Löhneysen, Phys. Rev. Lett. **74**, 1016, (1995).

46. H. G. Schlager *et al.*, J. Low Temp. Phys. **90**, 181, (1993).

47. J. D. Thompson *et al.*, J. Magn. Magn. Mater. **54-57**, 377, (1986).

48. S. Kawarazaki *et al.*, Physica B **206&207**, 298, (1995).

49. T. Graf *et al.*, Phys. Rev. Lett **78**, 3769, (1997).

50. D. E. MacLaughlin *et al.*, private communication, 1996.

51. F. Steglich *et al.*, Phys. Rev. Lett. **43**, 1892, (1979).

52. R. H. Heffner and M. R. Norman, Comments on Condensed Matter Physics **17**, 361, (1996).

53. G. R. Stewart *et al.*, Phys. Rev. Lett. **52**, 679, (1984).

54. Cooke *et al.*, Hyperfine Interactions **31**, 425, (1986); R. H. Heffner *et al.*, "Theoretical and Experimental Aspects of Valence Fluctuations and Heavy Fermions", Eds. L. C. Gupta and S. K. Malik, p. 319, (Plenum, 1987) .

55. G. Aeppli *et al.* Phys. Rev. Lett. **63**, 676, (1989); E. D. Isaacs *et al.*, Phys. Rev. Lett. **75**, 1178, (1995).

56. R. A. Fisher *et al.*, Phys. Rev. Lett. **62**, 1411, (1989).

57. G. Bruls *et al.*, Phys. Rev. Lett. **65**, 2294, (1990).

58. S. Adenwalla *et al.*, Phys. Rev. Lett. **65**, 2298, (1990).

59. M. Boukhny *et al.*, Phys. Rev. **B50**, 8985, (1994).

60. G. M. Luke *et al.*, Phys. Rev. Lett. **71**, 1466, (1993).

61. J. A. Sauls, J. Low Temp. Phys. **95**, 153, (1994) and Adv. Phys. **43**, 113, (1994).

62. S. M. Hayden *et al.*, Phys. Rev. **B46**, 8675, (1992).

63. P. Dalmas de Réotier *et al.*, Phys. Lett. **A205**, 239, (1995).

64. T. E. Mason *et al.*, Phys. Rev. Lett. **65**, 3189, (1990).

65. E. D. Isaacs *et al.*, Phys. Rev. Lett. **65**, 3185, (1990).

66. D. E. MacLaughlin *et al.*, Phys. Rev. **B37**, 3153, (1988).

67. E. A. Knetsch *et al.*, Physica B **186-188**, 300, (1993).

68. A. Yaouanc *et al.*, private communication, 1996.

69. M. B. Walker *et al.*, Phys. Rev. Lett. **71**, 2630, (1993).

70. D. F. Agterberg and M. B. Walker., Phys. Rev. **B50**, 563, (1994).

71. Y. J. Uemura *et al.*, Phys. Rev. **B39**, 4726, (1989).

72. G. M. Luke *et al.*, Phys. Rev. Lett. **73**, 1853, (1994).

73. R. Feyerherm *et al.*, Physica **B206-207**, 596, (1995); A. Amato, Physica **B199- 200**, 91, (1994).

74. N. Grewe and F. Steglich, "Handbook on the Physics and Chemistry of Rare Earths" **14**, 343, (1991).

75. H. R. Ott *et al.*, Phys. Rev. Lett. **50**, 1595. (1983).

76. Rauchschwalbe *et al.*, Europhys. Lett **3**, 751. (1987).

77. G. Schmiedeshoff *et al.*, Phys. Rev. **B48**, 16 417. (1993).

78. H. R. Ott *et al.*, Phys. Rev. **B31**, 1651. (1985).

79. B. Batlogg *et al.*, Phys. Rev. Lett. **55**, 1319. (1985).

80. R. H. Heffner *et al..* Phys. Rev B**40**, 806. (1989).

81. R. H. Heffner *et al..* Phys. Rev. Lett. **65**, 2816. (1990).

82. F. Kromer *et al..* Phys. Rev. Lett. **81**, 4476. (1998).

83. G. M. Luke *et al.*, Phys. Lett. **157**, 173. (1991).

84. R. H. Heffner *et al.*, Physica B230-232, 398, (1997).

85. W. Braunisch *et al.*, Phys. Rev. Lett. **68**, 1908, (1992).

86. David J. Thompson *et al.*, Phys. Rev. Lett. **75**, 529, (1995).

87. P. Kostic *et al.*, Phys. Rev. B**53**, 791, (1996); A. E. Koshelev and A. I. Larkin, Phys. Rev. B**52**, 13559, (1995); E. Zeldov *et al.*, Phys. Rev. Lett. **73**, 1428, (1994).

88. L. N. Bulaevskii, V. V. Kuzii and A. A. Sobyanin, Pis'ma Zh. Eksp. Teor. Fiz. **25**, 314, (1977); JETP Lett. **25**, 290, (1977).

89. Manfred Sigrist and T. M. Rice, J. Phys. Soc. Jap. **61**, 4283, (1992).

90. G. E. Volovik and V. P. Mineev, Zh. Eksp. Teor. Fiz. **81**, 989, (1981) [Sov. Phys. JETP **54**, 524 (1981)]; C. H. Choi and Paul Musikar, Phys. Rev. B**39**, 9664, (1989); Mario Palumbo and Paul Musikar, Phys. Rev. B**42**, 2681, (1990); Manfred Sigrist and Kazuo Ueda, Rev. Mod. Phys. **63**, 239, (1991).

4

NEUTRON SCATTERING FROM HEAVY FERMIONS

R. A. Robinson

Los Alamos National Laboratory
Los Alamos, NM 87545, U.S.A.

I. INTRODUCTION

Heavy fermion materials are intermetallic compounds exhibiting strongly enhanced low-temperature thermodynamic properties, particularly magnetic susceptibility and electronic specific heat. Indeed Stewart[1] has made the somewhat arbitrary definition that the term should be reserved for those materials that have linear terms in their low-temperature linear specific-heat coefficients $\gamma > 400$ mJ f-atom-mol^{-1}K^{-2}, though the term heavy fermion is often used for compounds with smaller Sommerfeld coefficients. All of these compounds contain an f-series metal component, typically Ce or U, but also in some cases Np, Yb or Pr. As regards other physical properties, there is no clear pattern in these compounds: some are superconductors, some are antiferromagnets, some are both and some are neither. Of the antiferromagnets, some have substantial well-developed moments, while others have tiny long-range-ordered moments that can barely be detected. Some antiferromagnets are commensurate, while others are incommensurate, and some heavy fermions exhibit both types of order within the same phase diagram. To this author's knowledge, none of these materials exhibit ferromagnetism, and there is good evidence from Curie-Weiss susceptibilities that antiferromagnetic interactions dominate in all cases. However, if a large enough magnetic field is applied to some heavy fermion systems, substantial magnetisations can be induced beyond a critical field B_c. In some cases, there is only one clear *metamagnetic* transition, while in the case of URu_2Si_2, at least, a series of transitions is observed. Very little work[2] has been performed to date with neutrons on these field-induced transitions.

The main contribution of neutron-scattering methods to this subject lies in two areas: diffraction studies of the long-rang-ordered magnetic states in these compounds, and inelastic scattering studies of the magnetic excitation spectra. Regarding the antiferromagnetically ordered heavy fermions, a major part of this chapter is devoted to cataloguing the magnetic structures of these compounds in a coherent way, and we attempt to include drawings of all the commensurate antiferromagnetic structures reported to date. The excitation spectrum on the other hand can be interpreted as representing the imaginary part of the generalised magnetic susceptibility $\chi''(\mathbf{Q},\omega)$. A dominant theme is that of very broad quasielastic or overdamped magnetic excitations, with a width corresponding to a characteristic energy scale (or Kondo temperature).

Neutron methods have made other contributions to this subject, particularly in the areas of trying to determine the valence of the f-electron species, via the form factor, characterising the phonons, and the studying the flux-line lattice in superconductors.

In Table I, which follows, we summarise work on the antiferromagnetic structures, as measured by neutron diffraction, while Table II summarises work on their spin dynamics. Only 50% of the materials listed here meet Stewart's 400 mJmol⁻¹K⁻² criterion. In Table I, the compounds are divided into four categories, superconducting antiferromagnets, regular antiferromagnets, regular superconductors and those that do not order into either of these ground states. Within each category, they are listed in order of decreasing crystallographic symmetry. In Table II, on the other hand, the compounds are categorised according to the active f-electron element (U, Ce, Pr and Yb).

Table I Crystallographic, Magnetic and Bulk Properties of Heavy Fermions

Material [Ref.(s)]	Structure Type	Space Group	Lattice Parameters (Å)	d_{ff} (Å)	γ (mJmol⁻¹K⁻²)	T_C(K)	T_N(K)	Ordered Moment (μ_B)	Low-T Magnetic Structure	Metamagnetic Field, B_C (T)	Critical Exponent β	Correlation Length ξ (Å)
Superconducting Antiferromagnets (*and related substitutions*):												
UPt_3 [3-7]	Ni_3Sn	$P6_3/mmc$	a = 5.764 c = 4.899	4.132	450	0.54	5	0.02	q = [1/2,0,0] $\mu \parallel a*$?	~20 (B ⊥ c)	1/2	ξ_{a*} = 280 Å ξ_{c*} = 500 Å
$U_{1-x}Th_xPt_3$ [8]	Ni_3Sn	$P6_3/mmc$	-	-	-	-	6.5	0.65	q = [1/2,0,0] $\mu \parallel a*$ assume single-q		« 1/2	
$UPt_{3-x}Pd_x$ [9-12]	Ni_3Sn	$P6_3/mmc$	-	-	-	-	5.8	0.6	q = [1/2,0,0] may be multi-q	13 (B ⊥ c)	« 1/2	
UNi_2Al_3 [13-15]	$PrNi_2Al_3$	P6/mmm	a = 5.207 c = 4.018	4.018	120	1	4.6	0.21	q = [1/2±τ,0,1/2] $\mu \parallel a$		0.34±0.03	
UPd_2Al_3 [12,16-21]	$PrNi_2Al_3$	P6/mmm	a = 5.365 c = 4.186	4.186	γ_p=210 γ_o=150	2	14	0.85	q = [0,0,1/2] $\mu \parallel a$	18 (B ⊥ c)	0.55±0.05 (Ref. 19) 0.33±0.02 (Ref. 20)	
URu_2Si_2 [22-27]	$ThCr_2Si_2$	I4/mmm	a = 4.121 c = 9.681	4.121	γ_p=180 γ_o=50	0.8	17.5	0.03	q = [100] $\mu \parallel c$	38 (B ∥ c)	1/2	∞ ?
$URu_{1.2}Re_{0.8}Si_2$ [28]	$ThCr_2Si_2$	I4/mmm	a = 4.126 c = 9.725	4.126	117	-	$T_C \sim 30$	0.5	ferromagnetic	-	-	ξ_{small}
Non-Superconducting Antiferromagnets (*and related substitutions*):												
$NpBe_{13}$ [29,30]	$NaZn_{13}$	$Fm\bar{3}c$	a = 10.233	5.11	900	-	3.4	μ_1=1.16 μ_2=0.94 $\mu \perp c$, noncoll.	q = [0,0,1/3]			
$CePb_3$ [31-33]	Cu_3Au	$Pm\bar{3}m$	a = 4.872	4.872	225	-	1.16	0.55	q = [δ_1, δ_2,1/2] $\mu \parallel$ [001]		0.30±0.05	
UCd_{11} [34]	$BaHg_{11}$	$Pm\bar{3}m$	a = 9.29	6.56	840	-	5		?	?		
UCu_5 [35-37]	$AuBe_5$	$F\bar{4}3m$	a = 7.038	4.977	86	-	15	1.55	q = [1/2 1/2 1/2] $\mu \parallel$ [111]			

Table I, continued.

Material [Ref.(s)]	Structure Type	Space Group	Lattice Parameters (Å)	d_{ff} (Å)	γ (mJmol^{-1}K^{-2})	T_C(K)	T_N(K)	Ordered Moment (μ_B)	Low-T Magnetic Structure	Metamagnetic Field, B_C (T)	Critical Exponent β	Correlation Length ξ(Å)
$UAgCu_4$ [35,38]	$AuBe_5$	$F\bar{4}3m$	a = 7.115	5.031	310		18					
$YbAgCu_4$ [39]	$AuBe_5$	$F\bar{4}3m$	a = 7.0696	4.999	245		0.6					
$YbPdCu_4$ [39]	$AuBe_5$	$F\bar{4}3m$	a = 7.0396	4.978	200		0.8					
YbBiPt [40]	MgAgAs	$F\bar{4}3m$	a = 6.595	4.663	8000		0.4	<0.1				
YbPdSb [41]	MgAgAs	$F\bar{4}3m$	a = 6.462	4.569	470		<2K?					
YbBiPd [41]	MgAgAs	$F\bar{4}3m$	a = 6.59	4.66	240		<2K?					
UPd_2Ga_3 [42-44]	BaB_2Pt_3	$P6_3/mmc$	a = 5.3015, c = 8.5112	4.2556	230	-	13	0.50	q = [0,0,1] μ⊥c	16.5	0.28±0.02	
$CePd_2Al_3$ [45-47]	$PrNi_2Al_3$	P6/mmm	a = 5.50816, c = 4.22668	4.2267	380	-	2.7	unknown	unknown μ⊥c	4.3 (B∥a)		
U_2Zn_{17} [48-50]	Th_2Zn_{17}	$R\bar{3}m$	a = 8.955, c = 13.124	4.375	100 (per U atom)		9.7	0.79	q = [000] μ∥a*(hex. coordinates)		0.36±0.02	
$CeRu_2Si_2$ [51,52]	$ThCr_2Si_2$	I4/mmm	a = 4.192, c = 9.78	4.192	385	-	1.5	~10^{-3} (μSR)		7.7 (B∥c)		
$CeRu_{2-x}Rh_xSi_2$ [53,54]	*$ThCr_2Si_2$*	*I4/mmm*	-		?	-	1.5	0.65	q = [0,0,0.42 - 0.45]			
$CeRu_{2-x}Pd_xSi_2$ [55]	*$ThCr_2Si_2$*	*I4/mmm*	-		?	-	~2.5	0.3	q = [0,0,0.38] μ∥c	1.5, 6		
$Ce_{1-x}La_xRu_2Si_2$ [56-58]	*$ThCr_2Si_2$*	*I4/mmm*	-		600	-	1-6	0-1.2	q = [0..31,0,0]			ξ = 200 Å
$CePd_2Si_2$ [59,60]	$ThCr_2Si_2$	I4/mmm	a = 4.24, c = 9.88	4.24	250 (p ~ 25 kbar)	0.2	9	0.62	q = [1/2,1/2,0] μ∥q	-		
UIr_2Si_2 [61,62]	$CaBe_2Ge_2$	P4/nmm	a = 4.0851, c = 9.8348	4.0851	γ_o=300, γ_p=105		6	0.10	q = [100] μ∥c	1.5 (B∥c)		

Superconducting Paramagnets:

Material [Ref.(s)]	Structure Type	Space Group	Lattice Parameters (Å)	d_{ff} (Å)	γ (mJmol^{-1}K^{-2})	T_C(K)	T_N(K)	Ordered Moment (μ_B)	Low-T Magnetic Structure	Metamagnetic Field, B_C (T)	Critical Exponent β	Correlation Length ξ(Å)
UBe_{13} [63]	$NaZn_{13}$	$Fm\bar{3}c$	a = 10.2607	5.13	1100	0.85	-	-	-	-		
$CeCu_2Si_2$ [64,65]	$ThCr_2Si_2$	I4/mmm	a = 4.105, c = 9.933	4.105	1000	1	-	-	-			

Vegetables (neither superconductivity nor antiferromagnetism) *and related substitutions:*

Material [Ref.(s)]	Structure Type	Space Group	Lattice Parameters (Å)	d_{ff} (Å)	γ (mJmol^{-1}K^{-2})	T_C(K)	T_N(K)	Ordered Moment (μ_B)	Low-T Magnetic Structure	Metamagnetic Field, B_C (T)	Critical Exponent β	Correlation Length ξ(Å)
$PrAg_2In$ [66]	Cu_2MnAl	$Fm\bar{3}m$	a = 7.075	5.003	6500							
$CeAl_3$ [67]	Ni_3Sn	$P6_3/mmc$	a = 6.541, c = 4.610	4.424	1620	-	-	-	-			
$CeNi_2Ge_2$ [68-70]	$ThCr_2Si_2$	I4/mmm	a = 4.150, c = 9.854	4.150	350 (for p > 15 kbar)	0.22	-	-	-	42		
U_2Pt_2In [71,72]	U_3Si_2	P4/mbm	a = 7.654, c = 3.725	3.701	425 (per U atom)		-	-				

Table I, continued.

Material [Ref.(s)]	Structure Type	Space Group	Lattice Parameters (Å)	$d_{f\text{-}f}$ (Å)	γ (mJmol^{-1}K^{-2})	T_C(K)	T_N(K)	Ordered Moment (μ_B)	Low-T Magnetic Structure	Metamagnetic Field, B_c (T)	Critical Exponent β	Correlation Length ξ(Å)
CePtSi [73]	LaPtSi	I4$_1$md	a = 4.202 c = 14.484	4.186	800	-	-	-				
Yb$_2$Ni$_2$Al [74,75]	Mo$_2$NiB$_2$ (=W$_2$CoB$_2$)	Immm	a = 8.224 b = 5.334 c = 4.117	3.372[a]	~750 (per Yb atom)							
CeCu$_6$ [76-80] (T> ~200K)	CeCu$_6$	Pnma	a = 8.112 b = 5.102 c = 10.162	4.839								
[81-83] (T< ~200K)		P2$_1$/c	a = 5.084 b = 10.1279 c = 8.0731 β = 91.442°	-	1670	-	-	-		2 (B $\parallel$ b$_{monoclinic}$ $\parallel$ c$_{orthorhombic}$)		
CeCu$_{5.8}$Au$_{0.2}$ [84,85]	*CeCu$_6$*	*Pnma*	-	-	-	0.25	0.02	$q = [\pm 0.79, 0,0]_{orthorhombic}$ $\mu \parallel c_{orthorhombic}$				$\xi_a = 150 - 250$ Å
CeCu$_{5.5}$Au$_{0.5}$ [86]	*CeCu$_6$*	*Pnma*	*a = 8.160* *b = 5.067* *c = 10.24*	-		1.00	~1	$q = [\pm 0.59, 0,0]_{orthorhombic}$ $\mu \parallel c_{orthorhombic}$			*(b)*	
CeCu$_{6-x}$Au$_x$ [87,88]	*CeCu6*	*Pnma*						$q = [\pm \delta_p, 0, \pm \delta_2]_{orthorhombic}$				
CeCu$_5$Au [89]	CeCu$_6$	Pnma	-		640	-	2.2	-		-	2.2, 3	

(a) Atom positions are not reported in Ref. 75, and the lattice parameters are reported in different order to the convention used in *Pearson's Handbook of Crystallographic Data for Intermetallic Phases*, P. Villars and L. D. Calvert, American Society for Metals, Metals Park, 1985. However, assuming the atom positions for W$_2$CoB$_2$, we derive the value for $d_{f\text{-}f}$ given in the table.

(b) These authors report the temperature variation of the order parameter in a somewhat unconventional manner as $I = I_o(1-(T/T_N)^{5/2})$, and if taken literally this would give $\beta = 1/2$.

Table II Dynamic Properties of Heavy Fermions

Material	γ (mJmol⁻¹K⁻²)	Quasielastic Width Γ_0 (meV) (Lorentzian HWHM)	Q-dep of Correlations	Gap (meV)	Crystal-field Energies (meV)	Comments on Excitations (p = polycrystal; s = single crystal pol. = polarised)	Crystal-Field Ground State	References
Uranium Compounds:								
UPt_3	450	9	powder avg.			p, pol.		[90,91]
		5	$q = (0,0,1)$			s		[92]
		0.2	$q = (1/2, 0,1)$			s		[4]
		0.5 ($q = 0.11$Å⁻¹)	$q \rightarrow 0$			p, $\Gamma \propto q$ (Fermi-liquid model)		[93]
URu_2Si_2	$\gamma_p = 180$	6.0 (T = 50K)	AF	5.5		p		[94]
	$\gamma_o = 50$	~6 (T>T_N)				s	singlet out of J=4	[23,95-97]
		<<1 (T<T_N)		1.8 (T<T_N)		s, longitudinal		
UBe_{13}	1100	13				p, pol.		[98]
		1.5	q-indep.	p		[99]		
U_2Zn_{17}	100	10	$q = (1,0,2)$			s, damped antiferromagnetic spin waves?		[49]
		8.9				p		[100]
UCu_5	86	9.8	antiferromagnetic			p		[100]
$UCu_{5-x}Pd_x$	NFL					p		[101-103]
UPd_2Al_3	$\gamma_p = 210$		n			s, damped antiferromagnetic spin waves?		[20]
	$\gamma_o = 150$		n			s, quadratic away from q_0 along c^*		[104]
		~0.3	$q_0 = (0,0,1/2)$			s, $\xi_c \sim 50$Å		[105]
			at q_0 0.35 (T<T_c)			s		[106,107]
Cerium Compounds and related substitutions:								
$CeAl_3$	1620				5.2,7.6	p	Γ_9 doublet	[108]
		0.5	$q = 0.1 - 1.2$ Å⁻¹			p		[109]
		1.2,13.0			7.6 only	p	Γ_7 doublet	[110]
$CePb_3$	225	<0.6			6.2	s		[33]
		1.0 ± 0.3			5.8	p	Γ_7 doublet	[111]
		0.17 ± 0.03				p	Γ_7 doublet	[112]
$CePd_2Al_3$	380	1.92			2.8		not known	[113]
							maybe Γ_7 $|\pm1/2>$	[47]
$CeCu_6$	1670	0.2 - 0.5				s, q-dependent		[114]
		0.5			5.5,11	p	doublet	[115]
		$\Gamma_{s-s} = 0.42$	$q_1 = (0,0,1)_{monoclinic}$			s, $\Gamma_{i-s} \sim \hbar\omega_0 = 0.2$ meV		[116]
			$q_2 = (0.85,0,0)_{monoclinic}$			$\xi_a = 9$Å; $\xi_c = 3.5$Å		
			7,13.8				doublet	[117]
$CeCu_{6-x}Au_x$	*NFL*					*s*		*[118,88]*
(x = 0.1)								

Table II, continued.

Material	γ (mJmol^{-1}K^{-2})	Quasielastic Width Γ_0 (meV) (Lorentzian HWHM)	Q-dep of Correlations	Gap (meV)	Crystal-field Energies (meV)	Comments on Excitations (p = polycrystal; s = single crystal pol. = polarised)	Crystal-Field Ground State	References
CeRu$_2$Si$_2$	385	$\Gamma_{s\text{-}s} = 2.0$	$\mathbf{q}_1 = (0.3,0,0)$ $\mathbf{q}_2 = (0.3,0.3,0)$			s, $\Gamma_{i\text{-}s} \sim \hbar\omega_0 = 1.2$ meV $\xi_\parallel = 8\text{Å}$		[116]
		1.2	corresponding to $\mathbf{q}_2$			s		[119]
		1.78 (T = 5K)				p		[120]
				none to 40 meV		p		[121]
		~ 2.2				p		[122]
Ce$_{1\text{-}x}$La$_x$Ru$_2$Si$_2$								
	600	*$\Gamma_{s\text{-}s} = 1.2$*				*s, $\Gamma_{i\text{-}s} = 0.2$ meV*		*[58]*
CeNi$_2$Ge$_2$	350	4					not known	[68]
CeCu$_2$Si$_2$	1000	~1 (@10K)			12.1,31.4	p	mainly $|\pm5/2\rangle$	[123,124]
		~1 (@2.5K)				p, pol.		[125]
CeCu$_{1.8}$Si$_2$	?	0.7			20.7,32.7	p	mainly $|\pm5/2\rangle$	[126]
CeCu$_{2.2}$Si$_2$	?	0.7			17.0,33.1	p	mainly $|\pm5/2\rangle$	[126]
CePd$_2$Si$_2$	250	0.76			18.4, 20.5	p	mainly $|\pm5/2\rangle$	[120,121]
		1.2			0.7 (T < T$_N$)	s		[127,128]
CePtSi	800	1.5			6.3,17.8	p	0.63$|\pm5/2\rangle$ +0.78$|\pm3/2\rangle$	[129]

Ytterbium Compounds:

Material	γ	Quasielastic Width Γ_0	Q-dep of Correlations	Gap	Crystal-field Energies (meV)	Comments on Excitations	Crystal-Field Ground State	References
YbBiPt	8000	0.2	n		1,2,6	p	Γ_7 doublet (and Γ_8 quartet)	[130]
YbBiPd	240	0.4			2,7.5	p		[131]
YbSbPd	470	1.4			6.8, 8.7	p		[131]
YbAgCu$_4$	245	6.45			~10	p		[132]
YbPdCu$_4$	200	0.09				p		[132]

Praesodymium Compounds:

Material	γ	Quasielastic Width Γ_0	Q-dep of Correlations	Gap	Crystal-field Energies (meV)	Comments on Excitations	Crystal-Field Ground State	References
PrAg$_2$In	6500	-				p (quadrupolar Kondo?)	Γ_3 doublet (nonmagnetic)	[66,133]

$\Gamma_{s\text{-}s}$ = single-site halfwidth (i.e. Q-independent)

$\Gamma_{i\text{-}s}$ = inter-site halfwidth (i.e. Q-dependent, and representing correlations between sites)

$\xi_\parallel$ = in-plane correlation length (i.e. within tetragonal basal plane)

NFL = non-Fermi-liquid behaviour (i.e. logarithmic or other divergence in C/T)

II. WHAT NEUTRON SCATTERING CAN TELL US

The neutron-scattering method has been a major tool in solid-state physics for roughly fifty years now and its strengths are well known. There are several excellent textbooks on the theory[134] and practice of neutron scattering,[135,136] as well as a series of proceedings from the triennial International Conferences on Neutron Scattering.[137] Likewise, there have been previous reviews[138,50] of the role neutron scattering has played in the heavy fermion field. The neutron is a weak scattering probe (unlike X-rays and electrons) and the first Born approximation applies. This means that reliable quantitative cross-sections can be extracted from the scattering experiments and one can make detailed quantitative comparisons with theory. The scattering process can be via the strong nuclear interaction (*nuclear* scattering) or via dipole-dipole magnetic scattering from magnetic moments typically in the electronic system of a material (*magnetic* scattering). The main contribution of neutron scattering to heavy fermion physics has been from magnetic scattering experiments, but there have also been a few important results from nuclear scattering. The technique can be further divided into *elastic* and *inelastic scattering*: the former term is often used loosely for total scattering with no energy analysis of the scattered beam or, put another way, neutron *diffraction*, on the grounds that the elastic cross-sections for scattering from solids are typically several orders of magnitude greater than the inelastic cross-sections. The nuclear diffraction cross-section is analogous to conventional x-ray diffraction, and given the relatively high cost of producing neutrons, as compared with x-rays, neutrons are only used to acquire information that one cannot obtain with x-rays. In the case of heavy fermions, the definitive contribution has been in structural studies to locate positions of light atoms in the presence the heavy *f*-electron element, or where the neutron contrast between constituent elements is greater than with x-rays. An example of the former is the case of determining beryllium atom positions in UBe_{13}, something that was first performed with neutrons more than 40 years ago.[139] The same subject was revisited in 1985 by Goldman *et al.*,[140] once it was known that UBe_{13} exhibited heavy fermion superconductivity, and these authors applied the Rietveld method,[141] a very powerful least-squares fitting technique for refinement of structures based on fitting the whole diffraction pattern at once. An example of using the different contrast properties of neutrons and x-rays is that of determining the structure of YbBiPt.[142] While the x-ray scattering power increases monotonically with atomic number as one proceeds down the Periodic Table of elements, the nuclear neutron scattering lengths have less rhyme and reason to their variation and different isotopes of the same element may have markedly different scattering properties.

In the case of Yb, Bi and Pt, the neutron contrast is roughly twice that of x-rays, and it has the opposite sign. Indeed with modern Rietveld refinement programs like GSAS,[143] one can even perform combined refinements of neutron and x-ray powder-diffraction data, and this was done in the case of YbBiPt. While the crystal structure of YbBiPt is very simple, with the face-centred-cubic half-Heusler or MgAgAs structure type, with only 3 atoms per primitive unit cell, and all sites having the same tetrahedral $\bar{4}3m$ point-group symmetry, one site is unique in that it has a full shell of nearest neighbours, while the other two sites have half-occupied nearest neighbour shells. As one might expect, based on the fact that Pt has the smallest atomic radius in the problem, it occupies the unique site. Of course, knowledge of the crystallographic structure of any solid is fundamental to the understanding any of its properties, and provides the basis from which to perform calculations of electronic structure, consideration of symmetry properties as to their effect on bulk measurements, analysis of NMR, μSR and Mössbauer experiments and so on. Other important contributions of neutron structural analysis have been the determination of the low-temperature monoclinic structure of $CeCu_6$,[81,82] whose parameters are given in Table I, and a study[144] of the Pd-Cu order in UCu_4Pd.

There has been a much larger contribution from magnetic neutron diffraction to the solution of ordered antiferromagnetic structures in heavy fermion compounds, and related compounds, and this subject forms the bulk of Section III. Aside from determining the structure, the antiferromagnetism has been used as a probe of the superconductivity in a number of the heavy fermion superconductors.

If one comes to nuclear inelastic scattering, this provides information on lattice vibrations (phonons) and this is the subject of Section V. There is a similar distinction between magnetic neutron diffraction and magnetic neutron inelastic scattering. The main contribution of magnetic neutron diffraction has been in determining the ground-state magnetic structures of antiferromagnetic heavy fermions, and this is the subject of section III. Magnetic inelastic scattering, on the other hand, provides the excitation spectrum in the material and this is directly related to its thermodynamic properties. In particular, the double-differential magnetic scattering cross-section is directly related to the scattering function $S(\mathbf{Q},\omega)$, which is in turn proportional to the imaginary part of the generalised magnetic susceptibility $\chi''(\mathbf{Q},\omega)$:

$$\frac{d^2\sigma}{d\Omega dE_f} = \frac{k_f}{k_i} S(\mathbf{Q},\omega) = \frac{k_f}{k_i} \langle 1 + n(\omega)\rangle \chi''(\mathbf{Q},\omega) \qquad (1)$$

where the suffices i and f refer to the incident and scattered (final) neutron beams respectively, $\hbar\mathbf{Q}$ and $\hbar\omega$ are the momentum and energy transferred to the sample in the neutron-scattering event,

$$\mathbf{Q} = \mathbf{k}_i - \mathbf{k}_f;$$
$$\hbar\omega = \frac{\hbar^2}{2m}\left(k_i^2 - k_f^2\right) \qquad (2)$$

Contributions to $\chi''(\mathbf{Q},\omega)$ may be due to spin waves, crystal-field excitations, spin fluctuations, quasiparticle-quasihole excitations and so on. If single crystal samples are available, the full vector $\mathbf{Q}$-dependence of the scattering can be measured, and this is typically done using a triple-axis spectrometer at a steady-state reactor. Some triple-axis spectrometers have polarised-beam and polarisation analysis capability. On the other hand, if single crystals of sufficient size are not available, it is better to use a constant-E_i chopper spectrometer either at a reactor or at a pulsed spallation source, in conjunction with the time-of-flight technique, because these instruments collect data over orders of magnitude more of the available solid angle. A powder average of $S(\mathbf{Q},\omega)$ is necessarily measured, though this may be more readily compared with thermodynamic quantities that are also integrated over the whole Brillouin zone. As large solid angle polarisers are not yet available, these time-of-flight measurements are always unpolarised.

One potential problem with neutron scattering as a probe is that certain isotopes, and therefore certain elements in their natural isotopic abundances, have very large cross-sections for neutron absorption. In particular ^{113}Cd, which has a natural abundance of around 12%, has a huge absorption cross-section, and this has precluded any neutron measurements on UCd_{11}, despite the fact that bulk measurements[34] indicate an antiferromagnetic ground state below 5K. Similar problems would occur with ^{10}B, ^6Li, ^{155}Gd and ^{157}Gd, if they were constituents of compounds discovered in the future. This problem can, in principle, be circumvented by making samples from source material isotopically enriched with the isotopes that do not absorb significantly.

In principle, the use of polarised neutron beams can give a wealth of extra information because the magnetic neutron scattering cross-sections are strongly spin dependent. The use of an incident polarised beam in conjunction with a magnetic field on the sample can give a lot of information on ferromagnetic moments, or in the paramagnetic state using the field to induce moments on the f-element site. This was done by Stassis *et al.21* for the superconductors UPt_3, UBe_{13} and $CeCu_2Si_2$, and all were found to possess typical f-shell magnetic form factors. Moreover, the

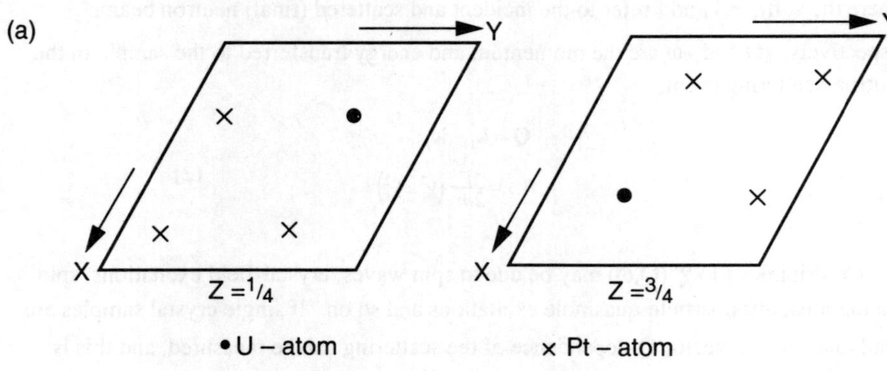

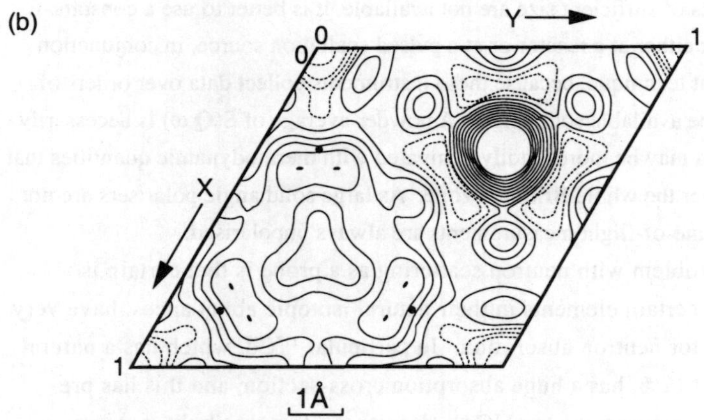

Fig. 1.   Induced magnetisation density within the basal plane in UPt$_3$ at 5K, as measured with polarised neutrons in an applied magnetic field of 4.6T (after Neumann *et al.*, Ref. 147). Panel (a) shows sections through the hexagonal unit cell at z = 1/4 and 3/4 respectively, while panel (b) shows the magnetization density for z = 1/4. The dotted lines indicate negative magnetisation density, while the solid lines indicate positive magnetisation density. There is clearly significant magnetisation density on the Pt sites and it is parallel to the uranium magnetisation. Presumably, there is some hybridisation with the platinum 5d electrons. All in all, the uranium 5f electrons account for only 50% of the magnetisation in the cell, the remainder being due to uranium 6d and 7s states and platinum 5d states. A similar experiment on UBe$_{13}$, reported in the same paper, showed no evidence for asphericity in that case.

flipping ratio (which can be related directly to the static magnetic susceptibility) was constant as a function of temperature, even into the superconducting state. These measurements were made on a limited set of crystallographically independent reflections (4 for UPt_3, 9 for UBe_{13} and 3 for $CeCu_2Si_2$[146]), and a more recent study of UPt_3 and UBe_{13} using 40 reflections has been made by Neumann *et al.21* In UPt_3, they found evidence for induced moments on the Pt sites in addition to those on the uranium site and parallel to the uranium moment, in addition to some hybridisation of the uranium $5f$-state with 6d and 7s states (see Fig. 1). On the other hand, in UBe_{13}, there was only evidence for purely f-character moments, and a similar study[148] of UPd_2Al_3 failed to find any induced moment on the Pd site in that compound. This technique is of less utility for antiferromagnets, and as those heavy fermions that order do so into antiferromagnetic ground states, the method has not been widely used. If one also analyses the scattered beam for its polarisation, however, much extra information can be extracted both regarding the antiferromagnetic structures and the magnetic excitations. The only problem is that one typically pays a huge penalty in intensity, because polarising monochromators and analysers (typically Cu_2MnAl Heusler crystals) are not as efficient as the unpolarised equivalents (typically pyrolytic graphite). As many of the experiments in this field are at the boundary of what can be done as regards signal-to-noise ratio, the vast majority of studies in this field have used unpolarised neutron beams.

To this author's knowledge, it is only the following few cases, in addition to those just mentioned above, that have used polarised beams with or without polarisation analysis:

In the case of URu_2Si_2, Walker *et al.21* have recently used polarisation analysis of a number of magnetic Bragg peaks to investigate the possibility of complex order parameters, as opposed to simple magnetic dipole ordering. In the earliest neutron study[125] of $CeCu_2Si_2$, a polarised beam was used in conjunction with magnetic field (up to 2T) to study inelastic scattering from a polycrystalline sample, without polarisation analysis. Apart from a fairly inconclusive study using full polarisation analysis by Johnson *et al.*,[150] subsequent work on $CeCu_2Si_2$ by other authors has used unpolarised neutrons. Likewise, the first inelastic neutron scattering experiments on UPt_3 were performed on a polycrystalline sample using polarised neutrons, this time in conjunction with polarisation analysis.[90,91] Again, apart from an experiment using polarisation analysis to check the magnetic nature of the antiferromagnetic Bragg peaks in Th-doped UPt_3,[9] all subsequent work on UPt_3, and there has been a lot of it, has been done with unpolarised beams. Another polarisation-analysis study was that by Goldman *et al.*[98] on UBe_{13}, though the more precise data on the quasielastic response in their paper are unpolarised.

III. ANTIFERROMAGNETISM IN HEAVY FERMIONS

In Table I, we list the antiferromagnetic ground states of a number of heavy fermion compounds, divided into four categories according to whether or not they superconduct and whether or not they have long-range-ordered antiferromagnetic ground states. There are compounds in all four categories, including some "vegetables" which neither superconduct nor order magnetically. Within each category, the compounds have been listed in the order that their space groups appear in the *International Tables for Crystallography*,[151] with higher symmetry space groups at the top. In some cases, compounds with low moments or no long-range magnetic order achieve more substantial magnetic ordering when one of the species is chemically substituted, and such substitutions are listed in italics along with the parent stoichiometric compound. There are two further distinctions that can be made in categorising the magnetically ordered states. First and most well known is that some of the antiferromagnets ($CeRu_2Si_2$, UPt_3, URu_2Si_2 and $YbBiPt$) have extremely small moments, some of which have eluded measurement by neutron diffraction, while others have moments of order $1\mu_B$ (most notably $NpBe_{13}$, UPd_2Al_3 and UCu_5). A second distinction is that between *commensurate* and *incommensurate* magnetic order. Both types of order can be characterised by a wave propagation vector $\mathbf{q}$. In the simplest picture, there is a sinusoidal modulation of the magnetisation density characterised by this wave-vector. Commensurate structures are those in which $\mathbf{q}$ is related by a rational number or numbers to the reciprocal lattice vectors of the material, and such structures can always be described within a magnetic unit cell, which will be at least as large as the crystallographic unit cell. Commensurate magnetic structures can always be categorised using the magnetic Shubnikov groups. On the other hand, incommensurate magnetic structures are characterised by $\mathbf{q}$-vectors that are related by at least one irrational number to the parent crystallographic reciprocal lattice. In general, one then sees extra satellite reflections at $\pm\,\mathbf{q}$ in the diffraction patterns. Incommensurate magnetic ground states have been observed in substitutionally doped forms of $CeRu_2Si_2$ and $CeCu_6$, as well as the pure compounds $CePb_3$ and UNi_2Al_3, while commensurate magnetic ground states are observed in $CePd_2Si_2$, $NpBe_{13}$, UIr_2Si_2, URu_2Si_2, UPd_2Al_3, UPd_2Ga_3, U_2Zn_{17}, UPt_3 and UCu_5. The $\mathbf{q}$-vectors associated with these states, both commensurate and incommensurate, are also listed in Table 1.

In both commensurate and incommensurate magnetic structures, one may have to consider whether the structure is single-$\mathbf{q}$ or multi-$\mathbf{q}$ (e.g. 2-$\mathbf{q}$, 3-$\mathbf{q}$, 4-$\mathbf{q}$, etc.). For a given $\mathbf{q}$-vector, the crystallographic symmetry may allow a number of different symmetry-equivalent modulations (e.g. [100], [010] and [001] in cubic crystals). If the magnetic

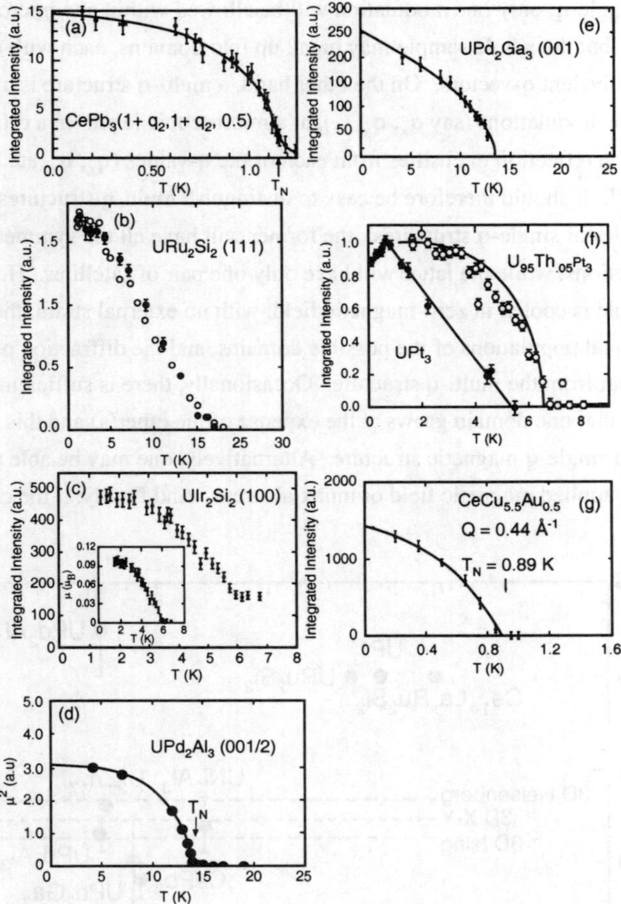

Fig. 2. Antiferromagnetic order parameters for various heavy fermion antiferromagnets: (a) $CePb_3$ (Vettier *et al.*, Ref. 33), (b) URu_2Si_2 (solid circles: neutron data from Mason *et al.*, Ref. 25; open circles: x-ray data from Isaacs *et al.*, Ref. 27), (c) UIr_2Si_2 (Vernière *et al.*, Ref. 62), (d)UPd_2Al_3 (Petersen *et al.* Ref. 20), (e) UPd_2Ga_3 (Süllow *et al.*, Ref. 42), (f) UPt_3 and Th-doped UPt_3 (Isaacs *et al.*, Ref. 7) and (g) Au-doped $CeCu_6$ (Chattopadhyay *et al.*, Ref. 190). Fits to Eqn. (3) are shown as solid lines in the following cases: (a) $CePb_3$, (d) UPd_2Al_3 and (e) UPd_2Ga_3 and all three materials give $\beta \sim 0.30$. Note that the commensurate phases in panels (d) and (e) have the same magnetic structure shown in Fig. 14 and the same Bragg reflection was used: the factor of 2 difference in indexing in the case of UPd_2Ga_3 is due to a small structural distortion (see details in Table I) which doubles its c-axis lattice parameter.

structure is single-**q**, only one modulation will be allowed within any microscopic region of the crystal, but the whole sample may break up into domains, each with one of symmetry-equivalent **q**-vectors. On the other hand, a multi-**q** structure is a coherent sum of the possible modulations (say $\mathbf{q}_A$, $\mathbf{q}_B$, ...), at a microscopic level: in a diffraction experiment, the reflection intensities from each of the **q**-values ($\mathbf{q}_A$, $\mathbf{q}_B$, etc.) will necessarily be equal. It should therefore be easy to distinguish multi-**q** structures from mono-domain samples of single-**q** structures: the former will have all the symmetry-equivalent satellite reflections, while the latter will have only one pair of satellites. However, if the single-**q** sample is cooled in zero-magnetic field, with no external strain, the sample will likely have equal populations of the possible domains, and the diffraction pattern will be identical to that from the multi-**q** structure. Occasionally, there is sufficient internal strain in the sample that one domain grows at the expense of the other(s) and this is clear evidence for a single-**q** magnetic structure. Alternatively, one may be able to grow one domain using applied magnetic field or uniaxial stress. And finally, if the coupling

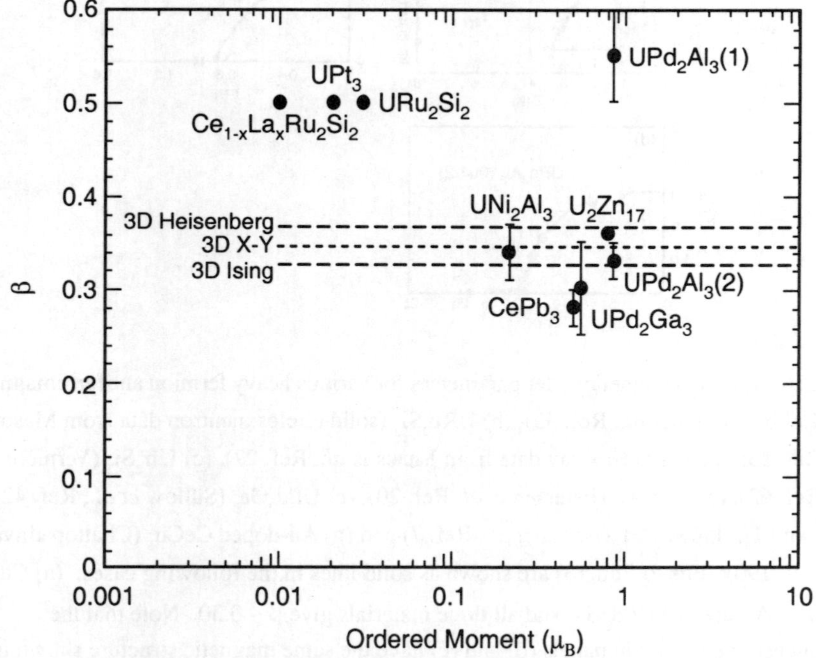

Fig. 3. Critical exponents β, for the antiferromagnetic order parameter following Eqn. (3), in various heavy fermion antiferromagnets, as a function of ordered moment in the ground state (from Table I).

between the magnetism and the lattice is sufficiently strong, one may observe magnetically driven structural distortions with the same symmetry breaking in single-**q** cases, while the highest level of multi-**q** structures preserve the crystallographic symmetry. Nevertheless, it is common for authors to assume the single-**q** magnetic structure, this being simpler to visualise than the multi-**q** case. In addition, multi-**q** structures are typically noncollinear. This may be an issue in the cases of UPt_3 and in UCu_5, both of which are described in more detail below.

As regards other parameters listed in Table I, $d_{f\text{-}f}$ is the closest spacing between f-electron elements in the crystal structure, the γ_p and γ_o refer to the specific heats in the paramagnetic and magnetically ordered phases respectively, and the critical exponent β is defined according to the normal convention:

$$\frac{\mu(T)}{\mu(0)} = \left(\frac{T_N - T}{T_N}\right)^\beta \qquad (3).$$

The exponents β characterise the dimensionality of the magnetic system, and the number of degrees of freedom that the moment has on each site.[152] $\beta = 0.326$ for a 3-d Ising system, 0.345 for a 3-d X-Y (easy-plane) system and 0.367 for a 3-d Heisenberg system. The order parameters for a variety of antiferromagnetic heavy fermion compounds are shown in Fig. 2, along with fits to Eqn. (3) as described in the caption. In Fig. 3, β-values for a range of heavy fermion antiferromagnets are plotted as a function of the low-temperature ordered moment. It is clear that the larger-moment systems have critical exponents in the normal range, but that the low-moment systems seem to have anomalously large β-values close to 0.5. This latter point is not presently understood. In addition, if the observed Bragg peaks are broader than the instrumental resolution, this is normally due to the fact that the scattering is from a relatively small volume, rather than an infinite crystal. If the profile is Lorentzian in Q, the width is inversely proportional to a correlation length ξ, which may be different in different crystallographic directions and may be temperature dependent. The resolution of typical neutron diffraction experiments is such that correlation lengths in the hundreds of Å, and less, can be measured. Some ξ-values from the literature are included in Table 1: whether these values are intrinsic or not is still a matter of debate, and there is good evidence in URu_2Si_2 that the correlation lengths vary with sample quality and are related to crystallographic defects. The same may be true in the other cases (UPt_3 and $CeCu_{5.8}Au_{0.2}$) reported here

In the following subsections, we give a more detailed description on the antifer-

214

romagnetic structures and neutron diffraction work on the heavy fermion compounds listed in Table I.

A. UBe$_{13}$ and NpBe$_{13}$

The face-centred-cubic superconductor UBe$_{13}$ does not order magnetically. However, the isostructural neptunium compound, which is also a substantial heavy fermion

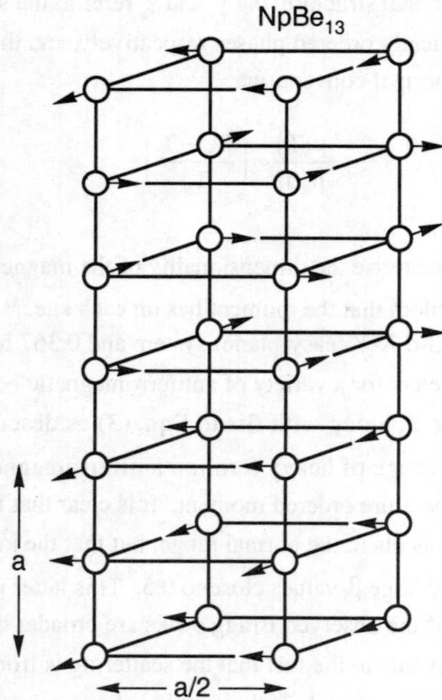

Fig. 4.   Antiferromagnetic structure of NpBe$_{13}$, as reported by Hiess et al.(Ref. 30). While the crystallographic structures of NpBe$_{13}$ and UBe$_{13}$ are face-centred cubic, there are two formula units per primitive unit cell, and the 5f element lies on a simple cubic lattice, with half the full crystallographic periodicity. The magnetic structure is noncollinear and commensurate with a 6-layer periodicity. There is a net magnetisation on each layer, and this magnetisation has the sequence + + + - - -. The figure shows equal moments on all Np sites, and this is a good approximation to the structure reported in Ref. 30: in detail Hiess et al. report that the moment magnitudes on the second and fifth layers are different from those on the other layers, but that may simply be a consequence of not seeing all the higher Fourier harmonics in their neutron diffraction experiment.

but not a superconductor, is an antiferromagnet below 3.4K. Its magnetic structure has recently been determined[30] and it forms in the 6-layer noncollinear structure shown in Fig. 4. The structure is close to having all Np moments identical, with moments rotating by 90° from layer to layer. Part of the motivation for studying $NpBe_{13}$ was to gain clues as to where, in reciprocal space, to look for enhanced inelastic scattering (spin fluctuations) in UBe_{13} itself.

B. Cerium and Uranium Compounds with the $ThCr_2Si_2$ and $CaBe_2Ge_2$ Structure Types

There are a number of heavy fermion compounds with the closely related tetragonal $ThCr_2Si_2$ and $CaBe_2Ge_2$ structures. In both cases, the f-element lies on a simple body-centred tetragonal lattice with $c > 2a$. In the $ThCr_2Si_2$ structure, the other constituent elements have inversion symmetry around the f-atom site and the whole structure is body centred with space group I4/mmm, while this inversion symmetry is missing in the $CaBe_2Ge_2$ structure, which has space group P4/nmm. Another way of thinking about this is that the transition-metal and Si/Ge layers are stacked in a different manner in the two structures, as shown in Fig. 5. In some compounds it is even possible to prepare the same material in either structure, depending on the heat treatment.[153] The most celebrated compound with this structure is $CeCu_2Si_2$, the first heavy fermion superconductor to be discovered.[64] Other compounds with the $ThCr_2Si_2$ structure include the antiferromagnetic superconductor URu_2Si_2 and the antiferromagnets $CeRu_2Si_2$ and $CePd_2Si_2$, while the antiferromagnet UIr_2Si_2 forms in the $CaBe_2Ge_2$ structure. If one derives the Shubnikov groups corresponding to either the $ThCr_2Si_2$ or the $CaBe_2Ge_2$ structure, there is only one possible antiferromagnetic arrangement without cell doubling: this arrangement has all moments along the c-direction, with the moment on the centre atom antiparallel to those at the corners, as is shown in Fig. 6(a), and it is exhibited by URu_2Si_2 and UIr_2Si_2. Their order parameters are shown in Figs. 2(b) and (c) respectively. While a number of Ce-compounds exhibit the same antiferromagnetic structure at low temperature, $CePd_2Si_2$ exhibits cell doubling with $\mathbf{q} = [1/2,1/2,0]$ as shown in Figure 6(b), with collinear moments parallel to $\mathbf{q}$, while substitutions of $CeRu_2Si_2$ are incommensurate, with both $\mathbf{q}$ and the moment along the c-axis (at least in the case of Rh-substituted $CeRu_2Si_2$). In the case of La-doped $CeRu_2Si_2$, long-range incommensurate magnetic order sets in beyond a critical La concentration of 7.5%.[56] At all concentrations, the wave vector is $\mathbf{q} = [0.31,0,0]$ at which peaked inelastic scattering is seen in the pure compound. Both the ordered moment and the Néel temperature rise rapidly reaching values of $1.2\mu_B$ and 6K respectively at 20% La doping. At the critical concentration,[58] there is some evidence for short-range-ordered incommensurate clusters (with a correlation length $\xi \sim 200$Å). The

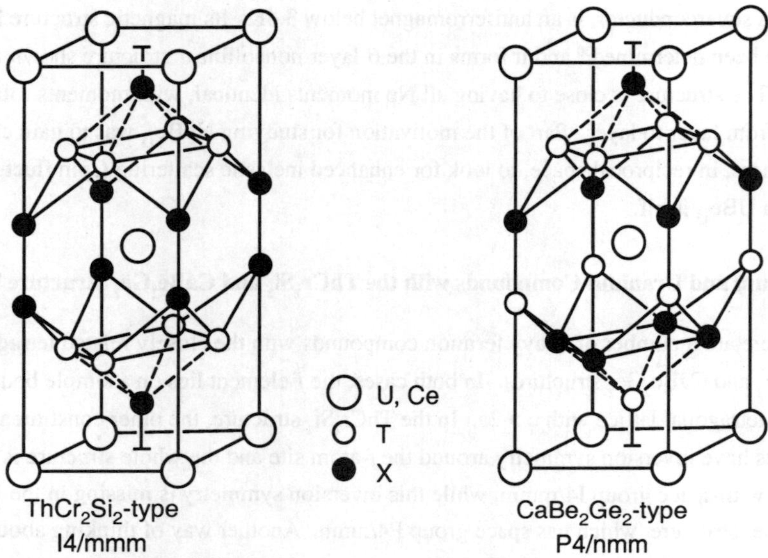

Fig. 5. Chemical structures of the ThCr₂Si₂ (body-centred tetragonal) and CaBe₂Ge₂ (primitive tetragonal) structure types.

moment associated with this state is very small, $\mu \sim 0.01\ \mu_B$, which is still substantially larger than the value of $0.001\mu_B$ inferred for the pure compound from μSR experiments.[51] For a fuller treatment of Ce-compounds forming in these structures, be they heavy fermions or not, see the studies by Grier *et al.*[59] and Loidl *et al.*21

A large amount of work has been performed on the antiferromagnetic superconductor URu₂Si₂, particularly with a view to understanding how the ordered magnetic state has such a small ordered moment (0.03 μ_B), why the spin-wave intensities are relatively so strong and why the specific heat jump at the superconducting transition ($T_C = 0.8$K) can be relatively so large. It is clear that the magnetic state is long-range ordered with the structure shown in Fig. 6(a), and that it coexists with superconductivity.[25] Nevertheless, the occurrence of very-small-moment antiferromagnetism below 17.5K has led to the suggestion that the ordered state might not be magnetic, but quadrupolar[155] or even more exotic.[156,157,158] This question has been thoroughly investigated by means of polarised-neutron diffraction in conjunction with symmetry analysis by Walker *et al.*[149]: polarization analysis of the fundamental (100) reflection shows that the order parameter for the 17.5K transition breaks time-reversal invariance and cannot therefore be quadrupolar. Of the five dipolar and octupolar order parameters allowed by symmetry, polarization

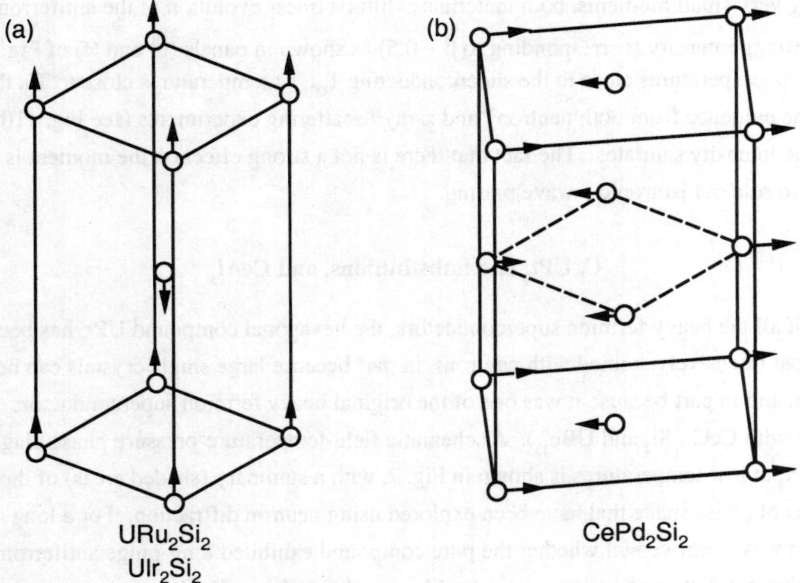

Fig. 6. Antiferromagnetic structures of tetragonal 1:2:2 heavy fermions: (a) the $\mathbf{q} =$ (0,0,1) structure exhibited by URu_2Si_2 and UIr_2Si_2 and (b) the $\mathbf{q} = (1/2,1/2,0)$ structure exhibited by $CePd_2Si_2$. Although URu_2Si_2 forms in the body-centred tetragonal $ThCr_2Si_2$ structure, and UIr_2Si_2 forms in the primitive tetragonal $CaBe_2Ge_2$ structure, in both cases the uranium ions form a body-centred tetragonal lattice and the magnetic strucure is anti-centred with moments along the c-axis. Both panels are drawn to the same scale and in the same crystallographic orientation: in (b) the dashed lines show a projection of the parent crystallographic unit cell.

analysis of the (111) reflection reveals that only the μ_z dipole moment orders at the transition. In other words, the magnetism looks normal, and there is no evidence for quadrupolar or more exotic ordering. In another study in magnetic fields up to 10T, Mason et al.21 have ruled out the possibility of multi-spin correlators. There has also been a small amount of work on rhenium-doped URu_2Si_2: the addition of rhenium rapidly suppresses the superconductivity and there is definitive evidence for ferromag-netism in both bulk and neutron scattering experiments,[28] though it is not clear that the ferromagnetism is truly long-range ordered.

Another question that has been pursued, by analogy with the work on UPt_3 described below, is the coupling between superconductivity and antiferromagnetism. In addition to having very small moments, both materials exhibit a linear evolution of the antiferromagnetic Bragg intensity (corresponding to $\beta \sim 0.5$) as shown in panels (b) and (f) of Fig. 2, down to temperatures close to the superconducting T_C. At temperatures close to T_C, there is some evidence from both neutron[26] and x-ray[27] scattering experiments (see Fig. 11(b)) that the intensity saturates. The fact that there is not a strong effect on the moment is taken to rule out isotropic s-wave pairing.

C. UPt$_3$ and Substitutions, and CeAl$_3$

Of all the heavy fermion superconductors, the hexagonal compound UPt$_3$ has been the most intensively studied with neutrons, in part because large single crystals can be grown, and in part because it was one of the original heavy fermion superconductors (along with CeCu$_2$Si$_2$ and UBe$_{13}$). A schematic field-temperature-pressure phase diagram for UPt$_3$ at low temperatures is shown in Fig. 7, with a summary (shaded areas) of those regions of phase space that have been explored using neutron diffraction. For a long time, it was controversial whether the pure compound exhibited long-range antiferromagnetism or not, but there is now a broad consensus that it does. The key experiment was a tour de force by Aeppli and co-workers[4] using a triple-axis spectrometer with the analyzer installed to reduce the background, and a moment of order $0.02\mu_B$ was observed for $\mathbf{q} =$ [1/2,0,0]. Substitutions on both the U and Pt sites, by Th and Pd respectively, also order with slightly higher Néel temperatures and substantially greater moments ($\sim 0.6~\mu_B$). In Th-doped UPt$_3$, Goldman *et al.*[8] assumed the magnetic structure to be single-$\mathbf{q}$, and derived the moment configuration in Fig. 8(a). On the other hand Frings *et al.*[9] report evidence in Pd-doped UPt$_3$ that the structure may be multi-$\mathbf{q}$, and possible 2-$\mathbf{q}$ and 3-$\mathbf{q}$ structures are shown in Figs. 8(b) and (c), and there has recently been some suspicion[160] that this may be the case in pure UPt$_3$. Recent work performed by de Visser *et al.*[11] has drawn the distinction between the small-moment antiferromagnetism in the pure compound and for compositions up to $x = 0.005$, and the large-moment antiferromagnetism beyond that. While the ordering vector is the same for both ordered states there are some notable differences, apart from moment magnitude. As can be seen in Fig. 9, the small-moment state has a concentration independent Néel temperature of $\sim 6K$, while the large-moment state orders at different temperatures depending x, with a maximum at $x \sim 0.05$. Furthermore, the temperature dependence of the moment is very different in the two states. The large-moment state exhibits normal behaviour, while the Bragg intensity in the small-moment state (which is proportional to μ^2) grows linearly as temperature is lowered below T_N (at least down to the superconducting T_C, as shown in Figs. 2 and 10). Stated another way, the critical magnetic behaviour is very different in pure UPt$_3$ (which has a critical exponent $\beta \sim 1/2$) as compared to either Pd- or Th-doped material (in which β is much smaller).

In another very interesting experiment, Aeppli *et al.21* studied the interaction between superconductivity and antiferromagnetism within the superconducting state. Most striking was the observation, shown in Fig. 11, that the (1,1/2,0) reflection is

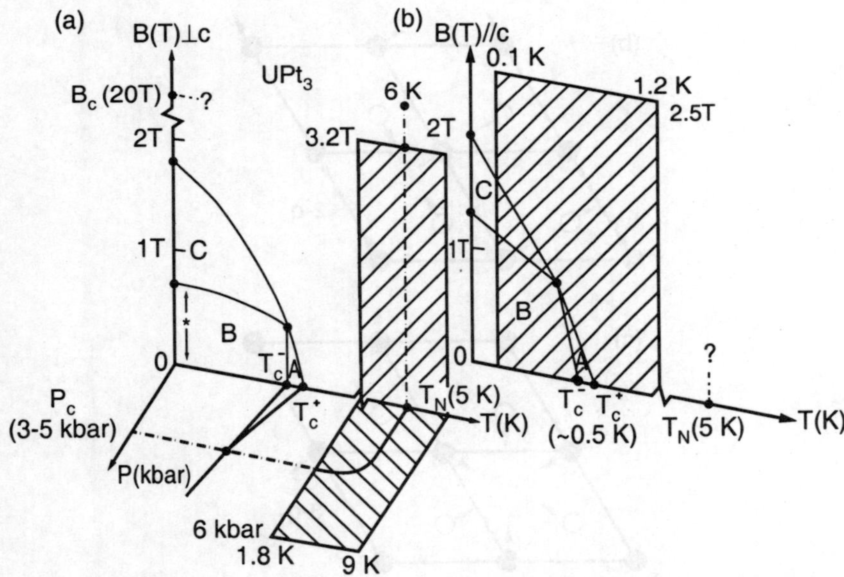

Fig. 7. Schematic field(B)-pressure(P)-temperature(T) phase diagram for UPt$_3$, (a) for magnetic fields perpendicular to **c**, and (b) for magnetic fields parallel to **c**, with shaded regions (and asterisk) to show where neutron diffraction experiments have been performed. A, B and C represent different superconducting phases. The solid circles represent transitions on the principal $B_\perp$, $B_\parallel$, P and T axes and critical points away from these axes, and the asterisk (with arrows) in (a) represents the point/temperature at which the flux-line lattice experiment of Kleiman *et al.* (Ref. 211) was performed. In (a), neutron experiments in the P-T plane are due to Hayden *et al.* (Ref. 5) and the $B_\perp$-T experiments are due to Lussier *et al.* (Ref. 160), while in (b) the $B_\parallel$-T experiments are due to Aeppli *et al.* (Ref. 161). Surprisingly, no experiments on the field dependence of T_N seem to have been reported, and less surprisingly, the field-induced transition at ~20T has yet to be studied with neutrons. For a fuller $B_\perp$-T phase diagram see Bogenberger *et al.* (Ref. 219), and for a fuller $B_\parallel$-T phase diagram see Taillefer *et al.* (Ref. 220).

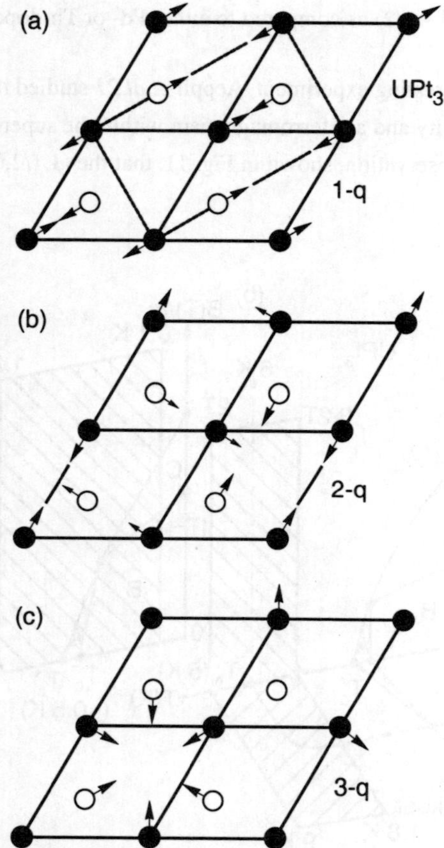

Fig.8.   Antiferromagnetic structures for UPt$_3$, and substitutions. While antiferromagnetism has been observed in pure UPt$_3$ below T$_N$ = 5K and with q = [1/2,0,0], the moment is so small that the detailed structure was not initially determined fully. However, Th-doped UPt$_3$, which has a moment of 0.65μ_B, was analysed assuming the single-q structure shown in panel (a), with moments parallel to **q**. On the other hand, work on Pd-doped UPt$_3$, which develops a similar-sized moment, claimed evidence that the structure might be multi-**q** (with the same **q** = [1/2,0,0]) and therefore noncollinear, and there is some independent support for a 3-**q** structure in pure UPt$_3$ by Lussier *et al.* (Ref. 160) Two possible related multi-**q** antiferromagnetic structures are shown in panels (b) and (c), and these necessarily imply different moments on different uranium sites: in the 3-**q** case, one site possesses identically zero moment. In all three panels, the uranium atoms shown as solid circles are at z = 0, while the open circles are at z = 1/2.

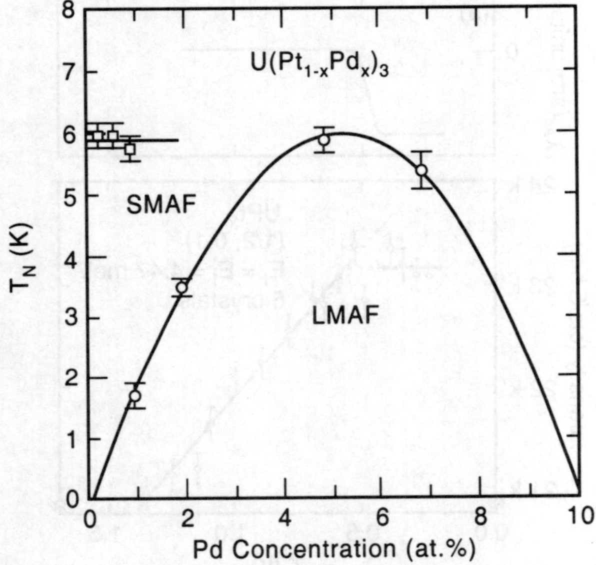

Fig. 9. The concentration variation of T_N for both the large-moment (LMAF) and small-moment (SMAF) magnetism in U(Pt$_{1-x}$Pd$_x$)$_3$ (from de Visser *et al.* Ref. 11).

suppressed by ~5% below a "phase" boundary stretching from 1.5 T to 0.4 K. This band corresponds to a set of phase transitions within the superconducting state, below H_{c2} and the moments may rotate below this range, suggesting that they are coupled to the superconducting order parameter. In a subsequent study as a function of pressure, Hayden *et al.*[5] showed that the same antiferromagnetic Bragg peaks disappear at a critical pressure of 5.4 ± 2.9 kbar (see Figs. 12 and 13), and that this is very close to the pressure at which two transitions seen in specific-heat measurements within the superconducting phase converge. This is taken as further evidence that the antiferromagnetism is intimately coupled to the superconductivity in UPt$_3$. The same linear temperature dependence in the magnetic Bragg peak intensity is seen at all pressures, and while the moments are strongly pressure dependent, the Néel temperature is pressure independent, as shown in Figs. 13(b) and 7. Isaacs *et al.*[7] conducted a careful study of the antiferromagnetic structure, using both resonant magnetic X-ray diffraction[162] and high-resolution neutron diffraction from 6 independent antiferromagnetic Bragg peaks. Samples were studied both before and after annealing - the annealed samples exhibit a 10% splitting in superconducting T_C - and the data are best analysed assuming no change whatsoever in the magnetic structure. This observation is evidence against the thesis that the

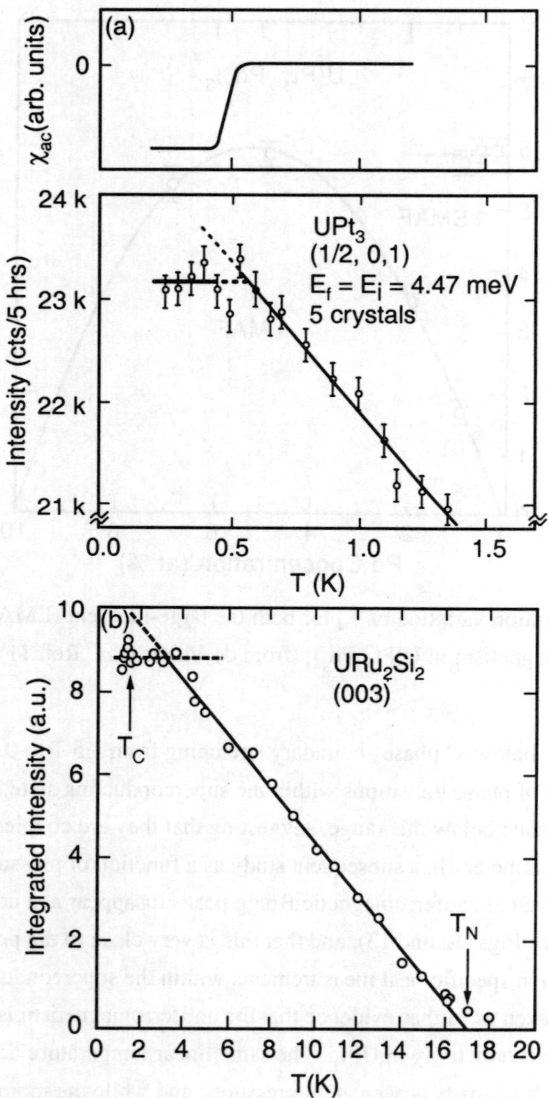

Fig. 10. Temperature dependence of antiferromagnetic reflection intensities in the vicinity of the superconducting transitions (a) in UPt_3 (From Aeppli *et al*, Ref.4), and (b) in URu_2Si_2 (from magnetic X-ray work by Isaacs *et al.*, Ref. 27). In (b), note that the (003) reflection would not be observable using neutron diffraction.

antiferromagnetism provides the symmetry-breaking field that drives the superconductivity. Rather the moment magnitude is the only varying parameter. These authors also extracted correlation lengths for the spatial extent of the antiferromagnetic order and came up with values of $\xi = 280 \pm 50$ Å along $\mathbf{a}^*$ and $\xi = 500 \pm 130$ Å along $\mathbf{c}^*$ in the annealed sample. Within error bars, the values are the same in the unannealed sample. They also examined the degree of homogeneity in the samples, with respect to an etched surface, using X-rays on the absorption edge, and found that there is no change in T_N or the temperature dependence of the magnetic order parameter on going from bulk to surface.

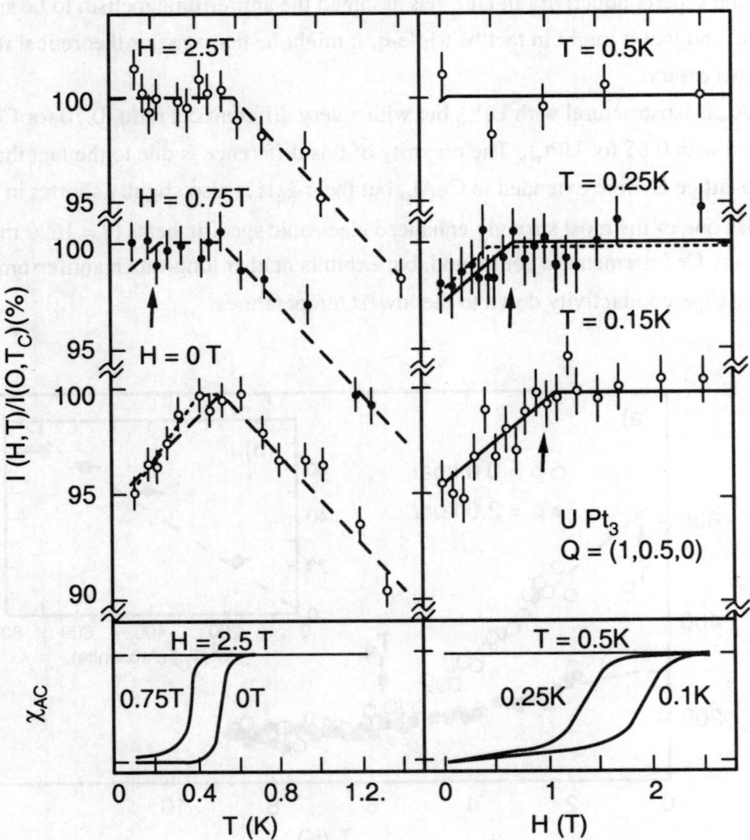

Fig. 11. Temperature and magnetic field dependences of the basal plane (1,1/2,0) antiferromagnetic reflection in UPt₃. Note the drop in intensity at the superconducting transition as manifest in the AC susceptibility χ_{AC} shown in the lower panels. This is evidence for coupling between the superconductivity and antiferromagnetism. From Aeppli *et al.*, Ref.161.

If one thinks of the antiferromagnetism in terms of a central peak, as in soft-mode phase transitions, this is very unusual. It is striking that the field dependence of the Néel temperature seems not to have been studied, nor has the temperature dependence of the field-induced transition at ~20T. It would be interesting to determine whether these two phase lines connect in any simple manner. Finally, a recent neutron diffraction experiment[160] has been performed with fields in the basal plane up to 3.2T. Such fields have no effect on the antiferromagnetism - the moments do not rotate, and it is not possible to grow one single-q domain at the expense of the others. Given that all theoretical work to date on the superconductivity in UPt$_3$ has assumed the antiferromagnetism to be single-q in nature, and that it might in fact be triple-q, it might be that a major theoretical reassessment is necessary.

CeAl$_3$ is isostructural with UPt$_3$, but with a very different c/a ratio (0.70 for CeAl$_3$ compared with 0.85 for UPt$_3$). The majority of this difference is due to the fact that the in-plane lattice is more extended in CeAl$_3$, but the c-axis is also slightly shorter in CeAl$_3$. CeAl$_3$ has one of the most strongly enhanced electronic specific heats (γ = 1620 mJmol^{-1}K^{-2}) of any Ce intermetallic compound, but exhibits neither long-range antiferromagnetic order nor superconductivity down to the lowest temperatures.

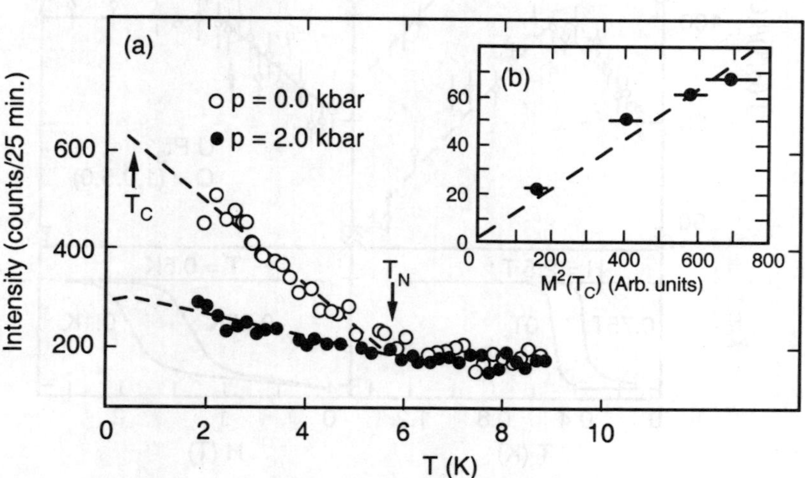

Fig. 12. The effect of pressure on the antiferromagnetic order parameter (intensity of the (1/2,1,0) reflection) in UPt$_3$ (from Hayden et $al.$, Ref. 5). Note that the temperature dependence of the intensity ($\propto \mu^2$) is linear at all pressures. The inset shows a correlation between the integrated intensity in the Bragg peak and the splitting (ΔT_c) between two superconducting transitions as a function of pressure (see Ref. 5).

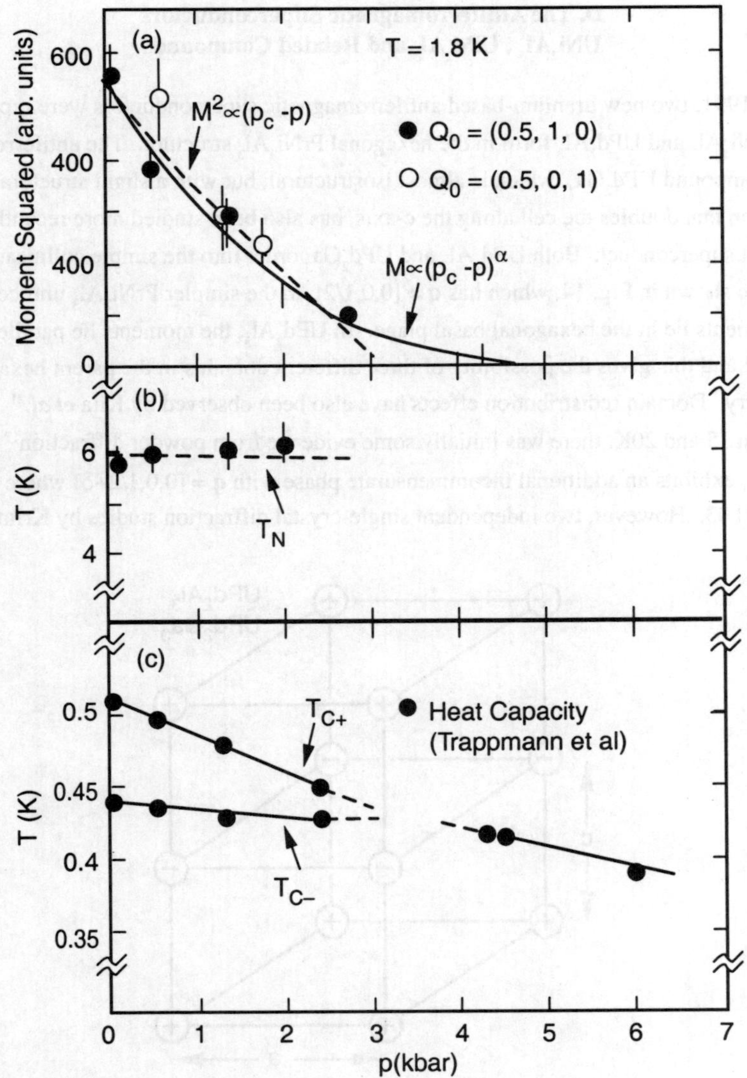

Fig. 13. The pressure dependence of (a) the antiferromagnetic Bragg peak intensity (closed circles: (1/2,1,0); open circles: (1/2,0,1)), (b) the Néel temperature T_N (both from Hayden *et al.*, Ref. 5) and the low-temperature specific-heat anomalies seen by Trappmann *et al.* (see Ref. 221) in UPt$_3$. In (a), the linear dashed line gives a critical pressure $p_c = 3.2 \pm 0.2$ kbar, while the solid-line fit to $\mu \propto (p_c - p)^\alpha$ yields $p_c = 5.4 \pm 2.9$ kbar and $\alpha = 2.6 \pm 1.9$. Note that in (b) T_N is pressure independent.

D. The Antiferromagnetic Superconductors
UNi$_2$Al$_3$, UPd$_2$Al$_3$ and Related Compounds

In 1991, two new uranium-based antiferromagnetic superconductors were reported: both UNi$_2$Al$_3$ and UPd$_2$Al$_3$ form in the hexagonal PrNi$_2$Al$_3$ structure. The antiferromagnetic compound UPd$_2$Ga$_3$, which is almost isostructural, but with a slight structural distortion that doubles the cell along the c-axis, has also been studied more recently, but it does not superconduct. Both UPd$_2$Al$_3$ and UPd$_2$Ga$_3$ order into the simple collinear structure shown in Fig. 14, which has **q** = [0,0,1/2], in the simpler PrNi$_2$Al$_3$ unit cell, and the moments lie in the hexagonal basal plane. In UPd$_2$Al$_3$, the moments lie parallel to the **a**-axis,[21] and this gives the possibility of three different domains in the parent hexagonal symmetry. Domain redistribution effects have also been observed by Kita *et al.*[21] Between 15 and 20K, there was initially some evidence from powder diffraction[17] that UPd$_2$Al$_3$ exhibits an additional incommensurate phase with **q** = [0,0,1/2+δ] where δ = 0.021± 0.03. However, two independent single-crystal diffraction studies by Krimmel *et*

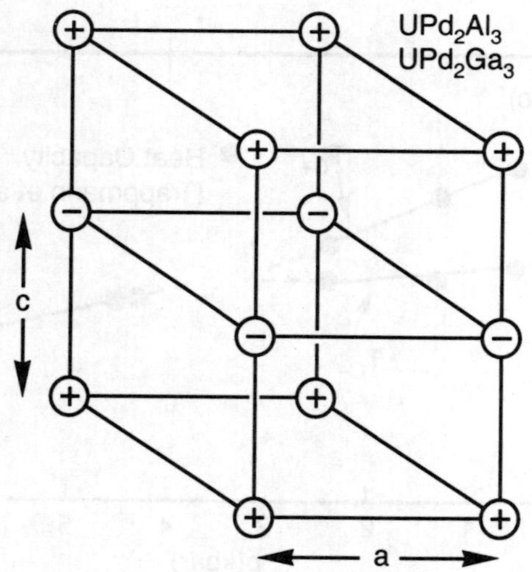

Fig. 14. Antiferromagnetic structure for UPd$_2$Al$_3$ and UPd$_2$Ga$_3$. Although UPd$_2$Ga$_3$ has a small lattice distortion making its unit cell twice as large in the c-dimension, in both cases, the uranium ions form a simple hexagonal lattic and the configurational symmetry of the magnetic order is as shown: the moments lie in the hexagonal basal plane, are ferromagnetically aligned within the (001) planes but alternate in direction as one proceeds along the c-axis. The moment direction within the basal plane is not yet known in UPd$_2$Ga$_3$, but in UPd$_2$Al$_3$ it lies along the hexagonal **a**-axis.

al.[19] and Petersen *et al.*[20] on different Czochralski-grown samples failed to find any evidence for this incommensurate phase, this difference being attributed to sample-preparation differences, and there must be considerable doubt whether the incommensurate phase is intrinsic or not. The single-crystal order parameter is shown in Fig. 2(e). We note that there is some discrepancy in the literature regarding the critical exponent β values: while Petersen *et al.* extract $\beta = 0.33 \pm 0.05$, Krimmel *et al.* obtain a value of 0.55 $\pm$ 0.05. The order parameter for UPd_2Ga_3, measured using the same Bragg reflection, is also shown in Fig. 2(e). The (0,0,1/2) antiferromagnetic reflection in UPd_2Al_3 has also been used by Metoki *et al.*[21] to probe the superconducting state in UPd_2Al_3 (below T_C = 2K), in much the same way as shown in Figs. 10 and 11 for UPt_3 and URu_2Si_2. In zero field, the intensity is depressed by ~1% below T_C, while the effect of superconductivity disappears completely by ~3T. Finally, in a polarised-neutron study, Paolasini *et al.*[21] have shown that there is no moment on the Pd sites, and that all the moment is attributable to the U sites.

On the other hand, the magnetic order in the superconductor UNi_2Al_3 is much weaker (by a fact of ~3 in both moment and Néel temperature), is incommensurate down to the lowest temperatures and evaded definitive detection for a long time. Krimmel *et al.*[17] initially reported powder-diffraction results showing no long-range magnetic order down to a sensitivity of ~$0.2\mu_B$, while μSR measurements[163] indicated moments of order $0.1\mu_B$. However, subsequent single-crystal neutron studies revealed incommensurate order with $\mathbf{q} = [1/2\pm\tau,0,1/2]$ and a moment amplitude of 0.24 μ_B. There is good evidence[15] that the moments lie in the hexagonal basal plane, rather than along the c-axis, and Lussier *et al.* claim that the moment is parallel to the hexagonal a-axis. However, the evidence for moments along $\mathbf{a}$, as opposed to $\mathbf{a}^*$, is not very strong. In addition, the same authors report that $\tau = 0.110 \pm 0.003$, and that $\beta = 0.34 \pm 0.03$ which is in agreement with other heavy fermion antiferromagnets with more substantial moments, and there is some coupling with the superconductivity, in that the antiferromagnetic moment is suppressed by ~3% below T_C.

$CePd_2Al_3$, which is also isostructural with UPd_2Al_3 and UNi_2Al_3, is also a heavy-electron antiferromagnet. However, its properties are highly sample dependent: a Néel temperature of 2.7K is obtained from bulk measurements on polycrystals, while single crystals do not order down to 0.3K.[113] This is thought to be due to small changes in aluminum stoichiometry, and this is also an issue in the case in UPd_2Al_3. There has been some neutron diffraction work on single crystals,[47] but there was no evidence for magnetic ordering down to 1.8K, and the nature of the antiferromagnetic state remains unknown. Nevertheless, bulk magnetisation measurements[47] indicate that the moments lie in the hexagonal basal plane, even if they do not order. Zero-field μSR measurements concur with the magnetic ordering scenario below 2.7K in polycrystals, and also indicate absence of magnetic order in single crystals. It is surmised that the Ce moments fluctuate in the basal plan, with a frequency greater than 10 MHz.

Finally, the neptunium analogue $NpPd_2Al_3$ has also recently been studied using neutron diffraction.[164] It orders with $\mathbf{q} = (1/3/1/3,0.36)$ at $T_N = 38K$, and additional commensurate Bragg peaks appear at $(1/3,1/3,1/2)$ below 25K. Both ordering wavevectors coexist, in both powder and single-crystal samples suggesting that there are two separate 5f responses. Similar ideas have been proposed for UPd_2Al_3.[165]

E. $CePb_3$

$CePb_3$ forms in the simple cubic Cu_3Au structure and it orders into an antiferromagnetic incommensurate state at 1.16K.[33] The ordering wave vector is close to the X-point $(1,1,1/2)$, but is displaced to $(1+\delta_1, 1+\delta_2, 0.5)$ where $\delta_1 = 0.135$ and is temperature independent up to T_N, and $\delta_2 = 1.056$ at the lowest temperature, but decreases continuously to a value of around 0.050 at the Néel point. Assuming that the structure is single-$\mathbf{q}$, there are 12 different domains, four of which were observed by Vettier *et al*.[33] There is fairly good evidence that the moment is parallel to the [001] direction, and that the amplitude at the lowest temperature is $0.55 \pm 0.10\ \mu_B$. The moment drops continuously on approaching T_N, with a critical exponent $\beta = 0.30 \pm 0.05$ (see Fig. 2(a)). There was no evidence for higher order harmonics (which would be evidence for squaring up) and one assumes the structure to be sinusoidal.

F. UCu_5 and Related Compounds

UCu_5, a moderately enhanced heavy fermion ($\gamma = 86$ mJmol^{-1}K^{-2}), forms in the face-centred cubic $AuBe_5$ structure and orders antiferromagnetically at 15K, into a commensurate structure with $\mathbf{q} = 1/2$ [111]. The structure was originally analysed[36] from neutron powder diffraction data assuming a single-$\mathbf{q}$ state, with moments lying along the [111] axis and parallel to $\mathbf{q}$ (see Fig. 15(a)). In the classification of possible face-centred cubic antiferromagnetic structures, this is known as Type II antiferromagnetism.[166] At the lowest temperature, the ordered moment is $1.55\ \mu_B$. There is a second hysteretic magnetic phase transition at ~1K,[35] and its nature has not yet been fully elucidated. Large single crystals are not yet available and all the neutron work has been performed on powders or polycrystalline samples. Schenck and co-workers[37] found no evidence in a powder experiment for any intensity changes or peak-position shifts on passing through the 1K transition, though their experiment only had a resolution of ~2%. Further work[167] at higher resolution, using the cold-neutron diffractometer IRIS[168] at the ISIS spallation source in England, has ruled out any shifts of the magnetic reflections to the level of 0.2%. However, NMR and NQR work due to Nakamura *et al.*,[169] has been interpreted to imply a 4-**q** magnetic structure between 1 and 15K, with the ground state below 1K being the single-**q** state. These two states are shown in Fig. 15, and it is clear that the 4-**q** state is noncollinear and preserves the cubic symmetry, while the single-**q** state is collinear but possesses the lower rhombohedral symmetry. In principle the 1-**q** state should have a corresponding rhombohedral structural distortion and this should be detectable in a high-resolution neutron or X-ray diffraction experiment. An argument against this scenario would be that there is a substantial specific-heat anomaly

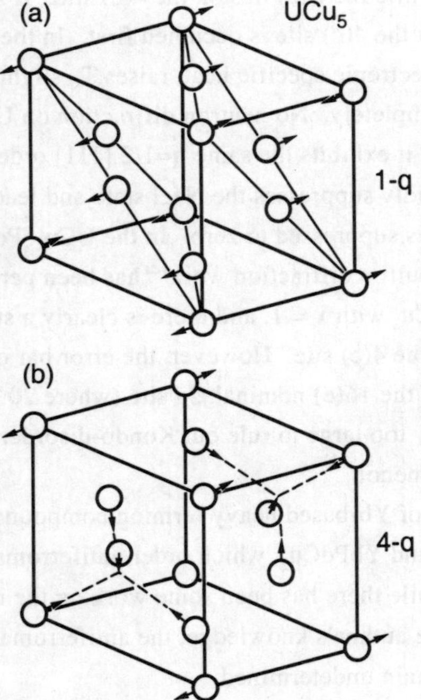

Fig. 15. Antiferromagnetic structure for UCu$_5$. The uranium atoms lie on a face-centred-cubic lattice, and neutron diffraction indicates that **q** = 1/2 [111], with moments along the [111] axis. It has been assumed that the magnetic structure is the single-**q** Type II structure shown in panel (a). However NMR and NQR studies indicate that UCu$_5$ may have the 4-**q** noncollinear structure shown in panel (b) between 1 and 15K, and transform to the 1-**q** structure below 1K (see Nakamura *et al.*, Ref. 169).

at 1K, and this should not be the case if the moments simply rotate. An alternative scenario for the 1K transition would be that UCu_5 is slightly incommensurate between 1 and 15K, but that it locks in to the $q=1/2$ [111] commensurate phase below 1K. While this picture might be supported by optical studies of DeGiorgi and co-workers,[170] it is ruled out by the high-resolution neutron diffraction experiment, at least at the 0.2% level. Of course, if single crystals of sufficient size can be grown, neutron diffraction experiments will probably reveal the true story very quickly: in the 1-q/4-q scenario, it should be possible to grow one domain at the expense of the others using applied magnetic field and the degree of commensurability (especially in transverse directions) is much more readily checked in single crystals.

Of the five copper atoms in UCu_5, four lie on the crystallographically equivalent 16(e) site at (5/8,5/8,5/8), while the fifth lies on the 4(c) site. If substitutions are made for Cu, it is likely that the 4(c) site is occupied first. In the case of $UAgCu_4$, this process enhances the electronic specific heat, raises T_N slightly to 18K and the 1K transition disappears completely. No neutron diffraction on $UAgCu_4$ has been reported, but it is likely that it exhibits the same $q=1/2$ [111] ordering. On the other hand, substitution of Pd rapidly suppresses the Néel state and leads to non-Fermi-liquid behaviour[171] once T_N is suppressed to zero. In the $UCu_{5-x}Pd_x$ system, this occurs for $x \sim 0.8$. Some neutron diffraction work[144] has been performed on the ideal stoichiometry of $UPdCu_4$ with $x = 1$, and there is clearly a strong preference for the Pd atoms to occupy the 4(c) site. However, the error bar on the measurement, ~4% Pd occupancy of the 16(e) nominal Cu site (where 20% would be completely random) is likely too large to rule out Kondo-disorder models[172,173,174] of the non-Fermi-liquid phenomenon.

There are also a couple of Yb-based heavy fermion compounds with the same $AuBe_5$ structure: $YbAgCu_4$ and $YbPdCu_4$, which order antiferromagnetically at 0.6 and 0.8K respectively.[39] While there has been some work on the inelastic scattering from both materials,[132] to the author's knowledge, the antiferromagnetic structures of these two compounds remain undetermined.

G. U_2Zn_{17}

U_2Zn_{17} forms in the rhombohedral Th_2Zn_{17} structure, with one formula unit (i.e. two uranium ions) within the primitive trigonal unit cell. Below $T_N = 9.7K$, Cox *et al.*[48] used neutron powder diffraction to show that these two uranium ions order so as to be antiparallel to each other. In other words $q = (0,0,0)$. Their directions are

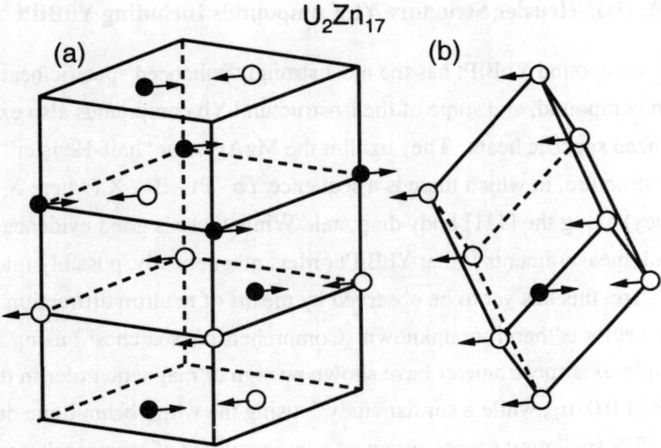

Fig. 16. Antiferromagnetic structure for U$_2$Zn$_{17}$. The crystallographic structure is rhombohedral and the magnetic unit cell is the same as the crystallographic cell, i.e. **q** = [000]. The moments lie in the basal plane, in hexagonal coordinates, and their direction within this plane is parallel to **a*** as shown. In (a), we show the structure in the hexagonal cell convention, while in (b) we show it in the primitive rhombohedral cell, which has only two uranium ions per cell aligned antiparallel to each other.

perpendicular to the unique 3-fold axis, and a subsequent single-crystal study[50] showed them to be aligned parallel to **a***. Note that, as the uranium ions lie on the 3-fold axis of symmetry, this magnetic structure necessarily breaks the trigonal

space-group symmetry and the structure is not a simple Shubnikov extension of $R\bar{3}m$. Stated another way, there should in principle be a corresponding symmetry breaking structural distortion, in the antiferromagnetic state. The antiferromagnetic structure is drawn in Fig. 16 in both hexagonal and rhombohedral axis conventions - the cell has three times the volume in the hexagonal coordinate system, and the antiferromagnetic structure appears very complicated. However it is very simple in the rhombohedral cell (See Fig. 16(b)). Doping with either Cu or Mn on the Zn site, at the 1 - 2 % level rapidly destroys the antiferromagnetic state, but leaves the specific-heat enhancement largely unchanged.[175] In a subsequent single-crystal study,[49] the critical exponent β for the antiferromagnetic transition was measured on the (102) reflection as β = 0.36 $\pm$ 0.02, which is close to the value for a 3-D short-range coupled x-y magnet.

H. Half-Heusler Structure Yb Compounds Including YbBiPt

The cubic compound YbBiPt has the most strongly enhanced specific heat of any heavy fermion compound, and some of the isostructural Yb compounds also exhibit strongly enhanced specific heats. They exhibit the MgAgAs or "half-Heusler" face-centred-cubic structure, in which there is a sequence Yb - Pt - Bi - X (where X is an ordered vacancy) along the [111] body diagonal. While there is good evidence from μSR[176] and bulk measurements[177] that YbBiPt orders magnetically, possibly into a spin-density-wave state, this has yet to be observed by means of neutron diffraction and the ordering wave vector is therefore unknown. Comprehensive searches[178] using a single crystal on a triple-axis spectrometer have shown no sign of magnetic order in the <hk0> zone to a level of $0.01\mu_B$, while a similar study[179] using the white-beam Laue diffraction method at the ISIS spallation source surveyed a large volume of reciprocal space and found nothing at the $0.1\mu_B$ level. On the other hand, YbPdSb and YbBiPd, if they order at all, do so at temperatures below 2K, the lowest temperature at which measurements have been made.[41]

I. UCd$_{11}$

While UCd$_{11}$ has one of the simplest of crystallographic structures, being cubic with one uranium atom per unit cell, and it has a clear antiferromagnetic transition at 5K, its magnetic structure is as yet unknown. This is largely because Cd is a strong neutron absorber. It may be that this question could be answered in the future, either using a sample enriched to avoid the strongly absorbing ^{113}Cd isotope, or by performing magnetic X-ray scattering.[162]

J. Tetragonal U$_2$T$_2$X Compounds

The tetragonal U$_2$T$_2$X compounds form a large group of interesting materials showing enhanced specific heats. U$_2$Pt$_2$In has the largest electronic specific heat in the series ($\gamma = 850$ mJmol^{-1}K^{-2}). There are two formula units, and therefore four uranium atoms, per unit cell and one needs to be careful when reading the literature to check whether thermodynamic quantities are reported as molar quantities or per uranium ion. The uranium atoms lie in a plane, and within one of two orthogonal mirror planes running at 45° to the cell axes. Most of the compounds in the series are noncollinear antiferro-magnets[180] though not all would be considered heavy fermions. Some of these com-pounds (e.g. U$_2$Pd$_2$In and U$_2$Pd$_2$Sn; see Ref. 181) order antiferromagnetically in the same crystallographic unit cell, while others exhibit cell doubling along the c-axis.[182,183] A major theme of the research on compounds with this structure has been a quest to understand the relationship between bonding and magnetic anisotropy at the uranium site: a phenomenological picture had been developed,[184,185] in which it seemed that the uranium moments lie perpendicular to directions of strong hybridisation. The physical idea behind this is based on theory by B. R. Cooper and co-workers,[186] who considered the magnetic anisotropy of cubic Ce compounds. But it also works well from a phenomenological viewpoint for uranium compounds, including compounds of symmetry lower than cubic. The interest in the U_2T_2X compounds arises from the fact that some compounds in the series have the shortest U-U link within the tetragonal basal plane $(d_\parallel)$, while in others it lies along the c-axis $(d_\perp)$. This thesis held up well in the cases U_2Pd_2In and U_2Pd_2Sn, which have $d_\perp < d_\parallel$ and moments in the tetragonal basal plane, while it breaks down in the case of U_2Rh_2Sn. Another interesting feature of these materials is that, in the case of U_2Pd_2Sn, electronic structure calculations[187,188] indicate that the in-plane magnetic anisotropy is much stronger than the conventional uniaxial magnetic anisotropy, with respect to the tetragonal c-axis.

K. CeCu$_6$ and Related Substitutions

$CeCu_6$ is orthorhombic at room temperature, but undergoes a slight distortion to monoclinic on cooling below ~200K.[81,82,83] The pure compound has one of the largest reported electronic specific heats ($\gamma = 1670$ mJmol^{-1}K^{-2}), but does not order into long-range antiferromagnetism, nor into a superconducting state. The reported transition temperatures for the structural distortion exhibit substantial variation, of order $\pm 50K$, and it is not clear whether the values were determined on heating or cooling, nor whether there is any hysteresis. The structural distortion can lead to confusion in interpreting other results from $CeCu_6$ and its substitutions, because the lattice parameters are permuted on changing from space group Pnma to P2$_1$/c, as listed in Table I. In particular, von Löhneysen's and Aeppli's groups adopt the orthorhombic notation throughout their work, while Rossat-Mignod's group adopt the monoclinic notation which is more correct from a formal viewpoint.

If a small amount of Au is substituted onto the Cu-site, antiferromagnetism can be induced.[189] Written as $CeCu_{6-x}Au_x$, this occurs for $x > 0.1$ giving $T_N = 2.2K$ at $x = 1$, whereupon the antiferromagnetism is rapidly depressed. The composition $x = 0.5$ has been studied extensively with neutrons. An initial powder diffraction study by Chattopadhyay $et\ al.$[190] found magnetic order for $x = 0.5$ below 0.89K and deduced a

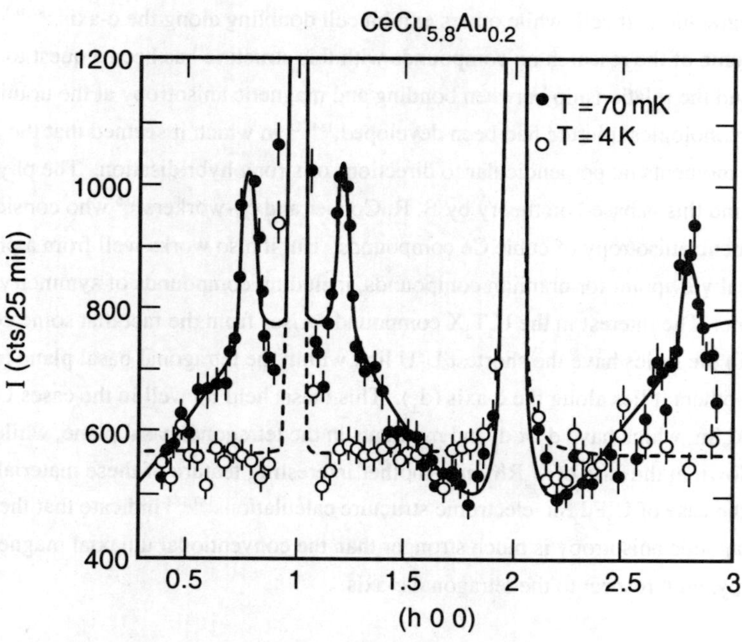

Fig. 17. Incommensurate antiferromagnetic reflections in CeCu$_{5.8}$Au$_{0.2}$, corresponding to $\mathbf{q}$ = (0.79,0,0) (from Stockert *et al.*, Ref. 84).

magnetic ordering vector in the a-c plane. However, subsequent single-crystal studies by collaborations including some of the same authors[86,84] demonstrated that this composition exhibits a transverse incommensurate structure with wave-vector along the orthorhombic **a**-axis and with moments polarised along **c** (see Fig. 17). Other compositions show short-range magnetic order, also with wave-vector along the orthorhombic a-axis, but long-range incommensurate order has now been observed[87,88] for all compositions between x = 0.2 and 0.4, with $\mathbf{q}$ = $(\delta_1, 0, \delta_2)$, where δ_1 varies between 0.605 and 0.625 and δ_2 varies between 0.22 and 0.27. The variations of q_a, the long-range-ordered antiferromagnetic moment and the Néel temperature as a function of x, extracted from these studies, are summarized in Fig. 18. In addition, Fig. 26(a) shows some of the diffraction data from Ref. 88 in the **a*-c*** plane. This develops into rods of quasielastic scattering at the critical concentration of x_c = 0.1, where non-Fermi-liquid behaviour occurs.

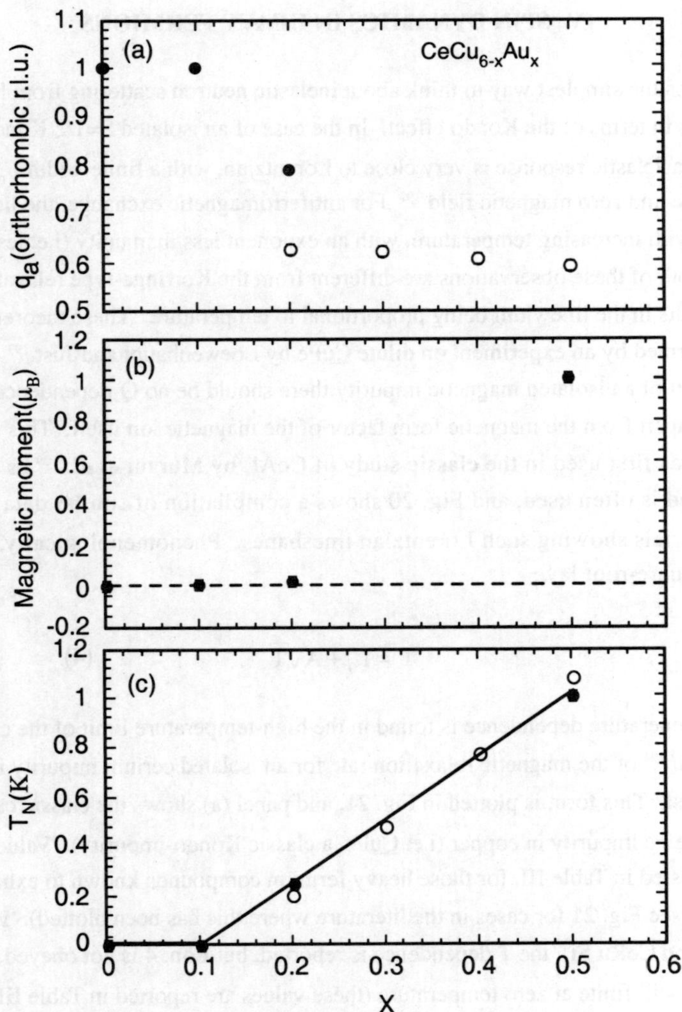

Fig. 18. The variation of (a) incommensurate ordering vector q$_a$, (b) ordered magnetic moment and (c) Néel temperature as a function of Au concentration x in the heavy fermion/non-Fermi-liquid system CeCu$_{6-x}$Au$_x$, as determined by neutron diffraction. The full circles represent neutron data from Refs. 84 - 86 and 116, while the open circles are data from Ref. 87.

IV. SPIN DYNAMICS IN HEAVY FERMIONS

Perhaps the simplest way to think about inelastic neutron scattering from heavy fermions is in terms of the Kondo effect. In the case of an isolated S=1/2 Kondo impurity, the quasielastic response is very close to Lorentzian, with a finite width Γ_0 at zero temperature and zero magnetic field.[191] For antiferromagnetic exchange, the linewidth increases with increasing temperature, with an exponent less than unity (i.e. less than linear). Both of these observations are different from the Korringa-type relaxation,[192] which results in the linewidth being proportional to temperature. These theoretical ideas were confirmed by an experiment on dilute $\underline{Cu}$Fe by Loewenhaupt and Just.[193] As the effect is that of an isolated magnetic impurity there should be no Q-dependence to the scattering apart from the magnetic form factor of the magnetic ion itself. This type of analysis was first used in the classic study of $CeAl_3$ by Murani $et\ al.$,[109] as shown in Fig. 19, and is often used, and Fig. 20 shows a compilation of similar data from other materials showing such Lorentzian lineshapes. Phenomenologically, Γ often obeys a square-root law:

$$\Gamma = \Gamma_0 + A\sqrt{T} \qquad (4)$$

and this temperature dependence is found in the high-temperature limit of the calculation by Cox $et\ al.$[194] of the magnetic relaxation rate for an isolated cerium impurity in a metallic host. This form is plotted in Fig. 21, and panel (a) shows the classic case of the s=3/2 dilute Fe impurity in copper (i.e. $\underline{Cu}$Fe, a classic Kondo-impurity). Values for Γ_0 and A are listed in Table III, for those heavy fermion compounds known to exhibit this behaviour (see Fig. 21 for cases in the literature where this has been plotted). In the cases of $CeCu_6$ and $CeRu_2Si_2$, the T dependence is reported, but Eqn. 4 is not obeyed, even though Γ is still finite at zero temperature (these values are reported in Table III as $\Gamma(0)$). It is striking that only rare-earth heavy fermion compounds, as opposed to uranium compounds, have been studied from this viewpoint. One would also expect that $\Gamma_0 \propto \gamma^1$, and this reasoning has been used by some authors to try to systematise the relationship and infer which part of the response is related to the heavy fermion behaviour in the specific heat: for instance, in their discussion of two or more energy scales in UBe_{13} and UPt_3, Lander $et\ al.$[99] derive $\gamma\,\Gamma_0 \sim 0.7$ meVJ mol^{-1} K^{-2} for two Ce compounds, but very different values for the two uranium compounds. In fact, simple arguments based on the R ln2 entropy associated with a doublet crystal-field ground state would lead one to the statement that $\gamma\,\Gamma_0 \sim 0.5$ J meV mol^{-1} K^{-2}. By way of comparison, a Lorentzian-broad-

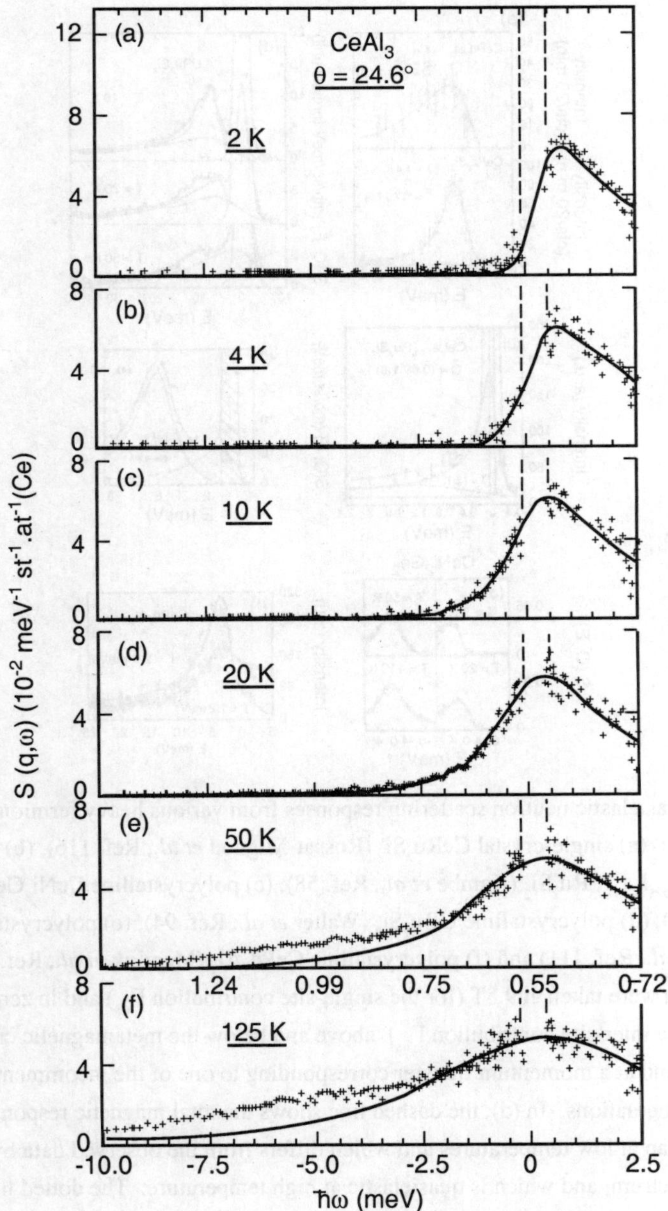

Fig. 19. The quasielastic response at various different temperatures as measured in CeAl$_3$ by Murani *et al.* (Ref. 109)

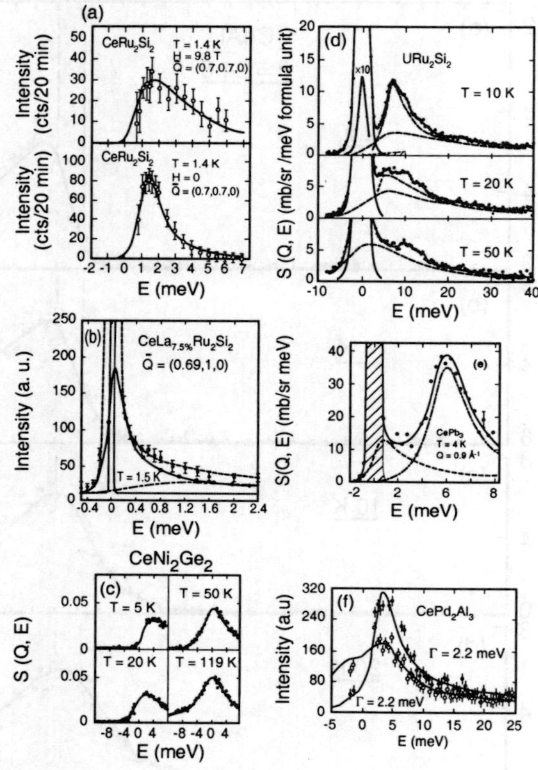

Fig. 20. Quasielastic neutron scattering responses from various heavy fermion compounds: (a) single-crystal $CeRu_2Si_2$ (Rossat-Mignod *et al.*, Ref. 116), (b) single-crystal $Ce_{0.925}La_{0.075}Ru_2Si_2$ (Kambe *et al.*, Ref. 58), (c) polycrystalline $CeNi_2Ge_2$ (Knopp *et al.*, Ref. 68), (d) polycrystalline URu_2Si_2 (Walter *et al.*, Ref. 94), (e) polycrystalline $CePb_3$ (Renker *et al.*, Ref. 111) and (f) polycrystalline $CePd_2Al_3$ (Mentink *et al.*, Ref. 113). In (a), the data were taken at 9.8T (for the single-site contribution Γ_{s-s}) and in zero magnetic field (for the inter-site contribution Γ_{i-s}), above and below the metamagnetic critical field $B_C = 7.7T$ and at a momentum transfer corresponding to one of the incommensurate magnetic fluctuations. In (d), the dashed line shows the total magnetic response, which exhibits a gap at low temperatures and which differs from the observed data by the phonon spectrum, and which is quasielastic at high temperature. The dotted line in (d) shows a putative quasielastic response with Γ scaled to the maximum in the dashed line, and the authors argue that there is no quasielastic scattering in the gap at 10K (i.e. below $T_N = 17.5K$). In (e) the dashed line is the quasielastic response, while the dot-dashed line is due to a crystal-field level at 5.8 meV.

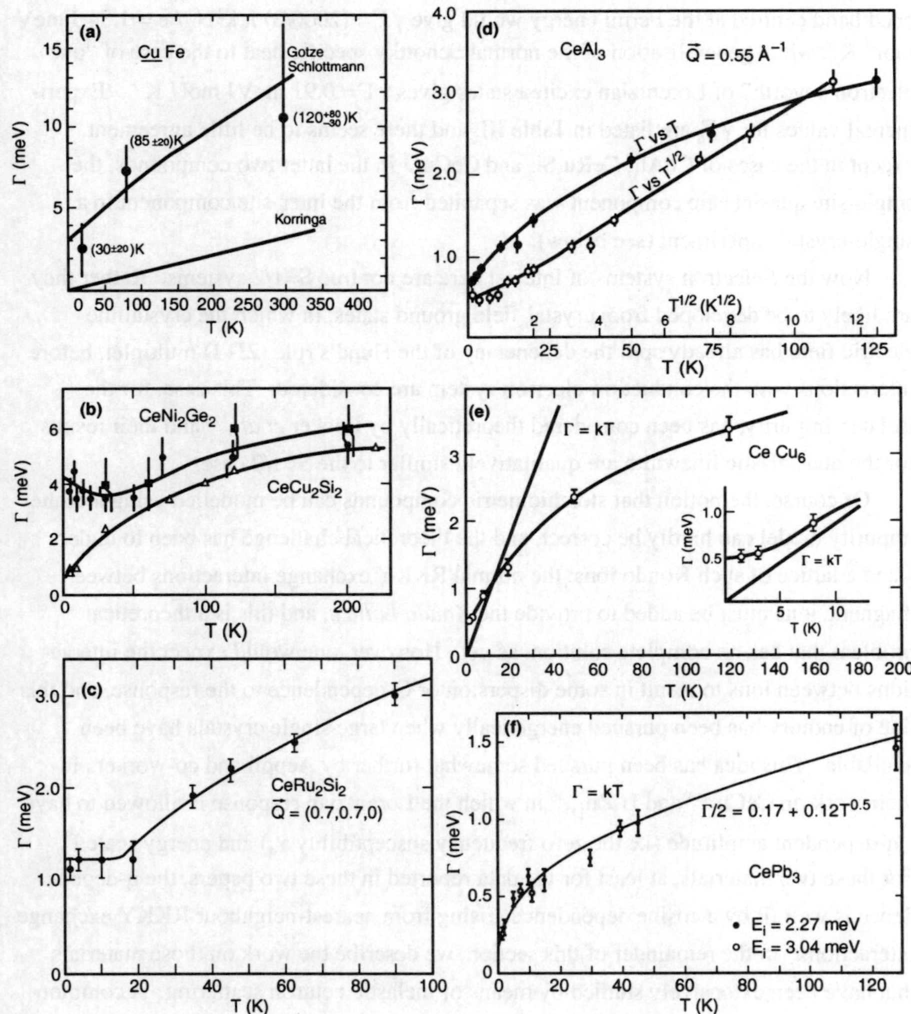

Fig. 21. Temperature variation of quasielastic halfwidths Γ for various compounds: (a) the classic dilute Kondo system $\underline{Cu}$Fe (from Loewenhaupt and Just, Ref. 193), (b) $CeCu_2Si_2$ and $CeNi_2Ge_2$ (Horn *et al.*, Ref. 124 and Knopp *et al.*, Ref. 68), (c) $CeRu_2Si_2$ (Regnault *et al.*, Ref. 119), (d) $CeAl_3$ (Murani *et al.* Ref., 109), (e) $CeCu_6$ (Walter *et al.*, Ref. 115) and (f) $CePb_3$ (Balakrishnan *et al.*, Ref. 112). It is worth comparing these data with Fig. 37 which shows qualitatively similar data for the narrow quasielastic component in some ytterbium compounds.

ened band centred at the Fermi energy would give $\gamma \Gamma = (2000/3) \pi k^2 N_A / e = 1.34$ J meV mol^{-1} K^{-2}, while generalisation of the normal Schottky specific heat to the case of "one electron's worth" of Lorentzian excited states gives $\gamma \Gamma = 0.91$ meVJ mol^{-1} K^{-2}. Experimental values for $\gamma \Gamma_0$ are listed in Table III, and there seems to be little agreement, except in the cases of CeAl$_3$, CeRu$_2$Si$_2$ and CeCu$_6$. In the latter two compounds, the single-site quasielastic component was separated from the inter-site component in a single-crystal experiment (see below).

Now the f-electron systems of interest here are not true S=1/2 systems. Rather they are likely to be developed from crystal-field ground states, in which the crystalline electric field has already split the degeneracy of the Hund's rule (2J+1) multiplet, before interactions with the conduction-electron system are considered. This case, for the isolated impurity, has been considered theoretically by Becker et $al.$,[195] and their results for the quasielastic linewidth are qualitatively similar to the S=1/2 case.

Of course, the notion that stoichiometric compounds can be modelled using a dilute-impurity model can hardly be correct, and the theoretical challenge has been to understand a lattice of such Kondo ions: the normal RKKY exchange interactions between magnetic ions must be added to provide the $Kondo$ $Lattice$, and this is a theoretical problem that has no complete solution, as yet. However, one would expect the interactions between ions to result in some dispersion or Q-dependence to the response, and this line of enquiry has been pursued energetically when large single crystals have been available. This idea has been pursued somewhat further by Aeppli and co-workers in their work on CeCu$_6$[114] and U$_2$Zn$_{17}$,[49] in which the Lorentzian response is allowed to have a q-dependent amplitude (i.e the zero frequency susceptibility χ_0) and energy scale Γ. For these two materials, at least for the data reported in these two papers, the q-dependence is well fit by a cosine dependence arising from nearest-neighbour RKKY exchange interactions. In the remainder of this section, we describe the work on those materials that have been extensively studied by means of inelastic neutron scattering. A common thread is that enhanced scattering is often seen at zone boundaries, or at incommensurate wave vectors, and this scattering is often referred to as $antiferromagnetic$ in character - enhanced scattering close to the zone center can be thought of as $ferromagnetic$. Though it is this $q \to 0$ ferromagnetic susceptibility that is formally equivalent to the bulk susceptibility (or that measured in other probes like NMR or μSR), the only study[93] to observe it has been in the case of the heavy fermion superconductor UPt$_3$. In fact UPt$_3$ is far and away the most intensively studied of all the heavy fermion materials. However, even in this case there is no complete understanding of the energy- and q-dependences of the scattering, as a function of temperature, let alone magnetic field or pressure. If such a complete knowledge of $\chi''(q,\omega)$ were to be available, one would be in a position to calculate all the thermodynamic quantities (susceptibility, specific heat, etc.) from it, in addition to making quantitative comparisons with de Haas-van Alphen results, for instance. In materials other than UPt_3, our knowledge is far less complete.

A cautionary note should be sounded in reading the literature, when one comes to interpret the Γ values in some papers, especially those of the Köln/Jülich/Darmstadt collaboration. While it is conventional in most circles to plot Lorentzian halfwidths (Γ), some papers consistently report $\Gamma/2$ values and use the terms linewidth and/or halfwidth rather loosely without defining it properly - following the papers by Loewenhaupt and Just[193] and Severing et al.[122] and the review by Holland-Moritz and Lander,[138] in which it is clear that their $\Gamma/2$ has the same meaning as the Γ used by other authors (i.e. the Lorentzian halfwidth), we will assume that this is the case whenever values for $\Gamma/2$ are reported. In Table III, these cases are shown with the § symbol.

Table III Quasielastic Linewidths for Selected Heavy Fermion Compounds Exhibiting $\sqrt{T}$ Dependence

Material	Γ_0 (mev)	$\Gamma(0)$ (meV)	A (mevK$^{-1/2}$)	γ (Jmol^{-1}K^{-2})	$\gamma\Gamma_0$ (JmeVmol^{-1}K^{-2})	$\gamma\Gamma(0)$	References
$CeAl_3$*		0.5		1.620		0.8	Murani et al.[109]
$CePb_3$§	0.17±0.03		0.12±0.01	0.225	0.038		Balakrishnan et al.[112]
$CeCu_2Ge_2$	0.12±0.01		0.115±0.004				Loidl et al.[154]
$CeNi_2Ge_2$	0	~4	0.454±0.009	0.350		~1.4	"
$CeAg_2Ge_2$	0.175±0.013		0.031±0.002				"
$CeRu_2Si_2$*		2.0[(a)]	0.32	0.385		0.77	Regnault et al.[119]
$CeCu_6$*		0.42	0.32	1.670		0.701	Walter et al.[115]
$CeCu_2Si_2$*§			0.38	1.000			Horn et al.[123,124]
$CePd_2Si_2$*§	0.9			0.16			Severing et al.[122]
$YbPdCu_4$§	0.18		0.25	0.200	0.009		Severing et al.[132]

* Values for $\Gamma(0)$ and A extracted from figures in literature.

§ Plots are in terms of $\Gamma/2$, and it is not entirely clear from literature whether the same definition of Γ (and therefore Γ_0 and A) is being used.

(a) There is also good evidence of inelasticity on this energy scale.

A. CeCu$_6$

Using the time-of-flight technique Walter,[115] and co-workers showed that, in addition to crystal field levels at 5.5 and 11 meV, there is strong quasielastic scattering in CeCu$_6$ (see Fig. 22(a)). It is strongly temperature dependent, increasing in both intensity and width from a halfwidth $\Gamma = 0.5$ meV at T=0 K roughly as $\sqrt{T}$ (see Fig. 21(e)). CeCu$_6$ has also been studied quite extensively, using the triple-axis technique,[114,116] in part because large high-quality single crystals have been available for a long time. See for instance contemporaneous single-crystal quasielastic scattering data taken in the **a*-c*** plane by Aeppli *et al.* in Figs. 22(b) and 23. Note that the product $\Gamma\chi_0$ remains more-or-less constant, and these authors used this evidence to speculate that $\chi''(\mathbf{q},\omega)$ might be **q**- and field-independent. Rossat-Mignod[116] and co-workers then showed that there is no

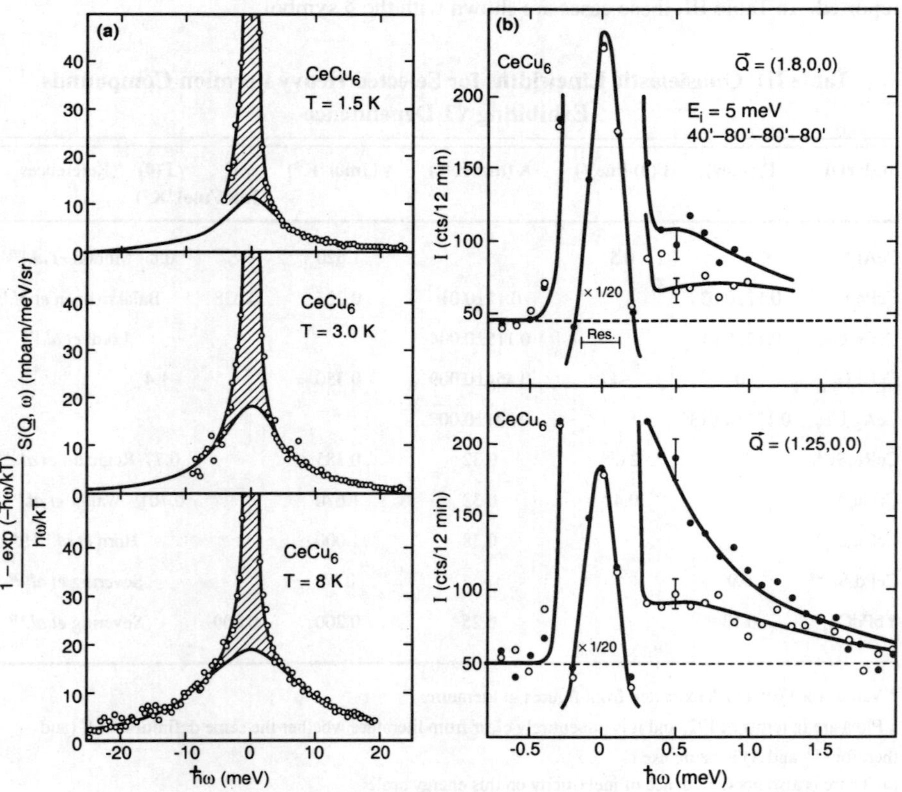

Fig. 22. Quasielastic neutron scattering response from CeCu$_6$: (a) polycrystalline (Walter *et al.*, Ref. 115), (b) single-crystal (Aeppli *et al.*, Ref. 114).

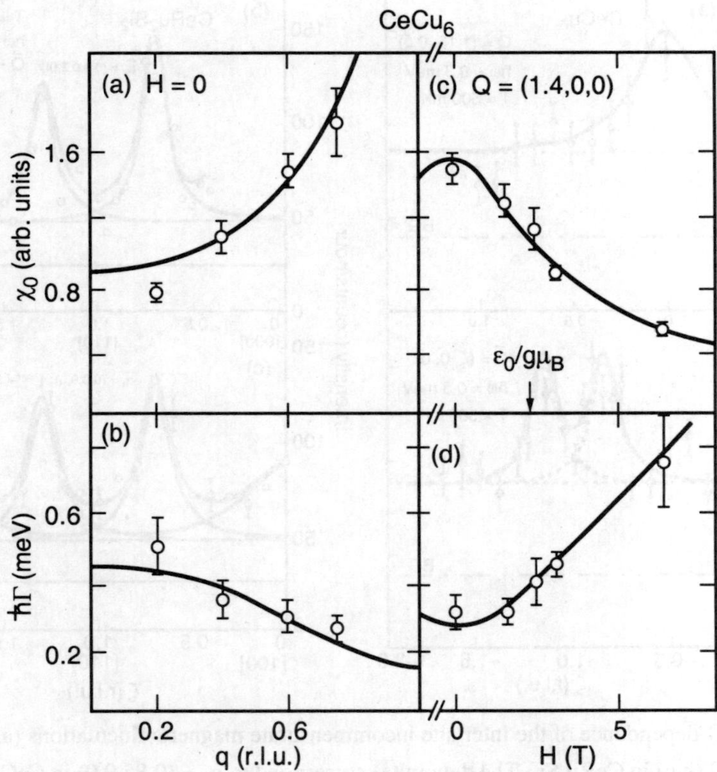

Fig. 23. Wavevector and magnetic field dependences of the quasielastic energy scale Γ and zero-frequency susceptibility χ_0 for CeCu$_6$ (from Aeppli *et al.*, Ref. 114), taken at a Q-value corresponding to single-site fluctuations (in Rossat-Mignod's parlance).

significant magnetic contribution for **Q** along the **b*** direction, indicating that ground-state doublet is Ising-like with uniaxial anisotropy along the **b**-direction, in agreement with conclusions from bulk magnetisation[196]. Using a cold-source triple-axis spectrometer with significantly better energy resolution than used by Aeppli *et al.*, they were able to separate the quasielastic scattering into two components: Q-independent single-site localised correlations and intersite correlations which have a strong Q-dependences. The Q-dependent part has strong correlations at $\mathbf{q}_1 = (0,0,1)$ and $\mathbf{q}_2 = (0.85, 0, 0)$ below 10K and 2T, as shown in Fig. 24(a). There is some evidence this Q-correlated response is slightly inelastic. The temperature dependences of the **a**- and **c**-axis magnetic correlation lengths are shown in Fig. 25. These results are consistent with those from

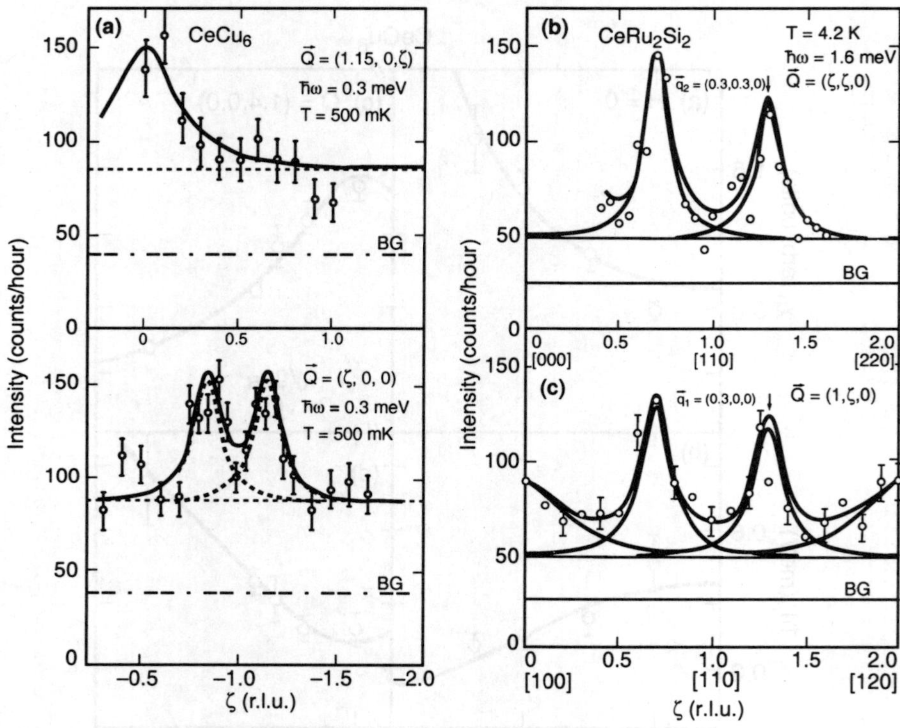

Fig. 24. Q-dependence of the inter-site incommensurate magnetic fluctuations (a) in $CeCu_6$ and (b,c) in $CeRu_2Si_2$. The data in (a) correspond to $\mathbf{q}_2 = (0.85,0,0)$ in $CeCu_6$, while the other two panels show the two modulations in $CeRu_2Si_2$: panel (b) shows $\mathbf{q}_2 = (0.3,0.3,0)$ and panel (c) shows $\mathbf{q}_1 = (0.3,0,0)$, both at an energy transfer of 1.6 meV and $T = 4.2K$ (from Rossat-Mignod *et al.*, Ref. 116).

time-of-flight experiments[115] on polycrystalline samples, which are not sensitive to the inter-site contribution as it has only 10% of the total weight in the q-integrated signal. In addition, the earlier work by Aeppli *et al.* was performed at **Q**-values well away from $\mathbf{q}_1$ and $\mathbf{q}_2$ (and with coarser energy discrimination). They therefore essentially measure the single-site contribution Γ_{s-s}, and interpreted this way, their results seem to be consistent with those of Rossat-Mignod *et al.*

Very recently some single-crystal inelastic scattering work has been performed on Au-doped $CeCu_6$. Beyond a critical concentration of $x_c = 0.1$, $CeCu_{6-x}Au_x$ develops long-range antiferromagnetic order, as described in the previous section. Right at the critical concentration, the phase transition is still thought to be 2nd-order, and there is a quantum

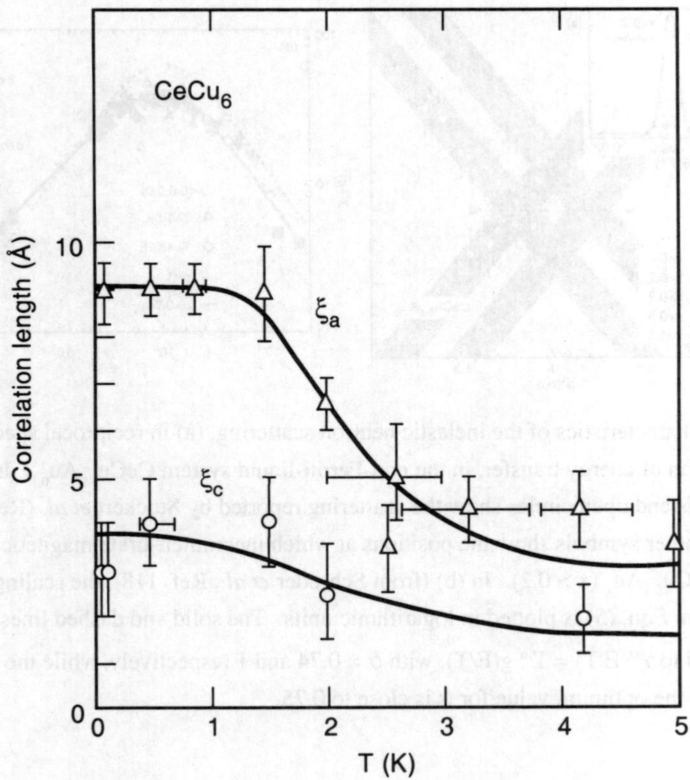

Fig. 25. Temperature variation of the a- and c-axis correlation lengths associated with the inter-site magnetic correlations in CeCu$_6$ as extracted from data like those shown in Fig. 24(a) (from Rossat-Mignod *et al.* Ref. 116).

critical point, and this gives rise to non-Fermi-liquid behaviour in thermodynamic properties like the specific heat.[197] The spin dynamics at this critical concentration have recently been studied in single-crystal samples in the vicinity of the $(100)_{orthorhombic} = (001)_{monoclinic}$ point. This corresponds to $\mathbf{q}_1$ in Rossat-Mignod's study[116] of pure CeCu$_6$, and it is also close to the ordering vectors observed for $x > x_c$. In fact, Stockert *et al.*[88] show that the scattering lies along rods in the a*-c* plane, as shown in Fig. 26(a), and they take these rods as evidence for two-dimensional fluctuations, possibly associated with next-nearest-neighbour Ce atoms. On the other hand, Schröder *et al.*[118] concentrate on the energy dependence of the scattering in the same q-region. As in polycrystalline samples of UPd$_x$Cu$_{5-x}$ (see Eqn. (6) below), they find E/T scaling, albeit with the slightly

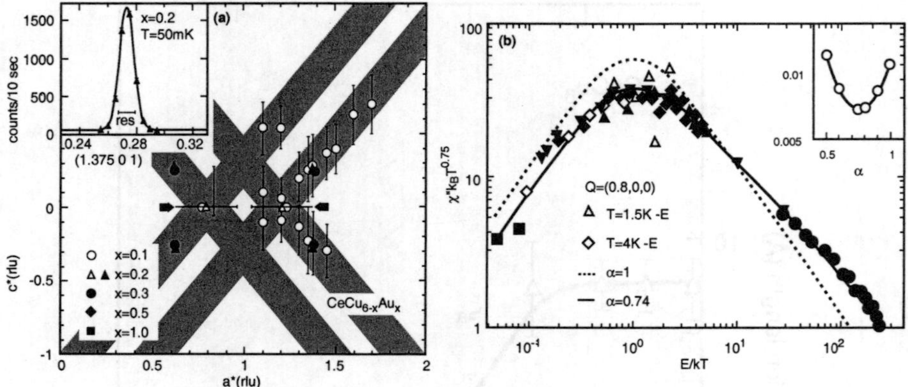

Fig. 26. Characteristics of the inelastic neutron scattering, (a) in reciprocal space and (b) as a function of energy transfer, in the non-Fermi-liquid system $CeCu_{5.9}Au_{0.1}$. In (a) the shaded rods and open circles show the scattering reported by Stockert *et al.* (Ref. 88), while the other symbols show the positions at which incommensurate magnetic order is seen in $CeCu_{6-x}Au_x$ (x > 0.2). In (b) (from Schröder *et al.*, Ref. 118), the scaling function g(E/T) from Eqn. (5) is plotted in logarithmic units. The solid and dashed lines correspond to $\chi''(E,T) = T^{-\alpha} g(E/T)$, with $\alpha = 0.74$ and 1 respectively, while the inset shows that the optimum value for α is close to 0.75.

different functional form

$$\chi''(\omega,T) \sim T^{-0.75} g(\omega / T) \tag{5}$$

where g is the universal scaling function shown in Fig. 26(b).

B. CeRu$_2$Si$_2$

One of the most impressive pieces of work in this field has been that performed on $CeRu_2Si_2$, using the single-crystal triple-axis method, by the Grenoble group.[119,116] In many ways, the behaviour of $CeRu_2Si_2$ is similar to that of $CeCu_6$ which is described above. Again the quasielastic scattering can be separated into a **q**-independent single-site contribution and a **q**-dependent correlated part (this time with $\mathbf{q}_1 = (0.3,0,0)$ and $\mathbf{q}_2 = (0.3,0.3,0)$): see Figs. 20(a) and 24. The in-plane correlation length for these fluctuations is $\xi_\parallel = 8\text{Å}$. The correlated part disappears at the 8-T metamagnetic transition just as the q-dependent scattering vanishes (at 2T) in $CeCu_6$ (see Fig. 27). They report[119,116] that the single-site and inter-site linewidths Γ_{s-s} and Γ_{i-s} in zero field increase from ~1.2 meV as $\sqrt{T}$, up to 100K (see Fig. 21(c)), but the work on polycrystalline samples measured up to 250K by Severing *et al.*[122] does not support this: rather, the temperature dependence is linear.

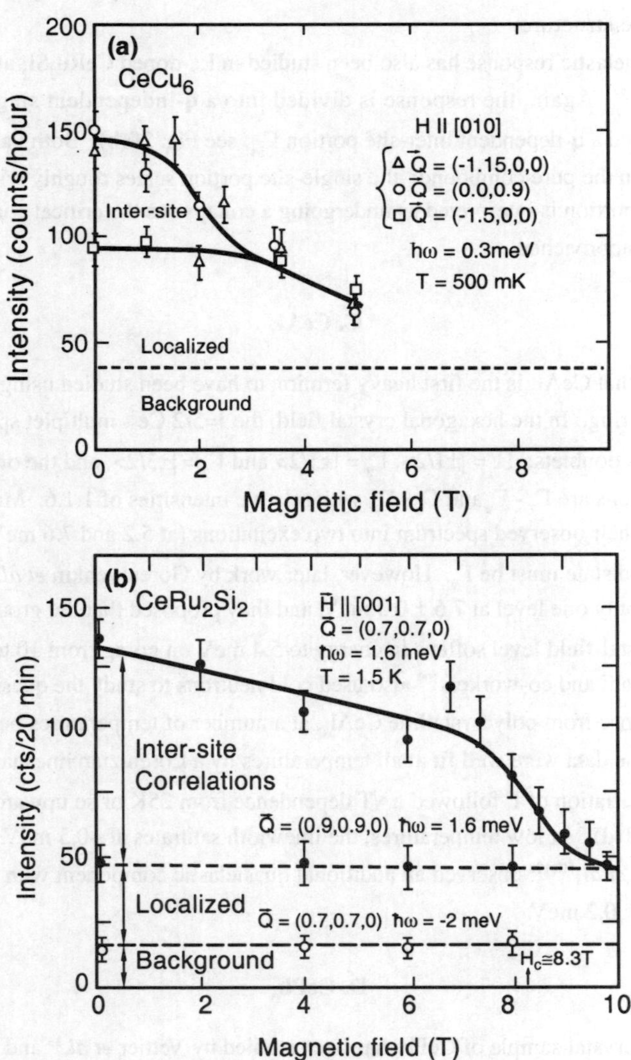

Fig. 27. Field dependences of the intensities of the inter- and single-site correlations (a) in CeCu$_6$ and (b) in CeRu$_2$Si$_2$ (corresponding to q_2 = (0.3,0.3,0)) (from Rossat-Mignod *et al.*, Ref. 116).

There was also some contemporaneous data taken on polycrystalline samples by Grier *et al.*[120] using the triple-axis technique and Severing *et al.*,[121,122] using the time-of-flight method. Grier's and Severing's values for Γ are in good agreement with the single-crystal result, while Severing *et al.* showed definitively that there are no clear crystal-field excitations up to 40meV, in contrast to the other Ce compounds with the same structure.

The quasielastic response has also been studied in La-doped $CeRu_2Si_2$ at the 7.5% doping level.[58] Again, the response is divided into a **q**-independent single-site portion $\Gamma_{s\text{-}s}$ and a q-dependent inter-site portion $\Gamma_{i\text{-}s}$: see Fig. 20(b). Both values are smaller than in the pure compound: the single-site portion scales roughly with $1/\gamma$, while the inter-site portion is interpreted as undergoing a collapse as the critical concentration ($x_c = 0.08$) is approached.

C. CeAl$_3$

It seems that $CeAl_3$ is the first heavy fermion to have been studied using inelastic neutron scattering. In the hexagonal crystal field, the J=5/2 Ce^{3+} multiplet splits into three Kramers doublets: $\Gamma_7 = |\pm1/2>$, $\Gamma_8 = |\pm5/2>$ and $\Gamma_9 = |\pm3/2>$, and the only allowed dipole transitions are $\Gamma_7 - \Gamma_9$ and $\Gamma_8 - \Gamma_9$, with relative intensities of 1:1.6. Murani *et al.*[108] decomposed their observed spectrum into two excitations (at 5.2 and 7.6 meV), implying that the ground state must be Γ_9. However, later work by Goremychkin *et al.*[154] found evidence for only one level at 7.6 ± 0.3 meV, and they proposed that the ground state is Γ_7. Their crystal-field level softens in energy to 5.4 meV on going from 10 to 77K. In addition, Murani and co-workers[109] also used cold neutrons to study the quasielastic neutron response from polycrystalline $CeAl_3$, at a number of temperatures between 60mK and 125K. The data were well fit at all temperatures by a Lorentzian lineshape, and the temperature variation of Γ followed a $\sqrt{T}$ dependence from 25K or so upwards, as can be seen in Fig. 21(d). At low temperatures, the linewidth saturates at ~0.5 meV. Again Goremychkin *et al.*[110,198] observed an additional quasielastic component with an energy scale of 13.0 ± 0.2 meV.

D. CePb$_3$

A single-crystal sample of $CePb_3$ has been studied by Vettier *et al.*[33] and a clear crystal-field level is seen at ~6 meV. Based on susceptibility measurements, it was assumed that the ground state is a Γ_7 doublet, and that this excitation corresponds to the $\Gamma_7 - \Gamma_8$ transition. The excitation is almost dispersionless, but there is some evidence for slight softening near the satellite positions for magnetic order (see above). The transition is broader than the instrumental resolution, by a factor of 3 or so. There is a crude estimate of the quasielastic linewidth which must be less than 0.6 meV, and this is substantially smaller than the width of the 6 meV transition. The next experiment was performed on a polycrystal by Renker et al.[111], using a triple-axis spectrometer with a resolution of 0.64 meV. They reached very similar conclusions, made an estimate that Γ ~ 1 meV, and definitively confirmed the crystal-field assignment that the ground state is the Γ_7 doublet. Some of Renker's data, showing both the quasielastic response and the Γ_7-Γ_8 crystal-field excitation are shown in Fig. 20(e). Finally, a high-resolution (46 - 90 μeV depending on E_i) time-of-flight measurement was made on a high-quality polycrystalline sample by Balakrishnan et al.,[112] who found the linewidth to be narrower still and temperature dependent with the square-root dependence of Eqn. (4) in which $\Gamma_0 = 0.17$ meV and A = 0.12 meVK^{-1} (see Fig. 21(f)).

E. CePtSi

A small amount of inelastic neutron scattering has been performed on a polycrystalline sample of CePtSi by Krimmel et al.[129] In addition to crystal-field levels at 6.3 and 17.8 meV, which are interpreted as the $|\pm1/2\rangle$ and $0.78|\pm5/2\rangle$-$0.63|\pm3/2\rangle$ Kramers doublets respectively, quasielastic scattering was also observed. At the lowest temperature of 5K, Γ ~ 1.5 meV, and the linewidth increases with increasing temperature, but the variation falls below a $\sqrt{T}$ dependence and appears to saturate at high temperatures. The Yb half-Heusler compounds described below seem to be qualitatively similar in this respect.

F. CeCu$_2$Si$_2$

One of the first heavy fermion compounds to be studied by means of inelastic neutron scattering was CeCu$_2$Si$_2$. Early work showed that the J=5/2 Ce^{3+} state is split into three Kramers doublets, the excited states lying at 12 and 31 meV respectively[123] (see Fig. 28). There is also strong quasielastic scattering resulting from the interaction of the 4f-moment with the conduction electrons, and this is strongly temperature dependent with the $\sqrt{T}$ dependence (see Fig. 21(b)). The T-dependence of Γ seems to correlated with the paramagnetic Curie-Weiss temperature θ and the whole data set can be well described with the phenomenological square-root form of Eqn. (4). There seems to be a linear correlation between Γ_0 and A when one compares with other Ce compounds.[124] This work was further supplemented with studies[126] of sub- and super-stoichiometric samples $CeCu_{1.8}Si_2$ and $CeCu_{2.2}Si_2$, part of the rationale being that sub-stoichiometric samples do not superconduct, while super-stoichiometric samples systematically exhibit superconductivity, and the ideal stoichiometry is very sensitive to sample history. The observed crystal-field levels and quasielastic linewidth results for both samples are qualitatively similar to the ideal stoichiometry, and are listed in Table II, although the detailed crystal-field energies differ somewhat. The same basic crystal field level scheme works for all three compositions: the tetragonal field splits the $J= 5/2$ multiplet into 3 Kramers doublets, the ground state being predominantly of $|\pm 5/2\rangle$ character, with the first excited state $|\pm 1/2\rangle$ and the second excited state predominantly $|\pm 3/2\rangle$. Some polarised-beam work[150] was performed both above and below T_C, and these experiments directly confirmed the magnetic nature of the scattering, and that it has a weak Q-dependence if any. Further polarised-beam work was also performed on poly-crystalline samples in fields up to $2T$[125], and these experiments give a value of $\Gamma = 0.7$ meV for the quasielastic linewidth.

G. Other Ce 1:2:2 silicides and germanides: $CeNi_2Ge_2$ and $CePd_2Si_2$

The isostructural compound $CeNi_2Ge_2$ has also been studied in polycrystalline form by Knopp et al.,[68] and it has been compared with other isostructural Ce compounds (including $CeCu_2Si_2$) by Loidl et al.[154] A Lorentzian lineshape is seen at all temperatures between 4 and 300K, as shown in Fig. 20(c). At high temperatures, the temperature dependence is consistent with a $\sqrt{T}$ dependence, but at low temperatures ($T < 50K$) there is significant deviation from this functional form, with saturation at $\Gamma \sim 4$ meV, and possibly an upturn with lowering temperature, as shown in Fig. 21(b). This is in marked contrast to other isostructural Ce compounds, some of which are shown in other panels of Fig. 21, that exhibit the $\sqrt{T}$ dependence down to much lower temperatures. The temperature at which the deviation from $\sqrt{T}$ dependence kicks in (~30 K) was taken as indicative of the Kondo temperature for this system.

Severing et al.[121] have studied a number of Ce 1:2:2 disilicides: $CePd_2Si_2$, $CeAg_2Si_2$, $CeAu_2Si_2$ and $CeRu_2Si_2$ (discussed above), and there was an earlier study using the triple-axis technique by Grier et al.[120] $CePd_2Si_2$ is often taken as a heavy fermion, even though its electronic specific heat seems never to have been measured in the paramagnetic state, presumably because it orders antiferromagnetically at 12K. Nevertheless the antiferro-magnetism can be suppressed by applying a pressure of 27 kbar or so.[199] At 12K, just above T_N, Grier et al. observed quasielastic neutron scattering with $\Gamma = 0.76$ meV and it broadens to 2.56 meV by 100K, the only other temperature measured. They also observed a single inelastic peak at ~20 meV, and interpreted this as a crystal-field

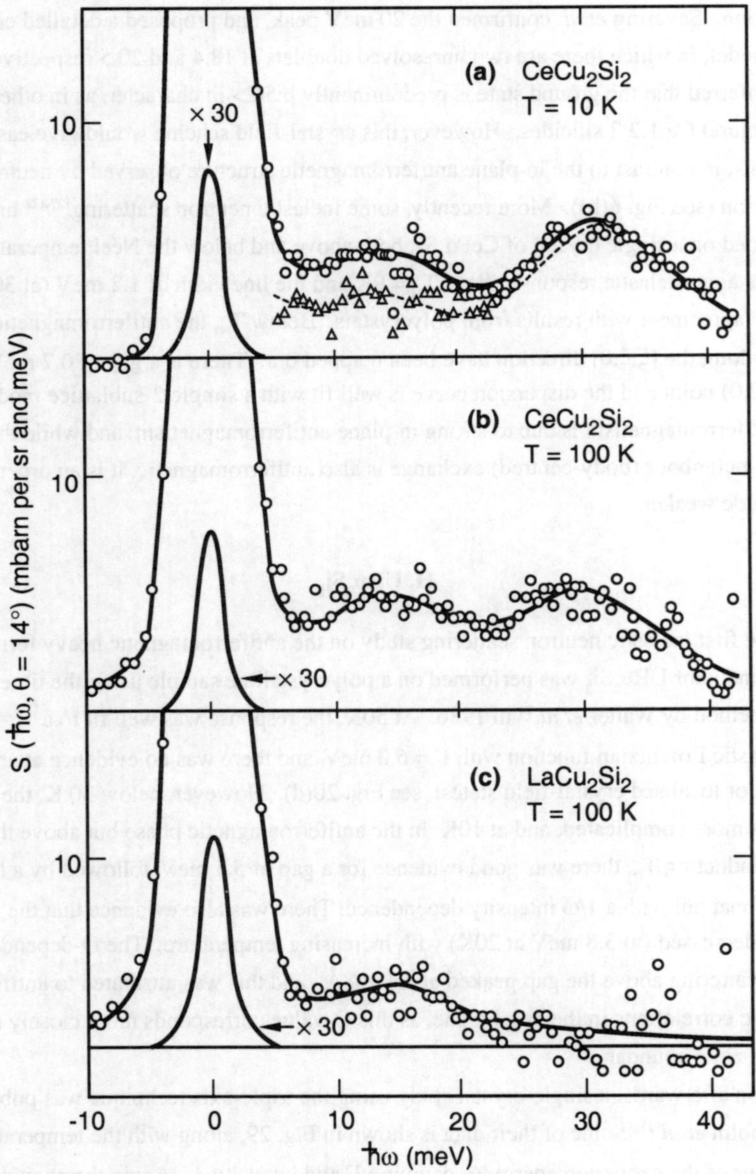

Fig. 28. Some of the original inelastic neutron scattering data (Horn *et al.*, Ref. 123) from polycrystalline $CeCu_2Si_2$, along with the "blank" analogue $LaCu_2Si_2$, showing crystal-field excitations at 12.1 and 31.4 meV. The triangles and dashed line in (a) show a weighted difference between the 10K $CeCu_2Si_2$, and the 100K $LaCu_2Si_2$ data.

excitation. Severing *et al.* confirmed the 20 meV peak, and proposed a detailed crystal-field model, in which there are two unresolved doublets at 18.4 and 20.5 respectively. They inferred that the ground state is predominantly $|\pm5/2\rangle$ in character, as in other isostructural Ce 1:2:2 silicides. However, this crystal-field scheme would give easy-axis moments, in contrast to the in-plane antiferromagnetic structure observed by neutron diffraction (see Fig. 6(b)). More recently, some inelastic neutron scattering[127,128] has been performed on a single crystal of $CePd_2Si_2$ both above and below the Néel temperature. There is a quasielastic response above $T_N = 9K$ and the linewidth of 1.2 meV (at 30 K) is in good agreement with results from polycrystals. Below T_N, the antiferromagnetic spin waves along the $[\xi,\xi,0]$ direction have been mapped out. There is a gap of 0.7 meV at the (1/2,1/2,0) point and the dispersion curve is well fit with a simple 2-sublattice model. The antiferromagnetism is due to strong in-plane antiferromagnetism, and while the next-nearest-neighbour (body-centred) exchange is also antiferromagnetic, it is an order of magnitude weaker.

H. URu_2Si_2

The first inelastic neutron scattering study on the antiferromagnetic heavy fermion superconductor URu_2Si_2 was performed on a polycrystalline sample using the time-of-flight method by Walter *et al.*[94] in 1986. At 50K, the response was well fit to a quasielastic Lorentzian function with $\Gamma = 6.0$ meV, and there was no evidence at any energy for localised crystal-field states: see Fig. 20(d). However, below 30 K, the line shape is more complicated, and at 10K, in the antiferromagnetic phase but above the superconducting T_C, there was good evidence for a gap of 5.5 meV, followed by a long relaxational tail with a $1/\omega$ intensity dependence. There was also evidence that the gap energy decreased (to 3.8 meV at 20K) with increasing temperature. The Q-dependence of the scattering above the gap peaked at $|Q| \sim .8\text{Å}^{-1}$ and this was attributed to antiferro-magnetic correlations in the basal plane, as this Q-value corresponds more closely to the in-plane zone boundary.

Soon afterwards, a single-crystal study using the triple-axis technique was published by Broholm *et al.*[23] Some of their data is shown in Fig. 29, along with the temperature variations of the excitation energy Δ, damping Γ and intensity A. Again the excitations are highly damped, if not overdamped, above T_N. Below T_N, the damping decreases substantially, and the scattering coalesces into sharp propagating magnons. There is good evidence that they are polarised along the c-axis, i.e. parallel to the ordered moment. There is also evidence that the ground state is developed from a singlet crystal-field state, possibly from the U^{4+} J=4 multiplet which would be split into two doublets and five

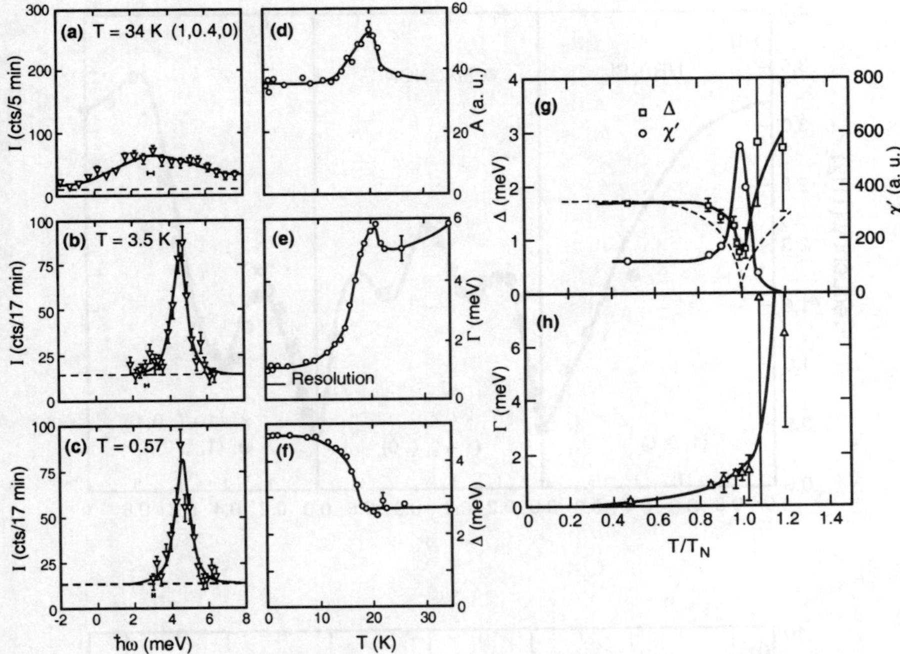

Fig. 29. The original single-crystal inelastic neutron scattering results from URu_2Si_2, due to Broholm *et al.*[23] at the (1,0.4,0) position in (a) the paramagnetic, (b) the antiferromagnetic and (c) the superconducting phases. Panels (d) - (e) show the temperature variations of the intensity A, damping Γ and gap Δ. Panels (g) and (h) show more recent results at the magnetic ordering wave vector ($\mathbf{q} = (001)$ and/or (100)) by Buyers[96], taken below and close to $T_N = 17.5K$.

singlets by the tetragonal crystal field. The total moment associated with these modes is 2.2 μ_B, roughly an order of magnitude larger than the ordered moment. More recent work by Buyers and co-workers[95,96] has looked at the critical behaviour close to T_N in more detail (see Figs. 29(g) and (h)), and their results are in good agreement with Broholm's. They conclude that the dynamic spin response below T_N is like that of a classic soft mode. However, the dynamics above T_N are less normal, consisting initially of very slow long-range commensurate correlations, before giving over to shorter-range incommensurate fluctuations above 20K. There is a much more complete report, by Broholm *et al.*,[97] of the work on URu_2Si_2, including that described above. In particular, they map out the magnon dispersion curve for the $(1,0,\zeta)$, $(1+\zeta,\zeta,0)$ and $(1,\zeta,0)$ directions as shown in

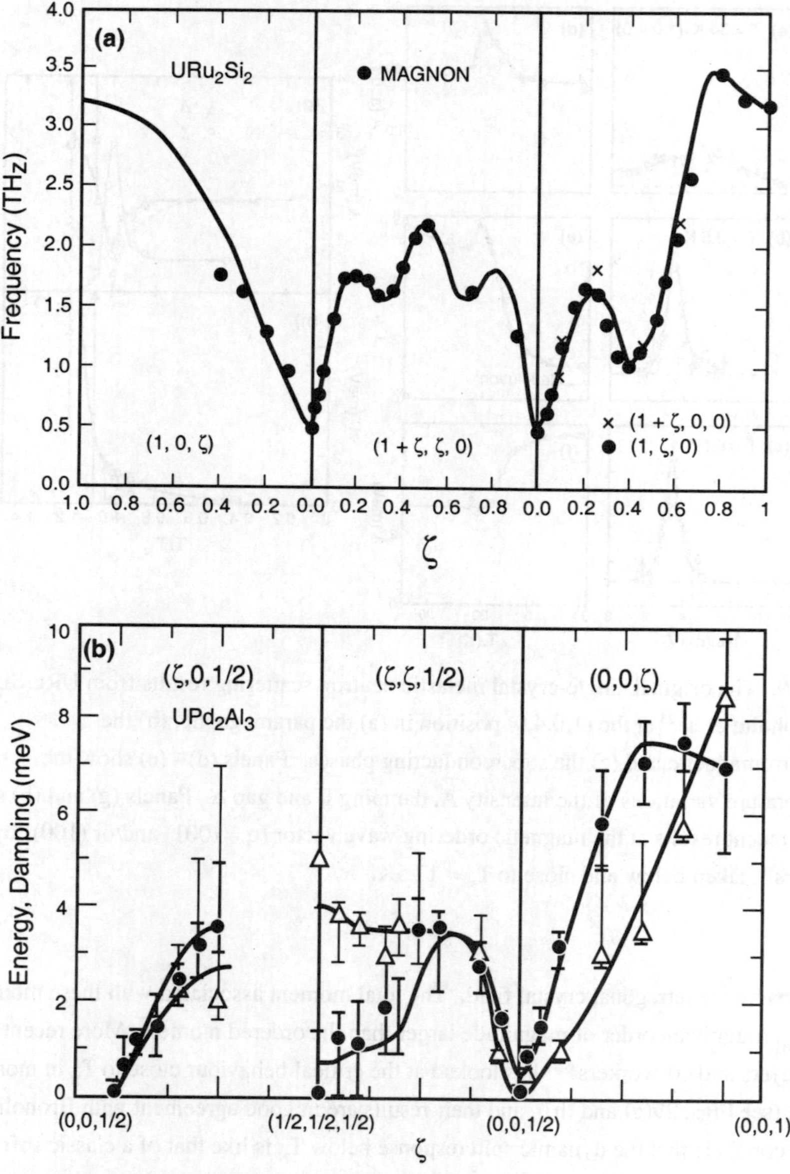

Fig. 30. The magnon dispersion curve for 3 principal symmetry directions (a) in URu$_2$Si$_2$ as measured at 1K (from Broholm *et al.*, Ref. 97), and (b) in UPd$_2$Al$_3$ (from Ref. 204). In (b), the filled circles represent excitation energies, while the open triangles represent the linewidth $\Gamma(\mathbf{Q})$. Note that the modes are overdamped for both in-plane propagation directions.

Fig. 30(a), and determine that a total moment of 1.2 μ_B is associated with the scattering in the ordered state. This behaviour is very different from that seen in other heavy fermions, even including U_2Zn_{17}, which is described below, and which orders on more or less the same temperature scale (T_N = 9.7 rather than 17.5K), has a similar γ (100 rather than 50 mJmol^{-1}K^{-2}, in the ordered state), but has better developed ordered antiferromagnetic moment (0.79 rather than 0.03 μ_B). Another observation is that it is rather unusual to observe nice propagating modes like those Fig. 30, in any uranium intermetallic magnet.[200,138]

I. UPt$_3$

As in the diffraction case, UPt$_3$ is the most intensively studied of all heavy fermion compounds by inelastic neutron scattering. The first inelastic neutron scattering experiments on UPt$_3$ were performed on a polycrystalline sample using polarised neutrons.[90] Spin-flip quasielastic scattering with $\Gamma = 9 \pm 2$ meV was observed in the normal-state phase at 1.3K (see Fig. 31), and the total moment obtained by integrating χ'' was in fairly

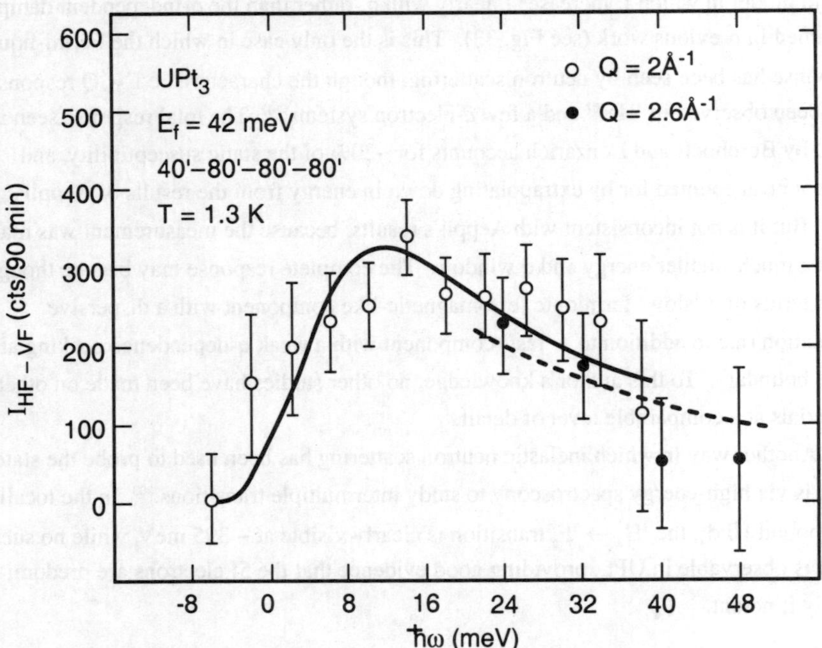

Fig. 31. Some of the original polarised-beam inelastic neutron scattering data (Aeppli *et al.*, Ref. 90) from polycrystalline UPt$_3$, showing the Lorentzian response with $\Gamma \sim 9$ meV.

good agreement with the moment deduced from the Curie-Weiss susceptibility. This work was followed up with a single-crystal study,[92] with better energy resolution, which revised the energy scale downwards somewhat to 5.0 ± 0.2 meV, and showed that there is substantial Q-dependence to the magnetic scattering (peaking at (001), (102), (202) and so on). This was attributed to antiferromagnetic correlations between the nearest uranium neighbours in adjacent hexagonal planes, each uranium ion having six such neighbours (see Fig. 32). There are also strong in-plane (next nearest neighbour) ferromagnetic correlations, and these persist up to 150K or so.[91] Now after the discovery of long-range antiferromagnetic order in Th- and Pd-doped UPt_3,[8,9] a further search[4] was made in the pure compound near the characteristic (1/2,0,1) ordering vector for the long-range order. This resulted in the observation of further structure in the inelastic scattering with a characteristic energy of 0.2 meV (4x the superconducting T_c).

Another study, focussed on the $\mathbf{q} \to 0$ ferromagnetic response, has been made using high-resolution time-of-flight spectroscopy by Bernhoeft and Lonzarich[93] using a high-quality polycrystalline sample. They extract the 10-K response, in the coherent state, by subtracting off a similar measurement made at 1.3K, and can fit their data using a Fermi-liquid model in which Γ increases linearly with q, rather than the q-independent damping assumed in previous work (see Fig. 33). This is the only case in which the Fermi-liquid response has been seen by neutron scattering, though the characteristic $\Gamma \propto Q$ response has been observed in ^3He[201] and a few d-electron systems.[202] The total response seen in UPt_3 by Bernhoeft and Lonzarich accounts for ~20% of the static susceptibility, and cannot be accounted for by extrapolating down in energy from the results of Aeppli et $al.$[92] But it is not inconsistent with Aeppli's results, because the measurement was made over a much smaller energy and q window. The complete response may best be thought of in terms of a 'slow' fermionic ferromagnetic-like component with a dispersive relaxation rate in addition to a 'fast' component with a weak q-dependence peaking at the zone boundary. To this author's knowledge, no other studies have been made on other materials at a comparable level of detail.

Another way in which inelastic neutron scattering has been used to probe the state of UPt_3 is via high-energy spectroscopy to study intermultiple transitions.[203] In the localised compound UPd_3, the $^3H_4 \to {}^3F_2$ transition is clearly visible at ~ 395 meV, while no such peak is observable in UPt_3, providing good evidence that the 5f electrons are predominantly itinerant.

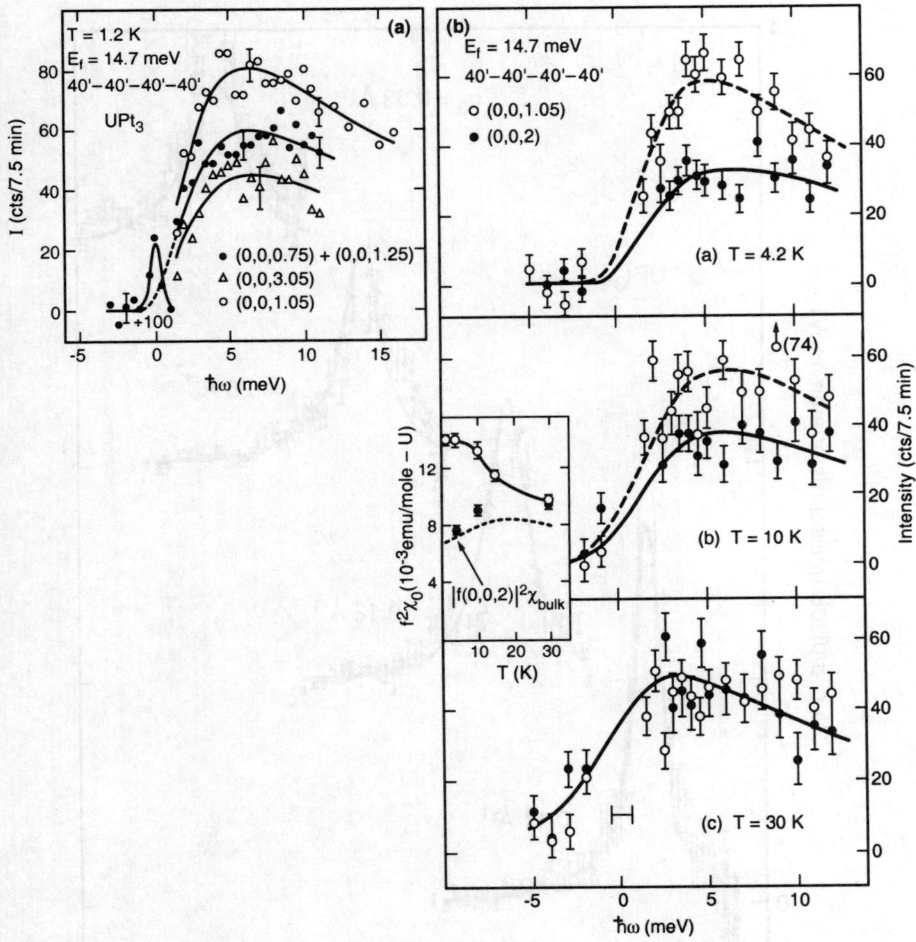

Fig. 32. The first single-crystal inelastic neutron scattering data from UPt$_3$, showing (a) the Q-dependence of the scattering, and (b) the temperature dependence of the antiferromagnetic correlations near the (001) zone boundary. In (a) the fact that the q$_z$ = 1.05 (open circles) is stronger than the sum of that for q$_z$ = 0.75 and 1.25, is taken as definitive evidence for antiferromagnetic correlations. The inset to (b) shows that the zero-frequency susceptibility corresponding to these antiferromagnetic fluctuations dies away with temperature above 10K or so (from Aeppli *et al.*, Refs. 91 and 92).

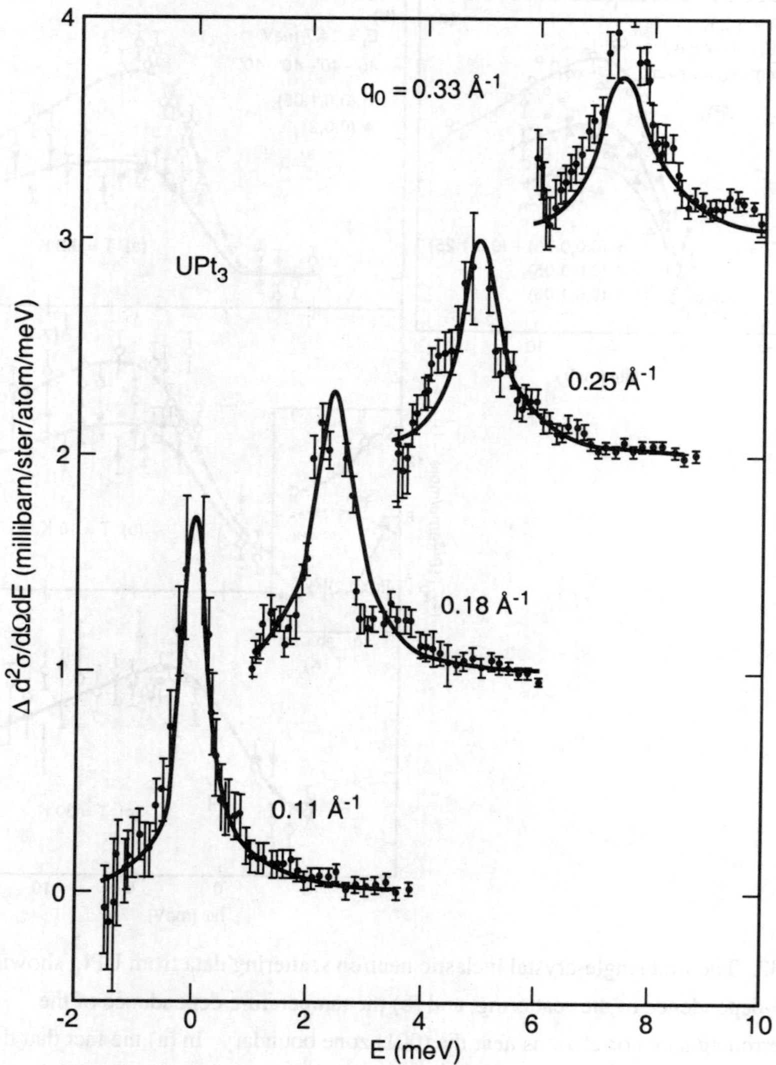

Fig. 33. The quasielastic Fermi-liquid response from UPt$_3$, measured at small Q-values in a polycrystalline sample using cold neutrons, by Bernhoeft and Lonzarch (Ref. 93). The origins are shifted by 2.5 meV and 1 mb sr^{-1} meV^{-1} atom^{-1} between each Q-value. Note that the linewidth increases linearly with Q.

J. UPd$_2$Al$_3$ and CePd$_2$Al$_3$

Some inelastic neutron scattering work has been performed in UPd$_2$Al$_3$ below T_N = 14.2 K, above and below the superconducting T_C (1.2K in the sample studied).[20] **a**- and **c**-axis spin waves were observed in the vicinity of the (0,0,1/2) point. They seem to be conventional antiferromagnetic spin waves, with no gap (down to the instrumental resolution of 0.35 meV), and resolution-corrected fits imply no gap whatsoever. Velocities of v_{100} = 17 meVÅ and v_{001} = 20 meVÅ were extracted for the two spin-wave propagation directions. No difference between the superconducting and normal states was observed, even at an energy transfer (0.3 meV) close to the superconducting gap

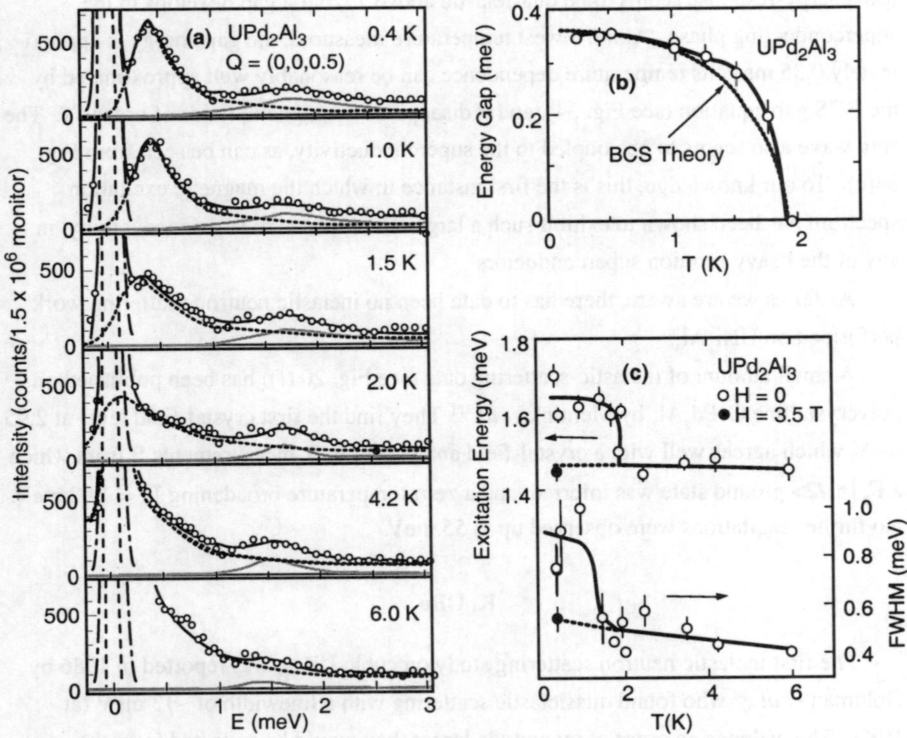

Fig. 34. The low-energy response of the heavy fermion superconductor UPd$_2$Al$_3$ in the vicinity of T_c = 2K (from Ref. 107). (a) shows a series of energy scans at the (0,0,1/2) point, with the gap excitation at 0.37 meV, and the spin-wave mode at ~1.5 meV. (b) shows the temperature evolution of the gap energy , while (c) shows the temperature evolution of both the spin-wave energy and its width at this point.

energy predicted using BCS theory. The whole dispersion curve (see Fig. 30 (b)) has subsequently been published by Mason and Aeppli,[204] and their work shows that the in-plane modes are overdamped, while the c-axis modes are much better defined. More recently, a higher resolution single-crystal study has been reported by Sato *et al.*[104,105] In addition to the dispersive spin wave modes shown in Fig. 30(b), they observe an additional narrower low-energy response that is confined close to the antiferromagnetic ordering vector $\mathbf{q}_0 = (0,0,1/2)$, with a correlation length of around 50Å (at 0.5K). At higher temperatures (10 - 20K), the two modes are indistinguishable and can be well fit with a Lorentzian lineshape in which Γ has quadratic dispersion away from $\mathbf{q}_0$. In a separate recent study, Metoki *et al.*[106,107] have concentrated on the response close to the superconducting $T_C = 2K$, and close to the (0,0,1/2) point. The additional low-energy-response seems to be quasielastic above T_C, but a gap develops in the superconducting phase. At the lowest temperature measured, the gap energy is approximately 0.35 meV, its temperature dependence can be reasonably well approximated by the BCS gap equation (see Fig. 34), and it disappears in magnetic fields of order 4T. The spin wave also seems to be coupled to the superconductivity, as can be seen from Fig. 34(c). To our knowledge, this is the first instance in which the magnetic excitation spectrum has been shown to exhibit such a large effect related to superconductivity, in any of the heavy fermion superconductors.

As far as we are aware, there has to date been no inelastic neutron scattering work performed on UNi_2Al_3.

A small amount of inelastic-scattering data (see Fig. 20 (f)) has been published on polycrystalline $CePd_2Al_3$ by Mentink *et al.*[113] They find the first crystal-field level at 2.95 meV, which agrees well with a crystal-field analysis of bulk measurements,[47] from which a $\Gamma_7 |\pm 1/2\rangle$ ground state was inferred, and a zero-temperature broadening $\Gamma_0 = 1.92$ meV. No further excitations were observed up to 55 meV.

K. UBe_{13}

The first inelastic neutron scattering study on cubic UBe_{13} was reported in 1986 by Goldman *et al.*,[98] who found quasielastic scattering with a linewidth of ~13 meV (at 10K). This value is an order of magnitude larger than would be expected from the specific heat γ, and in a follow-up study using high-resolution time-of-flight spectroscopy, Lander *et al.*[99] found a q-independent response below 1K with $\Gamma = 1.5$ meV. It seems that the magnetic response has at least two components, just as various different energy scales have been seen in UPt_3. Both studies of UBe_{13} used large polycrystalline samples. Now that long-range magnetic ordering has been seen in the isostructural heavy fermion compound $NpBe_{13}$,[30] with $\mathbf{q} = (0,0,1/3)$, it seems likely that a more detailed search of the excitations around this $\mathbf{q}$-value, in a single crystal of UBe_{13}, would be worthwhile.

L. U_2Zn_{17}

Single crystals of U_2Zn_{17} were available almost immediately and although there was one early time-of-flight polycrystalline study by Walter et al.,[100] almost all of our knowledge is derived from the contemporaneous single-crystal study by Broholm et al.[49] It exhibits spin dynamics characteristic of a short-range coupled x-y antiferromagnet, albeit with strong damping, and the phase transition is driven by a rise in the effective exchange interaction, rather than a change in the local single-ion susceptibility. This manifests itself in an excitation spectrum consisting of a vertical (parallel to $\hbar\omega$) ridge emanating from the antiferromagnetic Bragg point (e.g. 102), much as happens on a larger scale in the classic itinerant antiferromagnet Cr. This response is different to that seen in URu_2Si_2 and UPd_2Al_3, in both of which clear propagating modes were seen. The whole $S(\mathbf{Q},\omega)$ could be fitted at temperatures both above and below T_N using a model for a lattice of single f-like moments with a Kondo interaction, and with nearest neighbour isotropic RKKY interactions, but with temperature-dependent single-site susceptibility, exchange and Kondo damping. The same parametric model for $\chi''(Q,\omega)$ was used by Aeppli and co-workers[114] to analyse their data on $CeCu_6$. In U_2Zn_{17} and in contrast to normal magnets, the exchange increases as the temperature drops and this drives the transition to antiferromagnetism at 10K. The temperature dependence of Γ is very different to the $\sqrt{T}$ dependence described above for many Ce compounds, though this formalism emphasizes the inter-site response, and the $\sqrt{T}$ dependence is more normally thought of as a single-site response.

M. UCu_5

There is one report of inelastic scattering measurements from UCu_5 above and below $T_N = 15K$, by Walter et al.,[100] using a polycrystalline sample. They observed a quasielastic response of around 9 meV at both 10 and 50K, and its Q-dependence (below T_N) peaks broadly at $|Q|$ values corresponding to the $q = 1/2(111)$ long-range antiferromagnetic order. There has recently been increased interest in Pd-doped UCu_5 in recent years, because it exhibits power-law and logarithmic divergences in its thermodynamic and transport properties for Pd doping beyond $x = 0.8$. Inelastic neutron scattering studies[101,102] have been performed on polycrystalline samples of UPd_xCu_{5-x} (with $x = 1.0$

262

and 1.5). The magnetic response is very similar in both materials, even though UPdCu$_4$ is nominally an ordered compound.[144] Above 25 meV, the scattering is very similar to that in pure UCu$_5$, i.e. one sees the tail of a Lorentzian with halfwidth 8.8 meV. At lower energies, E/T scaling is observed with the form

$$\chi''(\omega, T) \sim \omega^{-\frac{1}{3}} Z(\omega / T) \qquad (6)$$

where Z is the universal scaling function shown in Fig. 35. Note that this form is similar to, but slightly different from, that given for another non-Fermi-liquid material CeCu$_{5.9}$Au$_{0.1}$ in Eqn. (5) above. In a more recent polarised-neutron study, Osborn et $al.$[103] found evidence of some Q-dependence (reminiscent of the 1/2 (111)-type long-range

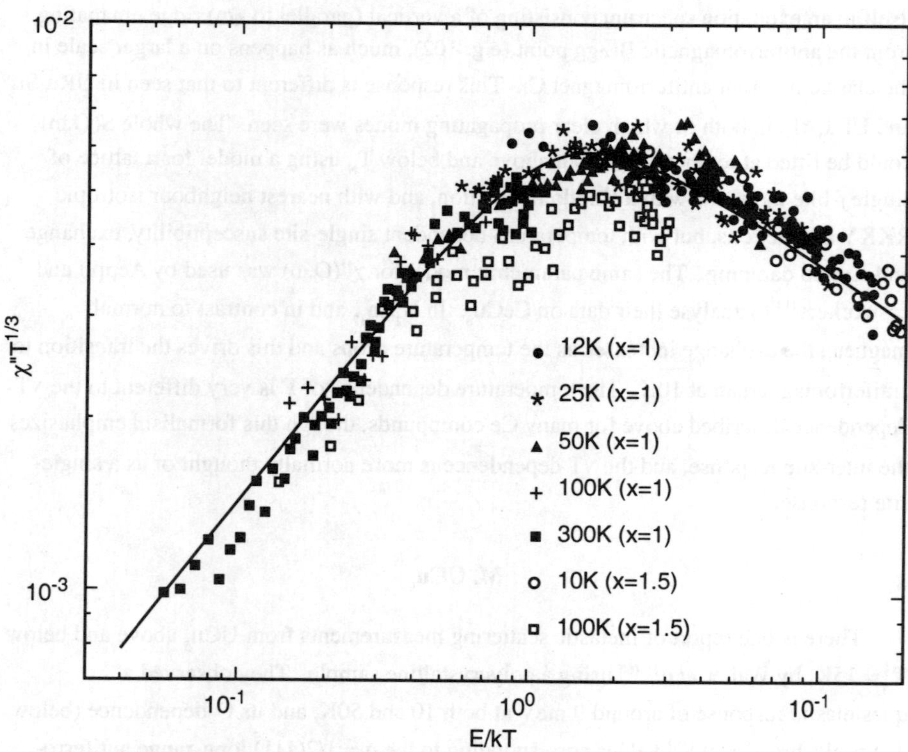

Fig. 35. The E/T scaling behaviour of the magnetic response in Pd-doped UCu$_5$, for two concentrations exhibiting non-Fermi-liquid behaviour in bulk properties UCu$_4$Pd (x = 1) and UCu$_{3.5}$Pd$_{1.5}$ (x = 1.5). The solid line has the form of Eqn. (6) with Z = tanh(ω/1.2T). (from Ref. 101)

order in pure UCu_5) in $UPdCu_4$ but not in $UPd_{1.5}Cu_{3.5}$, and interpreted this as a cross-over from the quantum critical regime to a low-temperature quantum-disordered regime.

N. $YbAgCu_4$ and $YbPdCu_4$

The cubic Yb compounds $YbAgCu_4$ and $YbPdCu_4$ have both been studied in poly-crystalline form, using time-of-flight spectroscopy, by Severing and co-workers.[132] No clear crystal-field levels were seen in either compound. In $YbPdCu_4$, the situation is complicated by the presence of impurity phases, but there is clear evidence for both Lorentzian quasielastic scattering and a broader inelastic response. In $YbAgCu_4$, there is much less quasielastic scattering, on an absolute intensity scale, at the lowest temperature (1.5K), but there is a very broad quasielastic response at high temperatures. On lowering the temperature, this develops into a strongly damped inelastic response, peaked at 10 meV or so.

O. YbBiPt

The cubic half-Heusler structure Yb compound YbBiPt has the largest electronic specific heat ($\gamma = 8$ Jmol^{-1}K^{-2}) yet observed in any heavy fermion compound, and there is an entropy of Rln5 by 20K, indicating a highly degenerate ground state. The excitations responsible for this specific heat have been observed in a polycrystalline sample using various inverted-geometry crystal-analyser spectrometers, in addition to a chopper spectrometer.[130] The response is quite complicated (see Fig. 36), there being a clear crystal-field excitation at 5.6 meV, and a more complicated low-energy response. At temperatures of 3.3K and above, there are at least two quasielastic components, which are better fitted with Gaussians (as opposed to Lorentzians). The broad component is roughly 5x broader than the narrow component, and it can be associated with an overdamped Γ_8 crystal-field quartet. In this scenario, which agrees fairly well with the specific heat results of Fisk *et al*. Ref. 40, the ground state consists of almost degenerate overdamped Γ_7 (doublet) and Γ_8 (quartet) states, with the 5.6 meV level being the remaining Γ_6 doublet. At lower temperatures, the broad quasielastic component shows clear inelastic structure, with discrete levels at ~1 and 2 meV. Single-crystal triple-axis experiments[205] showed that these levels do not disperse, and it may be that the Γ_8 quartet has split up into two Kramers doublets. However, this is not allowed within the parent cubic symmetry, and exhaustive searches for a distortion have failed to detect any structural deformation.[142] Likewise, the splitting can be seen clearly at 1.5K, and this effect cannot be correlated with the 400mK magnetic transition described above. The

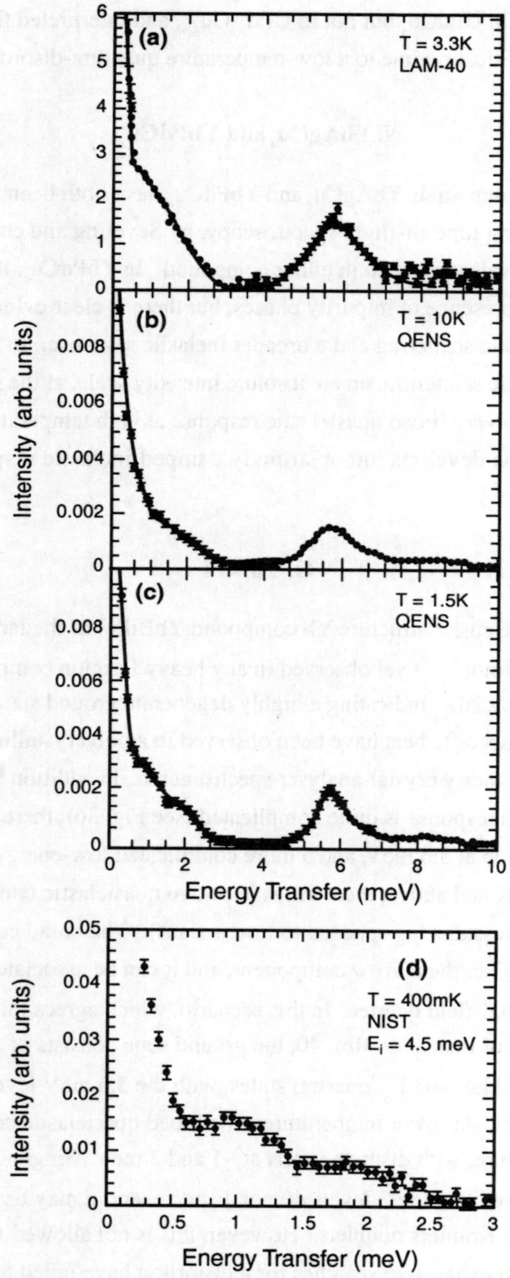

Fig. 36. Observed time-of-flight spectra from YbBiPt, taken at various temperatures on various different instruments (from Robinson *et al.*, Ref. 130)

temperature variation of the narrow quasielastic component is shown in Fig. 37, and while it bears some qualitative similarity to the $\sqrt{T}$ law (Eqn. (4) and see Fig. 21), there is evidence for saturation at high temperature. Note also that the low-temperature linewidth $\sigma = 0.2$ meV, and this would give $\gamma\,\sigma(0) = 1.6$ J meV mol^{-1} K^{-2}, by analogy with the $\gamma\,\Gamma(0)$ values listed in Table III.

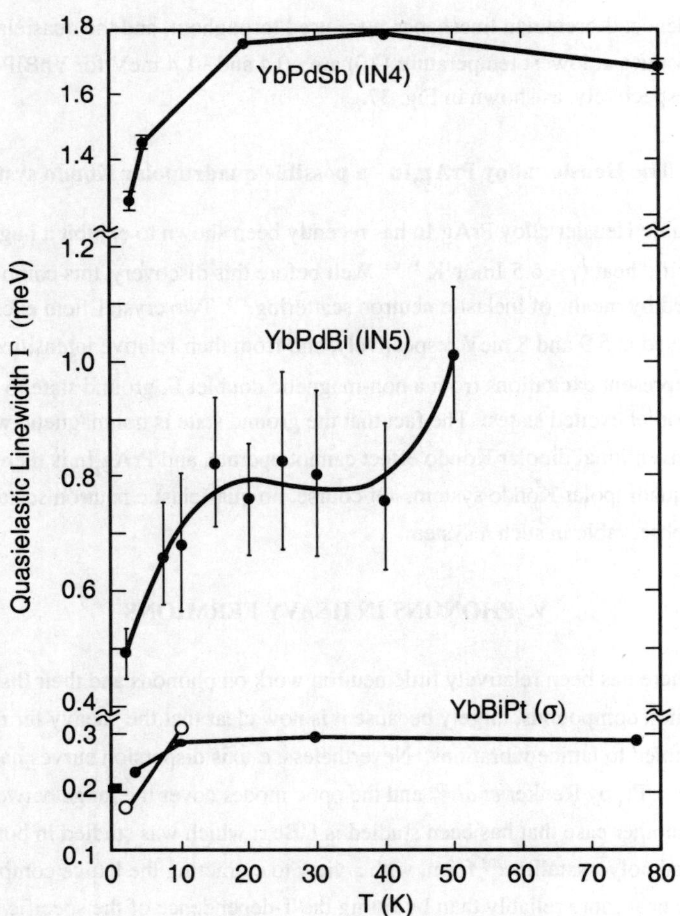

Fig. 37. The variation of the narrow quasielastic linewidths in YbBiPt (from Robinson *et al*. Ref. 130), YbPdSb and YbPdBi (from Marshall, Ref. 131), as a function of temperature. For YbBiPt, the data are gaussian σ's, and are shown using different symbols, were taken on three different spectrometers.

P. Other Yb-based Half Heuslers: YbBiPd and YbPdSb

The isostructural Yb compounds YbBiPd and YbPdSb have also been studied extensively by means of inelastic neutron scattering by Marshall.[131] The results are qualitatively similar to those from YbBiPt, though the electronic specific heat is not so high (see Tables I and II). Again a complicated low-energy response, including strongly damped crystal-field levels is observed. As in YbBiPt, and most rare-earth-based heavy fermions, the linewidths increase with increasing temperature, but not necessarily with a $\sqrt{T}$ dependence. Lorentzian lineshapes were used throughout, and the quasielastic linewidth values at lowest temperature $\Gamma(0)$ are ~0.4 and ~1.4 meV for YbBiPd and YbPdSb respectively, as shown in Fig. 37.

Q. The Heusler alloy PrAg$_2$In - a possible quadrupolar Kondo system

The cubic Heusler alloy PrAg$_2$In has recently been shown to exhibit a huge electronic specific heat ($\gamma = 6.5$ Jmol^{-1}K^{-2}).[66] Well before this discovery, this compound had been studied by means of inelastic neutron scattering.[133] Two crystal-field excitations were observed at 5.9 and 8 meV respectively, and from their relative intensities, it is clear that they represent excitations from a non-magnetic doublet Γ_3 ground state to Γ_4 and Γ_5 magnetic triplet excited states. The fact that the ground state is nonmagnetic would imply that the conventional dipolar Kondo effect cannot operate, and PrAg$_2$In is therefore a candidate quadrupolar Kondo system. Of course, no quasielastic neutron scattering should be observable in such a system.

V. PHONONS IN HEAVY FERMIONS

There has been relatively little neutron work on phonons and their dispersion in heavy fermion compounds, largely because it is now clear that the "heavy fermions" are weakly coupled to lattice vibrations. Nevertheless, c-axis dispersion curves have been reported for UPt$_3$ by Renker et $al.$[206] and the optic modes cover the range between 10 and 18 meV. Another case that has been studied is UBe$_{13}$, which was studied in both single-crystal[207] and polycrystalline[208] form, with a view to extracting the lattice component of the specific heat more reliably than by fitting the T-dependence of the specific heat itself. Robinson et $al.$[207] have published single-crystal elastic constants, and the only striking thing is that the c_{12} is very close to zero, an observation that may be interesting in the context that some cubic intermediate-valence compounds have large negative c_{12} values.[209,210] Using the time-of-flight method, Renker et $al.$[208] measured the phonon density of states and found the first peak at ~13meV, and this is presumably the energy for the mode in which the uranium ion vibrates within the relatively rigid beryllium cage.

VI. VORTEX LATTICES IN HEAVY FERMION SUPERCONDUCTORS

Another area in which neutron diffraction has contributed is in the study of vortex lines in superconducting UPt_3[211] (see Fig. 7(a) for the portion of the B-T phase diagram in which the measurements took place). The basic idea is due originally to de Gennes and Matricon,[212] who proposed that one should be able to diffract neutrons from the magnetic field distribution associated with the array of vortices, and the idea was verified experimentally in niobium by Cribier et $al.$[213] soon afterwards. Vortex lattices have now been seen and characterised in a wide range of type-II superconductors including the high-T_C superconductors and $NbSe_2$, in addition to UPt_3 and Nb.[214] In the simplest case, the Abrikosov vortex lattice should be simple hexagonal, but if the field is directed along the **a** axis in UPt_3 the flux-line lattice is distorted, due to the moderate anisotropy of the underlying crystal and electronic systems. This is consistent with anisotropic superconductivity, but does not constitute proof of it. Measurement of the rocking curve shows that the vortex lines are straight on a length scale of 3 μm or so. Using Ginzburg-Landau theory and the observed field variations of the obliqueness (deviation from perfect hexagonal) and the form factor H_1, Kleiman et $al.$[211] extract values for both the coherence length ($\xi = 111 \pm 2$ Å) and the London penetration depth ($\lambda = 6000 \pm 75$ Å). λ is very long compared with other superconductors,[215] while ξ is very short, though not to the degree exhibited by the high-T_C materials.

VII. CONCLUSIONS

Perhaps the most striking thing about heavy fermion compounds is the rich variety of ground states and excitation phenomena that can be exhibited (see Tables I and II). It is not even clear that the weak antiferromagnets have much in common with the materials with better developed moments. Neutron diffraction has been crucial to learning what we presently know about these antiferromagnets, and will likely play a key role making sense of the field in the future. Regarding dynamics, it is not clear that the Ce and U compounds are intrinsically similar. Clear crystal-field levels are typically seen in Ce and Yb heavy fermions, while they have never been observed definitively in uranium heavy fermions. In fact they are rarely observed in uranium intermetallics at all. And while one is clearly dealing with a one-f-electron system in Ce (or one hole in Yb), and the physics can be thought of as developing from a lattice of magnetic Kramers doublets, in uranium one is intrinsically dealing with a many-electron ion in the first place (f^3 for U^{3+}, f^2 for U^{4+}), whose valence is not known and whose notional non-interacting crystal-field ground states are also undetermined. On the experimental side, a striking observation is that Lorentzian behaviour has often been observed in Ce compounds, and the temperature dependence often obeys a $\sqrt{T}$ law at high temperatures. The corresponding measurements in uranium compounds have apparently not been made, and it is unclear whether the same $\sqrt{T}$ behaviour occurs in any uranium heavy fermions. Following this line of logic, the clearest inelastic neutron scattering experiments have been those by the Grenoble group on $CeRu_2Si_2$ and $CeCu_6$, in which the quasielastic response is divided up into a $\mathbf{q}$-independent single-site part (that is related to T_K) and a $\mathbf{q}$-dependent inter-site part that is manifest at special incommensurate positions. In both cases, long-range antiferromagnetic order sets in with one of the incommensurate wavevectors beyond a critical concentration of $x_c = 0.1$ for $CeCu_{6-x}Au_x$ ($\mathbf{q}_1 \sim (0,0,1)_{monclinic} = (1,0,0)_{orthorhombic}$) and $x_c = 0.08$ for $Ce_{1-x}La_xRu_2Si_2$ ($\mathbf{q}_1 = (0.3,0,0)$). It might be worthwhile to examine whether these special positions have any relationship to nesting of particular sheets of the Fermi surface, as measured by the de Haas-van Alphen effect.[216] This would entail calculating the Fermi surface and $\chi''(\mathbf{Q},\omega)$ in an electronic structure calculation and comparing to both experiments.

Another area of intense current interest is the nature of the magnetic excitations (and thermodynamics) in the vicinity of the *quantum critical points* at which the Néel temperature becomes zero as a function of pressure or composition. This is the point at which non-Fermi-liquid behaviour is manifest in thermodynamic and transport properties. To date, the phase diagrams for T_N (as a function of x) have only been measured with neutron diffraction in the $CeCu_{6-x}Au_x$ and $Ce_{1-x}La_xRu_2Si_2$, systems described in the previous paragraph. On the other hand, there have been careful studies of spin dynamics in the $UCu_{5-x}Pd_x$ and the $CeCu_{6-x}Au_x$ systems, and E/T scaling has been seen in both. Even more tantalising are the $CePd_2Si_2$ $CeNi_2Ge_2$ and $CeAl_3$ systems that also exhibit power-law behaviours in resistivity as the critical pressure is approached, and which then become superconducting.[70,60] There will certainly be attempts to study both the quantum critical behaviour and the superconductivity using neutrons in the coming years.

The uranium heavy fermion compound that has been most intensively studied is clearly UPt_3, and a whole hierarchy of energy scales is observed (see Table II). Even so, there is no unified picture of the spin dynamics of UPt_3, and it is not currently possible to sketch out a cartoon of its dynamics, accounting for all the observations. The relationship to superconductivity is even murkier, and this is surprising, given that antiferromagnetic spin fluctuations are widely thought to be the driving mechanism for Cooper-pair formation.

Another common thread, which can be seen in Table I, is that the crystallographic directions along which applied fields can induce metamagnetic transitions seem systematically to be along the ordered-moment directions in the antiferromagnetic heavy fermions. Perhaps this is simply saying that these are the soft magnetic directions in the material. If true, one could guess that any long-range antiferromagnetism in $CeRu_2Si_2$ will likely have its moment along the **c**-axis, because the 7.7T field-induced transition occurs for fields along **c**. This rule even applies to $CeCu_6$: while $CeCu_6$ itself does not order, Au doping leads to incommensurate structures with moments along the orthorhombic **c**-axis, while the 2-T metamagnetic transition occurs with field along the orthorhombic **c**-axis (which is the **b**-axis in monoclinic notation).

What is also clear is that there is a large number of loose ends that could be tidied up experimentally. For instance, what are the magnetic structures of the systems (e.g. $CeRu_2Si_2$, YbBiPt, etc.) in which μSR experiments imply very small ordered moments, but which have not yet been seen by neutron diffraction? Or, what is the field-dependence of T_N in UPt_3, which appears not to have been studied. Nor has the temperature dependence of B_C, though that will have to wait for new fast-repetition-rate high-field pulsed magnets,[217,218] expressly designed for use in diffraction experiments, are available.

Finally, it is also clear that more intense neutron sources, in conjunction with better large-solid-angle neutron polarisers, could eventually lead to a breakthrough - the scattering of interest is often quite diffuse and difficult to distinguish from background, phonon scattering and so on. Polarised beams, in conjunction with polarising analysers, can in principle allow unambiguous separation of magnetic from nuclear scattering. Relatively few experiments in this field, and even fewer of the important experiments, have used polarised beams, for the good reason that present polarisers are not very efficient, and a large intensity penalty is paid when using them.

ACKNOWLEDGMENT

I am very grateful to Garth Tietjen for help with the figures and to Joe Thompson and Thom Mason for a number of helpful discussions and to the research libraries at Los Alamos National Laboratory and the Institut Laue Langevin, where much of the effort in writing this chapter was expended. This work was supported in part by the Office of Basic Energy Sciences of the U.S. Department of Energy, under contract W-7405-ENG-36 with the University of California.

REFERENCES

1. G. R. Stewart, Rev. Mod. Phys. **56**, 755, (1984).

2. See work by Rossat-Mignod *et al.* on $CeCu_6$ and $CeRu_2Si_2$, which is described in more detail in Section IV.

3. G. R. Stewart, Z. Fisk, J. O. Willis and J. L. Smith, Phys. Rev. Lett. **52**, 679, (1984).

4. G. Aeppli, E. Bucher, C. Broholm, J. K. Kjems, J. Baumann and J. Hufnagl, Phys. Rev. Lett. **60**, 615, (1988).

5. S. M. Hayden, L. Taillefer, C. Vettier and J. Flouquet, Phys. Rev. B **46**, 8675, (1992).

6. P. H. Frings, J. J. M. Franse, F. R. de Boer and A. Menovsky, J. Magn. Magn. Mater. **31-34**, 240, (1983).

7. E. D. Isaacs, P. Zschack, C. L. Broholm, C. Burns, G. Aeppli, A. P. Ramirez, T. T. M. Palstra, R. W. Erwin, N. Stücheli and E. Bucher, Phys. Rev. Lett. **75**, 1178, (1995).

8. A. I. Goldman, G. Shirane, G. Aeppli, B. Batlogg and E. Bucher, Phys. Rev. B **34**, 6564, (1986).

9. P. Frings, B. Renker and C. Vettier, J. Magn. Magn. Mater. **63&64**, 202, (1987).

10. A. de Visser, A. Menovsky and J. J. M. Franse, Physica B **147**, 81, (1987).

11. A. de Visser, R. J. Keizer, A. A. Menovsky, M. Mihalik, F. S. Tautz, J. J. M. Franse, B. Fåk, N. H. van Dijk, J. Flouquet, J. Bossy and S. Pujol, Physica B **230-232**, 49, (1997).

12. J. J. M. Franse, F. R. de Boer, P. H. Frings and A. de Visser, Physica B**201**, 217, (1994).

13. C. Geibel, S. Thies, D. Kaczorowski, A. Mehner, A. Grauel, B. Seidel, U. Ahlheim, R. Helfrich, K. Petersen, C. D. Bredl and F. Steglich, Z. Phys. B**83**, 305, (1991).

14. J. G. Lussier, A. Schröder, B. D. Gaulin, J. D. Garrett, W. J. L. Buyers, L. Rebelsky and S. M. Shapiro, Physica B **199&200**, 137, (1994).

15. J. G. Lussier, M. Mao, A. Schröder, J. D. Garrett, B. D. Gaulin, S. M. Shapiro and W. J. L. Buyers, Phys. Rev. B **56**, 11749, (1997).

16. C. Geibel, C. Schank, S. Thies, H. Kitazawa, C. D. Bredl, A. Böhm, M. Rau, A. Grauel, R. Caspary, R. Helfrich, U. Ahlheim, G. Weber and F. Steglich, Z. Phys. B**84**, 1, (1991).

17. A. Krimmel, P. Fischer, B. Roessli, H. Maletta, C. Geibel, C. Schank, A. Grauel, A. Loidl and F. Steglich, Z. Phys. B**86**, 161, (1992).

18. K. Sugiyama, T. Inoue, K. Oda, T. Kumada, N. Sato, T. Komatsubara, A. Yamagishi and M. Date, Physica B**201**, 227, (1994).

19. A. Krimmel, A. Loidl, P. Fischer, B. Roessli, A. Donni, H. Kita, N. Sato, Y. Endoh, T. Komatsubara, C. Geibel and F. Steglich, Solid State Commun. **87**, 829, (1993).

20. T. Petersen, T. E. Mason, G. Aeppli, A. P. Ramirez, E. Bucher and R. N. Kleiman, Physica B **199&200**, 151, (1994).

21. H. Kita, A. Dönni, Y. Endoh, K. Kakurai, N. Sato and T. Komatsubara, J. Phys. Soc. Japan **63**, 726, (1994).

22. T. T. M. Palstra, A. A. Menovsky, J. van den Berg, A. J. Dirkmaat, P. H. Kes, G. J. Niewenhuys and J. A. Mydosh, Phys. Rev. Letters **55**, 2727, (1985).

23. C. Broholm, J. K. Kjems, W. J. L. Buyers, P. Matthews, T. T. M. Palstra, A. A. Menovsky and J. A. Mydosh, Phys. Rev. Lett. **58**, 1467, (1987).

24. K. Sugiyama, H. Fuke, K. Kindo, K. Shimohata, A. A. Menovsky, J. A. Mydosh and M. Date, J. Phys. Soc. Japan **59**, 3331, (1990).

25. T. E. Mason, B. D. Gaulin, J. D. Garrett, Z. Tun, W. J. L. Buyers and E. D. Isaacs, Phys. Rev. Lett. **65**, 3189, (1990).

26. T. E. Mason, H. Lin, M. F. Collins, W. J. L. Buyers, A. A. Menovsky and J. A. Mydosh, Physica B **163**, 45, (1990).

27. E. D. Isaacs, D. B. McWhan, R. N. Kleiman, D. J. Bishop, G. E. Ice, P. Zschack, B. D. Gaulin, T. E. Mason, J. D. Garrett and W. J. L. Buyers, Phys. Rev. Lett. **65**, 3185, (1990).

28. M. S. Torikachvili, L. Rebelsky, K. Motoya, S. M. Shapiro, Y. Dalichaouch and M. B. Maple, Phys. Rev. B **45**, 2262, (1992).

29. G. R. Stewart, Z. Fisk, J. L. Smith, J. O. Willis and M. S,. Wire, Phys. Rev. B **30**, 1249, (1984).

30. A. Hiess, M. Bonnet, P. Burlet, E. Ressouche, J. -P. Sanchez, J. Waerenborgh, S. Zwirner, F. Wastin, J. Rebizant, G. H. Lander and J. L. Smith, Phys. Rev. Lett. **77**, 3917, (1996).

31. J. R. Cooper, C. Rizzuto and G. L. Olcese, J. de Physique Colloque **32 C1**, 1136, (1971).

32. C. L. Lin, J. Teter, J. E. Crow, T. Mihalisin, J. Brooks, A. I. Abou-Aly and G. R. Stewart, Phys. Rev. Lett. **54**, 2541, (1985).

33. C. Vettier, P. Morin and J. Flouquet, Phys. Rev. Lett. **56**, 1980, (1986).

34. Z. Fisk, G. R. Stewart, J. O. Willis, H. R. Ott and F. Hulliger, Phys. Rev. B **30**, 6360, (1984).

35. H. R. Ott, H. Rudigier, E. Felder, Z. Fisk and B. Batlogg, Phys. Rev. Lett. **55**, 1595, (1985).

36. A. Murasik, S. Ligenza and A. Zygmunt, Phys. Stat. Solidi (a) **23**, K163, (1974).

37. A. Schenk, P. Birrer, F. N. Gygax, B. Hitti, E. Lippelt, M. Weber, P. Böni, P. Fischer, H. R. Ott and Z. Fisk, Phys. Rev. Lett. **65**, 2454, (1990).

38. A. M. Umarji, J. V. Yakhmi, N. Nambudripad, R. M. Iyer, L. C. Gupta and R. Vijayaraghaven, J. Phys. F **17**, L25, (1987).

39. C. Rossel, K. N. Yang, M. B. Maple, Z. Fisk, E. Zirngiebl and J. D. Thompson, Phys. Rev. B **35**, 1914, (1987).

40. Z. Fisk, P. C. Canfield, W. P. Beyermann, J. D. Thompson, M. F. Hundley, H. R. Ott, E. Felder, M. B. Maple, M. A. Lopez de la Torre, P. Visani and C. Seaman, Phys. Rev. Lett. **67**, 3310, (1991).

41. S. K. Dhar, N. Nambudripad and R. Vijayaraghavan, J. Phys. F **18**, L41, (1988).

42. S. Süllow, B. Ludolph, B. Becker, G. J. Niewenhuys, A. A. Menovsky, J. A. Mydosh, S. A. M. Mentink and T. E. Mason, Phys. Rev. B **52**, 12784, (1995).

43. S. Süllow, B. Ludolph, G. J. Nieuwenhuys, A. A. Menovsky and J. A. Mydosh, Physica B **223&224**, 208, (1996).

44. S. A. M. Mentink, T. E. Mason, S. Süllow, G. J. Nieuwenhuys, J. A. Mydosh and R. J. Donnaberger, Physica B **223&224**, 204, (1996).

45. H. Kitazawa, C. Schank, S. Thies, B. Seidel, C. Geibel and F. Steglich, J. Phys. Soc. Japan **61**, 1461, (1992).

46. S. A. M. Mentink, N. M. Bos, G. J. Niewenhuys, A. A. Menovsky and J. A. Mydosh, Physica B **186-188**, 497, (1993).

47. S. A. M. Mentink, N. M. Bos, G. J. Niewenhuys, A. Drost, E. Frikkee, L. T. Tai, A. A. Menovsky and J. A. Mydosh, Physica B **186-188**, 460, (1993).

48. D. E. Cox, G. Shirane, S. M. Shapiro, G. Aeppli, Z. Fisk, J. L. Smith, J. Kjems and H. R. Ott, Phys. Rev. B **33**, 3614, (1986).

49. C. Broholm, J. K. Kjems, G. Aeppli, Z. Fisk, J. L. Smith, S. M. Shapiro, G. Shirane and H. R. Ott, Phys. Rev. Lett. **58**, 917, (1987).

50. G. Aeppli and C. Broholm, in *Handbook on the Physics and Chemistry of the Rare Earths*, eds. K. A. Gschneidner, L. Eyring, G. H. Lander and G. R. Choppin, Vol. 19, Ch. 131, (Elsevier, Amsterdam, 1994).

51. A. Amato, C. Baines, R. Feyerherm, J. Flouquet, F. N. Gygax, P. Lejay, A. Schenck and U. Zimmermann, Physica B **186-188**, 276, (1993).

52. A. Amato, R. Feyerherm, F. N. Gygax, A. Schenk, J. Flouquet and P. J. Lejay, Phys. Rev. B **50**, 619, (1994).

53. S. Kawarazaki, Y. Kobashi, J. A. Fernandez-Baca, S. Murayama, Y. Onuki and Y. Miyako, Physica B **206&207**, 198, (1995).

54. P. Haen, M. J. Besnus, F. Mallmann, J. -M. Mignot and A. Meyer, Physica B **199&200**, 522, (1994).

55. J.-M. Mignot, F. Mallmann, P. Lejay and P. Haen, Physica B **223&224**, 319, (1996).

56. S. Quezel, P. Burlet, J. L. Jacoud, L. P. Regnault, J. Rossat-Mignod, C. Vettier, P. Lejay and J. Flouquet, J. Magn. Magn. Mater. **76-77**, 403, (1988).

57. R. A. Fisher, C. Marcenat, N. E. Phillips, P. Haen, F. Lapierre, P. Lejay, J. Flouquet and J. Voiron, J. Low Temp. Phys. **84**, 49, (1991).

58. S. Kambe, S. Raymond, H. Suderow, J. McDonough, B. Fåk, L. P. Regnault, R. Calemczuk and J. Flouquet, Physica B **223&224**, 135, (1996).

59. B. H. Grier, J. M. Lawrence, V. Murgai and R. D. Parks, Phys. Rev. B **29**, 2664, (1984).

60. N. D. Mathur, F. M. Grosche, S. R. Julian, I. R. Walker, D. M. Freye, R. K. Haselwimmer and G. G. Lonzarich, Nature **394**, 39, (1998).

61. A. J. Dirkmaat, T. Endstra, E. A. Knetsch, G. H. Niewenhuys and J. A. Mydosh, Phys. Rev. B **41**, 2585, (1990).

62. A. Vernière, S. Raymond, J. X. Boucherle, P. Lejay, B. Fåk, J. Flouquet and J. M. Mignot, J. Magn. Magn. Mater. **153**, 55, (1996).

63. H. R. Ott, H. Rudigier, Z. Fisk and J. L. Smith, Phys. Rev. Lett. **50**, 1595, (1983).

64. F. Steglich, J. Aarts, C. Bredl, W. Lieke, D. Maschede, W. Franz and H. Schäfer, Phys. Rev. Lett. **43**, 1892, (1979).

65. W. Rieger and E. Parthe, Monatsh Chemie **100**, 444, (1969).

66. A. Yatskar, W. P. Beyermann, R. Movshovich and P. C. Canfield, Phys. Rev. Lett. **77**, 3637, (1996).

67. K. Andres, J. E. Graebner and H. R. Ott, Phys. Rev. Lett. **35**, 1779, (1975).

68. G. Knopp, A. Loidl, R. Caspary, U. Gottwick, C. D. Bredl, H. Spille, F. Steglich and A. P. Murani, J. Magn. Magn. Mater. **74**, 341, (1988).

69. T. Fukuhara, K. Maezawa, H. Ohnuki, J. Sakurai, H. Sato, H. Azuma, K. Sugiyama, Y. Onuki and K. Kindo, J. Phys. Soc. Japan **65**, 1559, (1996).

70. S. J. S. Lister, F. M. Grosche, F. V. Carter, R. K. W. Haselwimmer, S. S. Saxena, N. D. Mathur, S. R. Julian and G. G. Lonzarich, Z. Phys. B **103**, 263, (1997).

71. L. Havela, V. Sechovsky, P. Svoboda, M. Divis, H. Nakotte, K. Prokes, F. R. de Boer, A. Purwanto, R. A. Robinson, A. Seret, J. M. Winand, J. Rebizant, J. C. Spirlet, M. Richter and H. Eschrig, J. Appl. Phys. **76**, 6214, (1994).

72. H. Nakotte, K. Prokes, E. Brück, N. Tang, F. R. de Boer, P. Svoboda, V. Sechovsky, L. Havela, J. M. Winand, A. Seret, J. Rebizant and J. C. Spirlet, Physica B **201**, 247, (1994).

73. W. H. Lee and R. N. Shelton, Phys. Rev. B **35**, 5369, (1987).

74. C. Geibel, U. Klinger, B. Buschinger, M. Weiden, G. Olesch, F. Thomas and F. Steglich, Physica B **223&224**, 370, (1996).

75. V. A. Romaka, Y. N. Grin', Y. P. Yarmolyuk, O. S. Zarechnyuk and R. V. Skolozdra, Phys. Met. Metall. **54**, no. 4, 58, (1982); Fiz. Metal. Metalloved. **54**, no. 4, 691, (1982).

76. G. R. Stewart, Z. Fisk and M. S. Wire, Phys. Rev. B **30**, 482, (1984).

77. Y. Onuki, Y. Shimizu and T. Komatsubara, J. Phys. Soc. Japan **53**, 1210, (1984).

78. A. Amato, D. Jaccard, J. F. Lapierre, J. L. Tholence, R. A. Fisher and S. E. Lacy, J. Low-Temp. Phys. **68**, 371, (1987).

79. Y. Onuki, Physica B **206&207**, 862, (1995).

80. D. T. Cromer, A. C. Larson and R. B. Roof, Acta Cryst. **13**, 913, (1960).

81. M. L. Vrtis, J. D. Jorgensen and D. G. Hinks, Physica **136B**, 489, (1986).

82. H. Asano, M. Umino, Y. Onuki, T. Komatsubara, F. Izumi and N. Watanabe, J. Phys. Soc Japan **55**, 454, (1986).

83. E. Gratz, E. Bauer, H. Nowotny, H. Mueller, S. Zemirli and B. Barbara, J. Magn. Magn. Mater. **63&64**, 311, (1987).

84. O. Stockert, H. von Löhneysen, A. Schröder, M. Loewenhaupt, N. Pyka, P. L. Gammel and U. Yaron, Physica B **230-232**, 247, (1997).

85. A. Rosch, A. Schröder, O. Stockert and H. von Löhneysen, Phys. Rev. Lett. **79**, 159, (1997).

86. A. Schröder, J. W. Lynn, R. W. Erwin, M. Loewenhaupt and H. von Löhneysen, Physica B **199&200**, 47, (1994).

87. H. Okumura, K. Kakurai, Y. Yoshida, Y. Onuki and Y. Endoh, J. Magn. Magn. Mater. **177-181**, 405, (1998).

88. O. Stockert, H. von Löhneysen, A. Rosch, N. Pyka and M. Loewenhaupt, Phys. Rev. Lett. **80**, 5627, (1998).

89. H. von Löhneysen, C. Paschke, G. Portisch, M. Ruck, H. G. Schlager, M. Sieck and C. Speck, Physica B **199&200**, 85, (1994).

90. G. Aeppli, E. Bucher and G. Shirane, Phys. Rev. B **32**, 7579, (1985).

91. A. I. Goldman, G. Shirane, G. Aeppli, E. Bucher and J. Hufnagl, Phys. Rev B **36**, 8523, (1987).

92. G. Aeppli, A. Goldman, G. Shirane, E. Bucher and M.-Ch. Lux-Steiner, Phys. Rev. Lett. **58**, 808, (1987).

93. N. R. Bernhoeft and G. G. Lonzarich, J. Phys.: Condens. Matter **7**, 7325, (1995).

94. U. Walter, C. -K. Loong, M. Loewenhaupt and W. Schlabitz, Phys. Rev. B **33**, 7875, (1986).

95. W. J. L. Buyers, Z. Tun, T. Petersen, T. E. Mason, J. -G. Lussier, B. D. Gaulin and A. A. Menovsky, Physica B **199&200**, 95, (1994).

96. W. J. L. Buyers, Physica B **223&224**, 9, (1996).

97. C. Broholm, H. Lin, P. T. Matthews, T. E. Mason, W. J. L. Buyers, M. F. Collins, A. A. Menovsky, J. A. Mydosh and J. K. Kjems, Phys. Rev. B **43**, 12809, (1991).

98. A. I. Goldman, S. M. Shapiro, G. Shirane, J. L. Smith and Z. Fisk, Phys. Rev. B **33**, 1627, (1986).

99. G. H. Lander, S. M. Shapiro, C. Vettier and A. J. Dianoux, Phys. Rev. B **46**, 5387, (1992).

100. U. Walter, M. Loewenhaupt, E. Holland-Moritz and W. Schlabitz, Phys. Rev. B **36**, 1981, (1987).

101. M. C. Aronson, R. Osborn, R. A. Robinson, J. W. Lynn, R. Chau, C. L. Seaman and M. B. Maple, Phys. Rev. Lett. **75**, 725, (1995).

102. M. C. Aronson, M. B. Maple, R. Chau, A. Georges, A. M. Tsvelik and R. Osborn, J. Phys.: Condens. Matter **8**, 9815, (1996).

103. R. Osborn, M. C. Aronson, B. D. Rainford, M. B. Maple, R. Chau and K. H. Anderson, Physica B **241**, 859, (1997).

104. N. Sato, N. Aso, G. H. Lander, B. Roessli, T. Komatsubara and Y. Endoh, J. Phys. Soc. Japan **66**, 2981, (1997).

105. N. Sato, N. Aso, G. H. Lander, B. Roessli, T. Komatsubara and Y. Endoh, J. Phys. Soc. Japan **66**, 1884, (1997).

106. N. Metoki, Y. Haga, Y. Koike, N. Aso and Y. Onuki, J. Phys. Soc. Japan **66**, 2560, (1997).

107. N. Metoki, Y. Haga, Y. Koike and Y Onuki, Phys. Rev. Lett. **80**, 5417, (1998).

108. A. P. Murani, K. Knorr and K. H. J. Buschow, in *Crystal Field Effects in Metals and Alloys*, p. 268, (ed. A. Furrer) Plenum New York, (1976).

109. A. P. Murani, K. Knorr, K. H. J. Buschow, A. Benoit and J. Flouquet, Sol. Stat. Commun. **36**, 523, (1980).

110. E. A. Goremychkin, I Natkaniec and E. Mühle, Sol. State Commun. **64**, 553, (1987).

111. B. Renker, E. Gering, F. Gompf, H. Schmidt and H. Rietschel, J. Magn. Magn. Mater. **63-64**, 31, (1987).

112. G. Balakrishnan, D. McK. Paul and N. R. Bernhoeft, Physica B **156&157**, 815, (1989).

113. S. A. M. Mentink, G. J. Niewenhuys, A. A. Menovsky, J. A. Mydosh, A. Drost, E. Frikkee, Y. Bando, T. Takabatake, P. Böni, P. Fischer, A. Furrer, A. Amato and A. Schenck, Physica B **199&200**, 143, (1994).

114. G. Aeppli, H. Yoshizawa, Y. Endoh, E. Bucher, J. Hufnagl, Y. Onuki and T. Komatsubara, Phys. Rev. Lett. **57**, 122, (1986).

115. U. Walter, D. Wohlleben and Z. Fisk, Z. Phys. B **62**, 325, (1986).

116. J. Rossat-Mignod, L. P. Regnault, J. L. Jacoud, C. Vettier, P. Lejay, J. Flouquet, E. Walker, D. Jaccard and A. Amato, J. Magn. Magn. Mater. **76&77**, 376 (1988).

117. E. A. Goremychkin and R. Osborn, Phys. Rev. B **47**, 14580 (1993).

118. A. Schröder, G. Aeppli, E. Bucher, R. Ramazashvili and P. Coleman, Phys. Rev. Lett. **80**, 5623, (1998).

119. L. P. Regnault, W. A. Erkelens, J. Rossat-Mignod, P. Lejay and J. Flouquet, Phys. Rev. B **38**, 4481, (1988).

120. B. H. Grier, J. M. Lawrence, S. Horn and J. D. Thompson, J. Phys. C **21**, 1099, (1988).

121. A. Severing, E. Holland-Moritz, B. D. Rainford, S. R. Culverhouse and B. Frick, Phys. Rev. B **39**, 2557, (1989).

122. A. Severing, E. Holland-Moritz and B. Frick, Phys. Rev. B **39**, 4164, (1989).

123. S. Horn, E. Holland-Moritz, M. Loewenhaupt, F. Steglich, H. Scheuer, A. Benoit and J. Flouquet, Phys. Rev. B **23**, 3171, (1981).

124. S. Horn, F. Steglich, M. Loewenhaupt and E. Holland-Moritz, Physica **107B**, 103, (1981).

125. Y. J. Uemura, C. F. Majkrzak, G. Shirane, C. Stassis, G. Aeppli, B. Batlogg and J. P. Remeika, Phys. Rev. B **33**, 6508, (1986).

126. E. Holland-Moritz, W. Weber, A. Severing, E. Zirngiebl, H. Spille, W. Baus, S. Horn, A. P. Murani and J. L. Ragazzoni, Phys. Rev. B **39**, 6409, (1989).

127. F. Hippert, B. Hennion, J.-M. Mignot and P. Lejay, J. Magn. Magn. Mater. **108**, 177, (1998).

128. N. H. van Dijk, B. Fåk, T. Charvolin, P. Lejay, J. M. Mignot and B. Hennion, Physica B **241-243**, 808, (1998).

129. A. Krimmel, A. Severing, A. Murani, A. Grauel and S. Horn, Physica B **180&181**, 191, (1992).

130. R. A. Robinson, M. Kohgi, T. Osakabe, F. Trouw, J. W. Lynn, P. C. Canfield, J. D. Thompson, Z. Fisk and W. P. Beyermann, Phys. Rev. Lett. **75**, 1194, (1995).

131. W. G. Marshall, Ph. D. thesis, Birkbeck College, University of London, (1994).

132. A. Severing, A. P. Murani, J. D. Thompson and C.-K. Loong, Phys. Rev. B **41** 1739, (1990).

133. R. M. Galera, J. Pierre, E. Siaud and A. P. Murani, J. Less-Common Met. **97**, 151, (1984).

134. W. Marshall and S. W. Lovesey, "Theory of Thermal Neutron Scattering", (Oxford 1971); G. L. Squires, "Introduction to the Theory of Thermal Neutron Scattering", (Cambridge 1978; reprinted by Dover, Mineola, 1997); S. W. Lovesey, "Theory of Thermal Neutron Scattering from Condensed Matter", (Oxford 1984).

135. For methods based at steady-state research reactors, see for instance, *Neutron Diffraction* 3rd Edition, G. E. Bacon (Oxford 1975).

136. For methods based at accelerator-driven pulsed spallation sources, see for instance, *Pulsed Neutron Scattering*, C. G. Windsor, (Taylor and Francis, 1981); *Neutron Scattering*, Los Alamos Science **19**, ed. N. G. Cooper, (LANL 1990).

137. Physica B **241-243** (1998); Physica B **213&214** (1995); Physica B **180&181** (1992); Physica B **156&157** (1989); Physica **136B** (1986); Physica **120B** (1983).

138. E. Holland-Moritz and G. H. Lander, "Handbook on the Physics and Chemistry of the Rare Earths", eds. K. A. Gschneidner, L. Eyring, G. H. Lander and G. R. Choppin, Vol. 19, Ch. 130, (Elsevier, Amsterdam, 1994).

139. W. C. Koehler, J. Singer and A. S. Coffinberry, Acta Cryst. **5**, 394, (1952).

140. A. I. Goldman, S. M. Shapiro, D. E. Cox, J. L. Smith and Z. Fisk, Phys. Rev. B **32**, 6042, (1985).

141. H. M. Rietveld, J. Appl. Cryst. **2**, 65, (1969).

142. R. A. Robinson, A. Purwanto, M. Kohgi, P. C. Canfield, T. Kamiyama, T. Ishigaki, J. W. Lynn, R. Erwin, E. Peterson and R. Movshovich, Phys. Rev. B **50**, 9595, (1994).

143. A. C. Larson and R. B. Von Dreele, Los Alamos National Laboratory Report, LAUR-86-748, (1986).

144. R. Chau, M. B. Maple and R. A. Robinson, Phys. Rev. B **58**, 139, (1998).

145. C. Stassis, J. Arthur, C. F. Majkrzak, J. D. Axe, B. Batlogg, J. Remeika, Z. Fisk, J. L. Smith and A. S. Edlestein, Phys. Rev. B **34**, 4382, (1986).

146. C. Stassis, J. D. Axe, C. F. Majkrzak, B. Batlogg and J. Remeika, J. Appl. Phys. **57**, 3087, (1985).

278

147. K. U. Neumann, P. J. Brown, H. Capellmann, K. R. A. Ziebeck, J. L. Smith, Z. Fisk, J. J. M. Franse and A. A. Menovsky, Physica B **171**, 273, (1991).

148. L. Paolasini, J. A. Paixao, G. H. Lander, A. Delapalme, N. Sato and T. Komatsubara, J. Phys. Condens. Matter **5**, 8905, (1993).

149. M. B. Walker, W. J. L. Buyers, Z. Tun, W. Que, A. A. Menovsky and J. D. Garrett, Phys. Rev. Lett. **71**, 2630, (1993).

150. S. M. Johnson, J. A. C. Bland, P. J. Brown, A. Benoit, H. Capellmann, J. Flouquet, H. Spille, F. Steglich and K. R. A. Ziebeck, Z. Phys. B **59**, 401, (1985).

151. *International Tables for Crystallography*, Vol. A, ed. Th. Hahn, (Reidel Dordrecht 1983).

152. See for instance, M. F. Collins, *Magnetic Critical Scattering*, Oxford, p.29, (1989).

153. A. Drost, Ph.D. thesis, Leiden, (1995).

154. A. Loidl, K. Knorr, G. Knopp, A. Krimmel, B. Caspary, A. Bohm, G. Sparn, C. Geibel, F. Steglich and A. P. Murani, Phys. Rev. B **46**, 9341, (1992).

155. A. P. Ramirez, P. Coleman, P. Chandra, E. Brück, A. A. Menovsky, Z. Fisk and E. Bucher, Phys. Rev. **68**, 2680, (1992).

156. L. P. Gor'kov, Europhys. Lett. **16**, 301, (1991).

157. L. P. Gor'kov and R. Sokol, Phys. Rev. Lett. **69**, 2586, (1992).

158. V. Barzykin and L. P. Gor'kov, Phys. Rev. Lett. **70**, 2479, (1993).

159. T. E. Mason, W. J. L. Buyers., T. Petersen, A. A. Menovsky and J. D. Garrett, J. Phys.: Condens. Matter **7**, 5089, (1995).

160. B. Lussier, L. Taillefer, W. J. L. Buyers, T. E. Mason and T. Petersen, Phys. Rev. B **54**, 6873, (1996).

161. G. Aeppli, D. Bishop, C. Broholm, E. Bucher, K. Siemensmayer, M. Steiner and N. Stüsser, Phys. Rev. Lett. **63**, 676, (1989).

162. see D. Gibbs, D. Harsman, E. D. Isaacs, D. B. McWhan, D. Mills and C. Vettier, Phys. Rev. Lett. **61**, 1241, (1988) and references therein.

163. A. Amato, C. Geibel, F. N. Gygax, R. H. Heffner, E. Knetsch, D. E. MacLaughlin, C. Schank, F. Steglich and M. Weber, Z. Phys. B **86**, 159, (1992).

164. A. Hiess, M. Bonnet, P. Burlet, E. Ressouche, J.-P. Sanchez, F. Boudarot, J. C. Waerenborgh, S. Zwirner, F. Wastin, J. Rebizant, G. H. Lander, E. Suard and J. L. Smith, Physica B**234-236**, 893, (1997).

165. R. Feyerherm, A. Amato, F. N. Gygax, A. Schenck, C. Geibel, F. Steglich, N. Sato and T. Komatsubara, Phys. Rev. Lett. **73**, 1849, (1994).

166. J. S. Smart, "Effective Field Theories of Magnetism", p.76, (Saunders, Philadelphia, 1966).

167. R. A. Robinson, G. Aeppli, D. Keen, W. Kagounya and D. Mandrus, unpublished data.

168. ISIS User Guide, Rutherford Appleton Laboratory Report RAL-88-030, (1988).

169. H. Nakamura, Y. Kitaoka, K. Asayama, Y. Onuki and M. Shiga, J. Phys.: Condens. Matter **6**, 10567, (1994).

170. L. DeGiorgi, H. R. Ott, M. Dressel, G. Grüner, C. Geibel, F. Steglich and Z. Fisk, Physica B **206&207**, 441, (1995).

171. see M. B. Maple, C. L. Seaman, D. A. Gajewski, Y. Daliaouch, V. B. Barbetta, M. C. de Andrade, H. A. Mook, H. G. Lukefahr, O. O. Bernal and D. E. MacLaughlin, J. Low-Temp. Phys. **95**, 225, (1994), and references therein.

172. O. O. Bernal, D. E. MacLaughlin, H. G. Lukefahr and B. Andraka, Phys. Rev. Lett. **75**, 2023, (1995).

173. E. Miranda, V. Dobrosavljevic and G. Kotliar, Physica B **230**, 569, (1996).

174. E. Miranda, V. Dobrosavljevic and G. Kotliar, J. Phys.: Condens. Matter **8**, 9871, (1996).

175. J. O. Willis, Z. Fisk, G. R. Stewart and H. R. Ott, J. Magn. Magn. Mater. **54-57**, 395, (1986).

176. A. Amato, P. C. Canfield, R. Feyerherm, Z. Fisk, F. N. Gygax, R. H. Heffner, D. E. MacLaughlin, H. R. Ott, A. Schenck and J. D. Thompson, Phys. Rev. **46**, 3151, (1992).

177. R. Movshovich, A. Lacerda, P. C. Canfield, J. D. Thompson and Z. Fisk, Phys. Rev. Lett. **73**, 492, (1994).

178. G. Aeppli, C. Broholm, T. Mason, R. A. Robinson and P. C. Canfield, unpublished data.

179. R. A. Robinson, G. Aeppli, D. Keen, V. Nield and P. C. Canfield, unpublished data.

180. L. Havela, V. Sechovsky, P. Svoboda, H. Nakotte, K. Prokes, F. R. de Boer, A. V. Andreev, K. Sugiyama, T. Kuroda and M. Date, J. Magn. Magn. Mater. **140-144**, 1367, (1995).

181. A. Purwanto, R. A. Robinson, L. Havela, V. Sechovsky, P. Svoboda, H. Nakotte, K. Prokes, F. R. de Boer, A. Seret, J. M. Winand, J. Rebizant and J. C. Spirlet, Phys. Rev. B **50**, 6792, (1994).

182. F. Bourée, B. Chevalier, L. Fournès, F. Mirambet, T. Roisnel, V. H. Tran and Z. Zolnierek, J. Magn. Magn. Mater. **138**, 307, (1994).

183. H. Nakotte, A. Purwanto, R. A. Robinson, K. Prokes, J. C. P. Klaasse, P. F. de Chatel, F. R. de Boer, L. Havela, V. Sechovsky, L. C. J. Pereira, A. Seret, J. Rebizant, J. C. Spirlet and F. Trouw, Phys. Rev. B **53**, 3263, (1996).

184. R. A. Robinson, A. C. Lawson, V. Sechovsky, L. Havela, Y. Kergadallan, H. Nakotte and F. R. de Boer, J. Alloys Compounds **213/214**, 528, (1994).

185. L. E. Delong, J. G. Huber and K. S. Bedell, J. Magn. Magn. Mater. **99**, 171, (1991).

186. B. R. Cooper, R. Siemann, D. Yang, P. Thayamballi and A. Banerjea, "Handbook of the Physics and Chemistry of the Actinides", eds. A. J. Freeman and G. H. Lander, Vol. 2, North-Holland, Amsterdam, Chap. 6, (1985).

187. L. M. Sandratskii and J. Kübler, Phys. Rev. Lett. **75**, 946, (1995).

188. L. M. Sandratskii and J. Kübler, Physica B **217**, 167, (1996).

189. see H. G. Schlager, A. Schröder, M. Welsch and H. von Löhneysen, J. Low-Temp. Phys. **90**, 181, (1993) and references therein.

190. T. Chattopadhyay, H. von Löhneysen, T. Trappmann and M. Loewenhaupt, Z. Phys. B **80**, 159, (1990).

191 W. Götze and P. Schlottmann, J. Low-Temp. Phys. **16**, 87, (1974).

192. J. Korringa, Physica **16**, 601, (1950).

193. M. Loewenhaupt and W. Just, Phys. Lett. **53A**, 305, (1975).

194. D. L. Cox, N. E. Bickers and J. W. Wilkins, J. Magn. Magn. Mater. **54-57**, 333, (1986); N. E. Bickers, D. L. Cox and J. W. Wilkins, Phys. Rev. B **36**, 2036, (1987).

195. K. W. Becker, P. Fulde and J. Keller, Z. Phys. B**28**, 9, (1977).

196. Y. Onuki and T. Komatsubara, J. Magn. Magn. Mater. **63&64**, 281, (1987).

197. H. von Löhneysen, T. Pietrus, G. Portisch, H. G. Schlager, A. Schröder, M. Sieck and T. Trappmann, Phys. Rev. Lett. **72**, 3262, (1994).

198. See Fig. 63 in E. Holland-Moritz and G. H. Lander, "Handbook on the Physics and Chemistry of the Rare Earths", eds. K. A. Gschneidner, L. Eyring, G. H. Lander and G. R. Choppin, Vol. 19, Ch. 130, (Elsevier, Amsterdam, 1994).

199. F. M. Grosche, S. R. Julian, N. D. Mathur, F. V. Carter and G. G. Lonzarich, Physica B **237 - 238**, 197, (1997).

200. W. J. L. Buyers and T. M. Holden, "Handbook on the Physics and Chemistry of the Actinides", eds. A. J. Freeman and G. H. Lander, Vol. 2, p. 239, (North-Holland, Amsterdam, 1985).

201. R. Scherm, K. Guckelsberger, B. Fåk, K. Sköld, A. J. Dianoux, H. Godfrin and W. G. Stirling, Phys. Rev. Lett. **59**, 217, (1987).

202. N. R. Bernhoeft, S. M. Hayden, G. G. Lonzarich, D. Paul and E. J. Lindley, Phys. Rev. Lett. **62**, 657, (1989); Y. Ishikawa, Y. Noda, C. Fincher and G. Shirane, Phys. Rev B **25**, 25, (1982); Y. Ishikawa, Y. Noda, Y. J. Uemura, C. F. Majkrzak and G. Shirane, Phys. Rev B **31**, 5884, (1985); N. R. Bernhoeft, G. G. Lonzarich, P. W. Mitchell and D. McK Paul, Phys. Rev B **28**, 422, (1983); N. R. Bernhoeft, S. A. Law, G. G. Lonzarich and D. McK. Paul, Physica Scripta **38**, 191, (1988).

203. M. J. Bull, K. A. McEwen, R. Osborn and R. S. Eccleston, Physica B **223&224**, 175, (1996).

204. T. E. Mason and G. Aeppli, Matematisk-fysiske Meddelelser **45**, 231, (1997), (Munksgaard, Copenhagen).

205. C. Broholm, R. A. Robinson, G. Aeppli and T. Mason, unpublished data.

206. B. Renker, F. Gompf, E. Gering, P. Frings, H. Rietschel, R. Felten, F. Steglich and G. Weber, Physica **148B**, 41, (1987).

207. R. A. Robinson, J. D. Axe, A. I. Goldman, Z. Fisk, J. L. Smith and H. R. Ott, Phys. Rev. B **33**, 6488, (1986).

208. B. Renker, F. Gompf, W. Reichardt, H. Rietschel, J. -B. Suck and J. Beuers, Phys. Rev. B **32**, 1859, (1985).

209. H. A. Mook, R. M. Nicklow, T. Penney, F. Holtzberg and M. W. Schafer, Phys. Rev. B **18**, 2925, (1978).

210. H. Boppart, A. Treindl, P. Wachter and S. Roth, Solid State Commun. **23**, 483, (1980).

211. R. N. Kleiman, C. Broholm, G. Aeppli, E. Bucher, N. Stücheli, D. J. Bishop, K. N. Clausen, K. Mortensen, J. S. Pederson and B. Howard, Phys. Rev. Lett. **69**, 3120, (1992).

212. P. G. de Gennes and J. Matricon, Rev. Mod. Phys. **36**, 45, (1964).

213. D. Cribier, B. Jacrot, L. Madhav Rao and B. Farnoux, Phys. Lett. **9**, 106, (1964).

214. for instance, see the set of 5 papers in J. Appl. Phys. **76**, starting with J. Appl. Phys. **76**, 6772, (1994), and references therein.

215. See for instance, C. Kittel, "Introduction to Solid State Physics", 5th edition, p.376, (Wiley, New York 1976).

216. G. G. Lonzarich, J. Magn. Magn. Mater. **76&77**, 1, (1988).

217. M. Motokawa, H. Nojiri, J. Ishihara and K. Ohnishi, Physica B **155**, 39, (1989); H. Nojiri, M. Uchi, S. Watamura, M. Motokawa, H. Kawai, Y. Endoh and T. Shigeoka, J. Phys. Soc. Japan **60**, 2380, (1991).

218. R. A. Robinson, Y. M. Eyssa, H. J. Schneider-Muntau and H. J. Boenig, in "Proceedings of Conference on Physical Phenomena at High Magnetic Fields II", Tallahassee FL, May 6-9, 1995, eds. Z. Fisk, L. Gor'kov, D. Meltzer and R. Schrieffer, World Scientific, Singapore, p. 735, (1996).
219. B. Bogenberger, H. von Löhneysen, T. Trappmann and L. Taillefer, Physica B **186-188**, 248, (1993).
220. L. Taillefer, J. Flouquet and G. G. Lonzarich, Physica B **169**, 257, (1991).
221. T. Trappmann, H. von Löhneysen and L. Taillefer, Phys. Rev. B **43**, 13714, (1991).

5

f-ELECTRONIC AND MAGNETIC BEHAVIOR BETWEEN ATOMICLIKE AND FULLY ITINERANT: RANDOM LOCALIZED MAGNETISM AND HEAVY FERMIONS

Bernard R. Cooper

Department of Physics
West Virginia University
Morgantown, WV 26506-6315, U.S.A.

I. OVERVIEW AND SUMMARY

Heavy fermion systems[1-4] are characterized by their extremely enhanced low-temperature electronic specific heat and Pauli susceptibility. If categorized by the nomenclature of Landau Fermi liquid theory, this behavior can be summarized as corresponding to an electronic mass enhanced by factors of many hundred, or indeed a thousand, compared to the free electron mass. Such systems can be superconducting, magnetically ordered, sometimes both, or "vegetables" showing neither superconductivity nor magnetic ordering. If there is magnetic ordering, it typically involves a very small ordered moment. The overall heavy fermion behavior is characterized by the Wilson-Sommerfeld ratio of the low-temperature susceptibility to low-temperature specific heat. The magnetically ordered systems typically have distinctly higher values of the Wilson ratio.

The extremely enhanced effective masses characteristic of heavy fermion behavior cannot be obtained from ab initio Local Density Approximation (LDA), i.e. one-electron-dynamics band theory, calculations for these materials. Such LDA calculations simulate the consequences of the true many-electron dynamics through an exchange-correlation potential based on the consequences of exchange and correlation effects for an electron gas. Given the inability adequately to capture the many-electron dynamics through LDA calculations incorporating such an exchange-correlation potential, heavy fermion systems are classified as strongly-correlated-electron (SCE) systems.

The nature of heavy fermion phenomenology, especially the severe reduction or total absence of magnetic ordering, is indicative of a magnetic-nonmagnetic (singlet) "competition" engendered by the electron correlation effects. At an early stage in the theoretical development, this led to attempts to understand heavy fermion behavior by utilizing the existing theory for the Kondo Effect,[5,6,6a] especially that for the isolated impurity for which the theory[6] is highly developed. This gave rise to the Kondo Impurity (KI) model and the consequent Kondo resonance theory[7] for heavy fermion behavior. This theory provides a nondispersive heavy fermion state with a characteristic temperature dependence for the spectral density behavior to be observed experimentally by spectroscopy techniques such as photoemission. For several years there has been an experimental controversy[8-10] as to whether or not there are at least some heavy fermion systems where this characteristic Kondo resonance behavior is not observed, and where indeed there is direct experimental observation of extremely-narrow-band dispersed behavior.[11] From the theoretical side, one can hope to settle this question only through extending the ab initio materially-predictive capability beyond that of LDA to include an adequate treatment of correlation effects, i.e. going sufficiently beyond one-electron dynamics. The thrust of the present article is to present the present state and future prospects of fulfilling this objective.

Heavy fermion materials have $4f$ or $5f$ transition-shell electronic behavior in the regime of chemical environments such that the transition-shell electrons are not significantly influenced by direct overlap with transition-shell electrons from other sites. In this situation, the transition-shell f electron role in the overall solid-state electron dynamics is determined by the hybridization of each such f electron with band electrons of non-f-transition-shell atomic origin, as constrained by coulomb exchange interaction with those band electrons and as diminished by coulomb repulsion with the other on-site transition-shell f electrons. As it happens, the constraint imposed by the band-f exchange (electron permutation symmetry) provides the greatest difficulty, but also plays a major role in controlling the physics of interest, in this situation. As described below, there are two many-electron terms in the ab-initio-based model hamiltonian used to treat the physics of interest. In fact, these are both two-electron terms, viz, the on-site f-f coulomb repulsion (correlation) and the band-f exchange. This situation, together with the nature of the band-f hybridization interaction leads us to adopt an ansatz of treating correlation effects at the level of two-electron dynamics.

This ansatz provides a situation between the two extremes of (1) LDA theory, where there are not true correlation effects, but only an attempt to mimic them via a one-electron potential, and of (2) the Kondo impurity problem, where one can solve exactly to include all correlation effects. As discussed in detail below, this ansatz leads us to a relatively simplified "energy inverted-helium-atom" picture that is applicable to a class of uranium compounds. This considerably simplifies the treatment of the physics, and also, for sufficiently weak hybridization allows us to perform ab-initio-based absolute predictions of Curie temperatures and ferromagnetic ordered moments in a class of uranium compounds.[12-16] Interestingly, this treatment of the physics also leads to the recognition that there may be fundamental differences between cerium-based and uranium-based heavy fermion systems. This difference in behavior seems to have at least tentative support from the nature of similarities and difference between the thermally-driven structural transitions in elemental plutonium and elemental cerium.[16] For that reason among others, we include a discussion of that behavior below.

The conclusion we are drawn to, for at least a certain class of uranium compounds, is that for sufficiently weak hybridization there are two types of $5f$ electrons occupying randomly distributed sites in a solid-solution-like situation providing high entropy. At sufficiently high temperature, the free energy advantage gained through this high entropy gives randomly located localized $5f$ sites, and this localization temperature can be substantially lowered by the occurrence of strong magnetic ordering. As hybridization increases through alloying and/or pressure, there is a total f-electron delocalization phase transition (which, in at least the magnetically ordered systems, is first order). This phase transition is to a correlated narrow band state and occurs when the $5f$ bonding energy more than compensates for the loss of free-energy lowering from the combined solid-solution-like entropy and magnetic-ordering energy in the randomly localized phase.[16] A very interesting materials system, with experimental behavior[17] consistent with this picture, is the cubic alloy $U(In_{1-x}Sn_x)$ where the indium-rich end shows strong antiferro-magnetism; there is an abrupt transition to non-magnetically-ordered and enhanced mass behavior as the tin concentration increases; and the specific heat enhancement within the tin-rich regime reaches values characterized as heavy fermion-like.

The experimental evidence and theoretical underpinning for dispersed very narrow band behavior in at least some heavy fermion systems does not necessarily preclude there being behavior for other systems at the very narrow correlated band limit that is observationally indistinguishable from that provided by the Kondo Impurity-Kondo Resonance model.[6,7] This may be especially true of the more atomiclike transition-shell f electron behavior in the cerium materials. In treating the response functions of many-electron systems, broadening and renormalization of the low lying excitation (quasi particle) levels go hand in hand,[18] for an example of the application of such theory in comparison with experiment see Reference 19. Thus if the "bare" dispersion is narrow enough, the broadening "covers" the dispersion; and the thermodynamic behavior would be that of a broadened impurity level, presumably indistinguishable from a broadened level obtained by resonant scattering broadening of a localized state. The answer to whether or not the same overall picture and theory yields both observationally narrow-band dispersed behavior and Kondo-resonance-like behavior depending on the specific material parameters can be answered once fully ab initio calculational results are available. Hopefully this answer will be provided by calculations performed using the technique described in Section IV B.

Thus, as discussed above, for strongly-correlated-electron (SCE) systems, one has to treat at least two-electron dynamics[20] and to include both p/d band-f hybridization and p/d band-f coulomb exchange. However, the two great simplifications possible when considering *magnetic ordering* for *uranium* SCE systems, discussed below, have enabled us to make absolute material-specific predictions of alloying or high-pressure effects using Local Density Approximation (LDA) input into many-electron dynamics with only one *non-ab-initio* number.[12-16] Experimentally, the alloying effects that must be predicted to validate the theory can be dramatic, e.g. in $U_xLa_{1-x}S$ strong ferromagnetic ordering abruptly disappears at about 55% uranium.[21-24] The theory is quite successful in its detailed absolute predictions, and this success and the general understanding gained has important implications for the overall understanding of electronic behavior in SCE systems including heavy fermion systems. The key conclusion is that the hybridization process, *as kinematically constrained by exchange symmetry*, leads to a chemical-environment-dependent sharp threshold for a drastic restructuring of the transition-shell-electron behavior in the equilibrium lattice state in SCE systems, i.e., *a phase transition*, with dramatic observable consequences. For light actinide systems, this restructuring is from a phase with coherent highly-correlated very narrow band behavior to a phase with a random distribution of transition-shell-atom sites with either localized stable, $5f$ valence or fluctuating $5f$ valence (Random-Localized-Fluctuating-Site, RLFS, solid-solution-like phase). This phase transition occurs as the solid-state chemical environment is made less hybridizing by alloying or as temperature increases. It is driven[16] by the magnetic-ordering enhanced, or indeed for fcc Pu self-induced, Anderson localization (i.e., local-ization caused by strong scattering from a random distribution of imperfections) of the transition-shell electrons (f electrons for light actinide and cerium-based SCE materials, d electrons for transition-metal-oxide-based SCE materials). To minimize the free energy, i.e., maximize the entropy, once the localization is initiated, it is favorable to localize the transition-shell electrons at half of the lattice sites occupied by the transition-shell atoms (or more than half if magnetic ordering or another energy lowering mechanism ensues). The one *non-ab-initio* number used in the predictive calculation is the baseline (i.e., for unalloyed material at ambient pressure) percentage division between transition-shell atom (e.g., uranium) sites having localized (magnetic) $5f$ electrons and those providing delocalized (nonmagnetic) $5f$ electrons.[12] Recent muon spin rotation/relaxation (μSR) measurements[24] on $U_xLa_{1-x}S$, showing a moment collapse at 55% uranium, from a ferromagnetic moment of well over $1\mu_B$ to an antiferromagnetic moment of about $0.1\mu_B$, support this picture of the phase transition. Furthermore, the ab-initio based absolute calculations described in Section III give the correct prediction of ferromagnetic moment and critical concentration for the phase transition.

In a broader sense, the physical picture developed here shows that hybridization treated via the Coqblin-Schrieffer[25] resonant-scattering point of view provides the *connection (under certain conditions described herein, an actual phase transition) between localized (i.e., coupled-magnetic-ions) magnetism and itinerant-electron magnetism in solids*. The remainder of this chapter is devoted to elucidating this physics and its consequences. In Section II, we describe the exchange-constrained hybridization process. The conceptual development of the methodology to treat the magnetic ordering of uranium systems in the weakly hybridizing subregime is given in Section III A. Then in Sections III B and III C we describe the calculations and results for the low-temperature ordered moment and Curie temperature of the strongly magnetically ordered and extremely anisotropic ferromagnetic NaCl-structure uranium monochalcogendies, US, USe, and UTe. Both the effects of high hydrostatic pressure (to ~20Gpa) and dilution alloying of the uranium by La or Y are calculated absolutely and compared to experiment. In addition, in Section III D we discuss possible experimental work to further validate and quantify the existence of two types of light actinide $5f$ site occupation. In Section IV we discuss the correlated narrow-band (heavy fermion) behavior in the strongly hybridizing subregime,[26] and in Section IV B we present a summary of a method[20] to perform ab initio electronic calculations including two-electron correlations for that correlated narrow-band behavior. In the concluding section (Section V) we present a discussion of the broader implications of our recognition of the nature of the exchange-constrained hybridization process and its consequences for solid-state behavior.

II. DESCRIPTION OF EXCHANGE-CONSTRAINED HYBRIDIZATION PROCESS

The need to include f-"ligand" (p/d) as well as f-f correlations, makes the development of *ab initio* (materially-predictive) theory for electronic and magnetic behavior difficult for Strongly Correlated Electron (SCE) systems. This is because of the computational complexity of going beyond refinements of LDA (one-electron dynamics) theory. One has to treat at least two-electron dynamics and to include both ligand-f hybridization and ligand-f coulomb exchange.

Starting from the first principles hamiltonian embodying the dynamics of the ~10^{25} f-transition-shell and p/d valence electrons for the lattice,

$$H = \sum_i \nabla_i^2 + V_0(r_i) + \frac{1}{2}\sum_{i \neq j} \frac{1}{r_{ij}}$$

(1)

(where V_0 is the periodic potential from the fixed background of the nuclei and filled-shell electrons), one can derive[27] a model hamiltonian describing f electrons interacting with non-f band electrons, where all the quantities in it can either be matched to the output of a LMTO (linear combination of muffin-tin orbitals) LDA (local density approximation) band calculation or be calculated separately using the information given by a LMTO band calculation.

$$H = H_0 + H_1,$$

(2a)

$$H_0 = \sum_k \varepsilon_k b_k^\dagger b_k + \sum_{Rm} E_f c_m^\dagger(R) c_m(R)$$

$$+ \frac{U}{2} \sum_{R,m \neq m'} n_m(R) n_{m'}(R),$$

(2b)

$$H_1 = \sum_{km\sigma R} [V_{km\sigma} e^{-ik \cdot R} b_{k\sigma}^\dagger c_{m\sigma}(R) + H.c.]$$

$$- \sum_{kk'} \sum_{mm',R} J_{mm'}(k,k') e$$

$$\times b_k^\dagger b_{k'} c_m^\dagger(R) c_{m'}(R).$$

(2c)

The nature of the terms and the sources of the parameters in this five-term Hamiltonian are summarized as follows. The first term is the non-f band energy and comes directly from the LDA calculation; the second term is the f-state energy, which is not a LDA energy but can be calculated indirectly from the LDA; the third term is the intra-atomic f-f coulomb interaction (correlation energy) (two-electron interaction) which can be calculated indirectly from the LDA; the fourth term is the LDA hybridization; and the fifth term is the band-f exchange coulomb interaction (two-electron interaction, see Ref. 27 for treatment in almost atomiclike limit).

What makes it extremely formidable to find the ground state of this hamiltonian are the two two-electron terms, the f-f coulomb interaction and the band-f coulomb exchange. As formidable as the problem is because of the presence of the on-site f-f coulomb interaction, the difficulty is much further increased by taking into account the two-electron nature of the coulomb exchange interaction.

An excellent and succinct presentation of the salient features of heavy fermion and valence fluctuation physics has been provided by Lawrence and Mills.[3] They emphasize the close relationship of heavy fermion and valence fluctuation physics. We briefly recapitulate a number of important points. In doing this, we will emphasize the changes in the theoretical framework that occur on recognizing that in addition to the qualitatively important role of f-f on-site coulomb repulsion (correlation effects) in conditioning the nature and consequences of band-f hybridization, the (typically ignored) importance of the kinematic constraint provided by band-f coulomb exchange interaction (electron permutation symmetry) also must be properly treated.

Lawrence and Mills[3] identify the central issue of the physics of heavy fermion compounds as being the understanding of how the coherent ground state evolves from the high temperature regime, and that, "very diverse ground states are realized in practice". They point out that the "Kondo Lattice Standard Model—starts from a localized description of the f electron" (non-interacting Kondo impurities), which "pays the price that it is then very difficult to describe the coherent band-like ground state".

In our own point of view as reviewed herein, we emphasize considering the central physics from the perspective of the change in behavior on going from the low-temperature coherent state toward the high-temperature regime, i.e., what happens as temperature increases. In providing an alternative to the standard Kondo model for the heavy fermion state and the transition to the high-temperature regime, we observe that a number of features identified as providing strong support for the Kondo model are, in fact, generic features expected for any model based on correlated-hybridization-driven behavior. We focus on the role of exchange and the resultant fine structure in phenomena often attributed to the Kondo effect. For comparison with the Kondo point of view, we briefly address the "Kondo" volume collapse transition, as occurs with increasing temperature in elemental cerium and plutonium, driven by the sensitivity of hybridization to cell volume. (We will discuss this topic further in Section V). In addressing the comparison to the Kondo volume collapse picture, we briefly address the twin issues of spectral density and entropy generation.

We then seek an entropy-generating mechanism that more than compensates for any loss of cohesive (bonding) energy, caused by exiting from the low-temperature coherent state, and which by thereby lowering the free energy as temperature increases drives a phase transition. Allen and Martin[28] began their consideration of the Kondo volume collapse model for the γ to α transition in elemental cerium by pointing out the difficulty of explaining the large atomic volume change (~15%) involved, while at the same time experimental probes find form factors and other behavior that is atomiclike for both phases. (We note that the α-to-δ atomic volume change in elemental plutonium is about twice that size.)[29] The central point of our model[16,26] of the heavy fermion state is that there is a fine structure of nonmagnetic (para) and magnetic (ortho) subbands imposed on the overall $5f$ spectral density by the kinematic constraint on the correlated-electron dynamics imposed by exchange (electron permutation) symmetry. We expect a degree of this fine structure to also be present in the α low-temperature phases of cerium and plutonium. Local probes then see the para (or ortho) atomiclike (i.e., narrow in energy) spectral density, but the atomic volume changes are governed by the cohesive energy associated with the overall $5f$ spectral width. The conventional entropy difference, for the Kondo model[28] or lattice-periodic Mott transition[30], depends on the difference in electronic entropy between an almost-localized and a fully-localized lattice-periodic state. In our model the transition from the low-temperature coherent state to the high-temperature state is driven by the additional entropy of mixing associated with having a solid-state-solution of para (fluctuating configuration) and ortho (stable, i.e. localized, configuration) f-sites, a perfect mixture having homogeneous lattice disorder. (As discussed below, the transition to this solid-solution-like state is driven by Anderson localization which is physically equivalent to Mott localization for a disordered system.)[30a] This entropy of mixing per site is significant compared to the difference in electronic entropy between lattice-periodic states with almost-localized and fully-localized behavior. (Depending on the degree of f-localization, e.g., cerium vs uranium in the same chemical environment, exchange tends to homogenize the para and ortho behavior, and we anticipate there being considerable homogenization for elemental cerium.)

The existence of the fine structure automatically deals with the question of there being a large energy scale associated with the primary electronic interactions (Fermi energy, i.e., occupied $5f$ spectral width, f-f correlation energy U, and the one-electron hybridization potential, and the small energy scale provided by the universality of the thermodynamics. The ortho/para fine structure provides the small energy scale, and in past work the results of absolute calculations have been published[31,32] as quoted below, providing the correct energy scale.

Given this view of the central features of the electronic behavior and their consequences, here we focus on the electronic behavior in chemical environments such that the transition-shell electrons are neither too atomiclike nor significantly influenced by direct overlap with transition-shell electrons from other sites. In the situation where direct overlap is not significant, the transition-shell electron role in the overall solid-state electron dynamics is solely determined by hybridization, as constrained by exchange, with band electrons of non-transition-shell atomic origin (where, in the case of cerium $4f$ and light actinide $5f$ electrons, the band electrons for hybridization do include those of $5d$ and $6d$ atomic origin, respectively). We first discuss the situation for cerium compounds when the 4f behavior is relatively atomiclike and exchange has to be treated by including the fifth term in the hamiltonian of Equation (2). Then we will go on to discuss the situation for uranium compounds where the 5f behavior is not too atomiclike.

Some thirty years ago, to deal with the Kondo impurity problem for cerium, Coqblin and Schrieffer[25] developed a resonant scattering type technique for dealing with a hamiltonian consisting of the first four terms of the hamiltonian in Equation (2) (i.e. without the coulomb exchange term), but for a single f-electron impurity rather than a lattice of such atoms. The validity of this treatment depended on the hybridization (first term in Equation (2c)) being sufficiently weak. Over a period of years, work by the present author and colleagues, culminating in the 1994 Physical Review B paper by Sheng and Cooper,[27] extended this treatment to a periodic lattice of cerium atoms (in compounds with sufficiently weak f with non-f-band hybridization so that the 4f electron behavior is rather atomiclike) and including the coulomb exchange interaction. (For cerium in the compounds where this treatment is appropriate, the f-transition-shell configuration is close to 4f.)[1] Earlier work by Wills and Cooper,[31] as discussed in detail by Sanchez-Castro, Cooper, and Bedell,[33] had shown that, without the coulomb exchange term, the hamiltonian of Equation (2) treated for a periodic lattice in the weak hybridization limit provides two main types of cerium-cerium interactions. These are a strongly anisotropic ferromagnetic, "generalized Coqblin-Schrieffer interaction", and an antiferromagnetic kinetic superexchange interaction. Judging by the results of detailed calculations[27] for CeSb and CeTe, the effect of the coulomb exchange is to enhance substantially the antiferromagnetic superexchange. For the cerium compounds studied, this provides an overall antiferromagnetic ordering brought about by a weak antiferromagnetic coupling of planar units within which there is very strong ferromagnetic coupling.

Then the question arises as to how this treatment can be changed to treat light actinide systems in the regime of chemical environments such that the transition-shell electrons are neither significantly influenced by direct overlap with transition-shell electrons from other sites nor too atomiclike. It turns out that by adopting ad hoc approximations appropriate to the "not too atomiclike" limit for the f-transition-shell electron behavior, the calculations become sufficiently simplified so that the physics can be made much more transparent. In this not-too-atomiclike limit, the effects of exchange can be treated in a much simplified manner. Since this is a very substantial simplification both for calculation and for conceptualization, we will briefly summarize the underlying physics, and how past work led to recognizing the conceptually simplified way in which we treat the exchange effects based on two-electron dynamics. For reasons summarized below, this limit, and these approximations with the consequent relatively simple physics, are applicable to certain uranium compounds (the uranium monochalcogenides) with strong highly anisotropic ferromagnetism.[13-16] (It is also pertinent to elemental plutonium and its stabilization into the fcc structure thermally or by trivalent additives such as some gallium.[16] This same physics is probably relevant to the more localized $4f$ electron behavior in elemental cerium,[34] but in a more complicated way,[16] and to some antiferro-magnetic Cu3Au-structure uranium compounds[17,35,36] such as UIn_3 and UGa_3.) In this not-too-atomiclike limit, as long as the hybridization is treated as acting between properly exchange-symmetrized two-electron wave functions, the effects of exchange can be incorporated adequately by its inclusion in the one-electron exchange-correlation potential. (Exchange [electron permutation] symmetry in a many-electron system requires having a many-electron wave function that is antisymmetric [changes sign] on interchange of any two electrons. Thus a properly exchange-symmetrized two-electron wave function consists of one of two possibilities. It must be either a product of a symmetric [para] two-electron orbital state and a two-electron spin singlet [since a two-electron spin singlet is antisymmetric on exchange of electrons] or a product of an antisymmetric [ortho] two-electron orbital state and a two-electron spin triplet [since a two-electron triplet is symmetric on exchange of electrons].) This is a very substantial simplification both for calculation and for conceptualization; and this is what has allowed the development for uranium compounds and elemental plutonium of calculational technique, incorporating what was learned from the work of Wills and Cooper[31] and Sheng and Cooper[27] on weakly hybridizing more-atomiclike cerium compounds. This conceptualization and calculational technique has been extraordinarily successful in predicting the magnetic ordering behavior of the NaCl-structure uranium chalcogenides under high pressures to about 20Gpa and under certain uranium-dilution-alloying changes described below. This extraordinary predictive power and its prediction of a phase transition involving the destruction of strong ferromagnetism then validates our physical picture of a transition for strengthening, but still relatively weak, hybridization to a correlated very-narrow-band (i.e. enhanced mass) phase which provides heavy fermion phenomenology.

The band-f hybridization term in the hamiltonian, the first term in Equation (2c), consists of a linear combination of terms that destroy one f electron and create one band electron or vice versa. Thus if we want to include those corrections to the one-electron dynamics that treat the correlated dynamics of one band and one f electron (exact instantaneous effects of motion coupled via the interelectronic coulomb interaction), the calculation of the necessary two-electron matrix elements requires the use of properly exchange-symmetrized two-electron wave functions as, for example, is done in treating the helium atom. Given this situation, the hybridization process can be correctly and usefully pictured by thinking of a lattice of energy-level-inverted helium (two-electron) atoms. We can describe the $5f$ behavior in weakly hybridizing actinides as if we had a lattice of interacting atoms with atomic ground states that have both s-$5f$ para singlet and s-$5f$ ortho triplet two-electron components and an excited state that is a doubly-s occupied para singlet. (A detailed calculation justifying this procedure is given in Sec. IV of Ref. 20.) In fact, the s-like part of these two-electron wave functions comes from the virtual occupation of this state by the itinerant band electrons which originate from p and d atomic states of atoms on other sites and which have their orbital angular momentum quenched[37] by itinerancy.

From Equation (2) we can see that hybridization conserves spin (does not involve spin) and thus acts only between the excited para singlet and the ground state s-$5f$ para singlet. Thus to lowest order, hybridization drives the singlet component of the $5f$ contribution to the two-electron spectral density to fluctuate between configurations differing by one $5f$ electron (to become itinerant); and the triplet component remains localized (has a stable configuration). This description, in fact, corresponds to the resonant scattering physics of the Coqblin-Schrieffer treatment[25] of hybridization.

As hybridization strengthens at fixed temperature within the intermediate delocalization regime of transition-shell-electron (e.g., $5f$) behavior, we find there are two subregimes of behavior. (1) In the weaker hybridization subregime, described in the preceding paragraph, the $5f$ behavior viewed locally at a given site is of one of two types: (a) either totally localized (ortho), and therefore capable of having a large (free-ion-like) orbital contribution to an ordered magnetic moment (and because of strong spin-orbit coupling, large spin ordering), or (b) of fluctuating configuration (para), providing an itinerant 5-f component and thereby providing a mechanism for passing a hybridization-mediated message of orbital magnetic polarization between the localized-$5f$ sites. Thus in the more weakly hybridizing subregime of behavior, magnetic ordering with a large orbital contribution can occur. A key aspect of this behavior is that the *localized sites are randomly distributed on the lattice*. This disorder supplies a source of entropy that, at sufficiently high temperature, stabilizes this Random-Localized-Fluctuating Site (RLFS) para-ortho solid-solution-like phase (i.e., gives lower free energy $F = U - TS$ than when the $5f$ electrons become bonding) even in the absence of magnetic ordering. In this subregime (the RLFS phase), the $5f$ electronic contribution to the entropy is analogous to the configurational entropy in a random alloy. As the hybridization strengthens, (2) there is an abrupt delocalization to a narrow correlated $5f$-band phase as the $5f$ bonding (banding) contribution to lowering the free energy more than compensates for the loss of the *combined lowering of free energy coming from the magnetic ordering acting in concert with the electronic entropy from the random distribution of the two types of 5f local behavior.*

Thus, viewed in reverse as hybridization weakens, we have a *magnetic-ordering-enhanced Anderson localization of the 5f electrons.*[16] This process of 5f localization can occur in a pure material (with or without magnetic ordering present), i.e., can be wholly self-induced, as described in Section V for elemental plutonium, at a temperature sufficiently high to develop the necessary free energy lowering via the configurational entropy. However, for alloyed uranium compounds such as the NaCl-structure monochalcogenides, e.g. $U_xLa_{1-x}S$, the randomness provided by the alloying itself may nucleate this process. Whether or not magnetic ordering occurs upon localization depends sensitively on the bonding energy, e.g., 5f, that must be overcome since this determines the temperature at which the combination of configurational entropy and magnetic ordering "wins". For compounds such as the uranium monochalcogenides the spacing provided by the anion provides an environment where there is less bonding energy to overcome than in the pure light actinide elements; hence the phase transition occurs at a temperature where magnetic ordering can be present in practice.

It is very important to recognize, that once Anderson localization occurs, it is *energetically favorable to develop the maximum entropy.* Thus, in the absence of magnetic ordering, there is an *abrupt localization*, at a sufficiently high temperature to have a phase transition, *at half the lattice sites occupied by the light actinide.* If magnetic ordering occurs, this abrupt localization can occur at *more than half* these lattice sites.

In the delocalized hybridizing 5f subregime, coherent behavior develops, and one has very narrow correlated bands of nonmagnetic para (singlet) and magnetic ortho (triplet) character. (However, any ordered magnetism will be much weaker than in the "solid solution" ("random alloy") weakly hybridizing phase. In this correlated very narrow band phase, depending on how narrow and well defined the singlet or triplet 5f subbanding is, qualitatively a variety of behavior is possible. One extreme, as described in Reference 26 is behavior characterizing the heavy fermion regime; and as shown in Ref. 20 and Ref. 26, the characteristic enhanced electronic specific heat and Pauli susceptibility behavior in our treatment of this regime occurs with the correct energy (temperature) scale. In addition, as shown in Ref. 26 and shown in Section IV herein, the division into nonmagnetic and magnetic subbands, which may be overlapping or nonoverlapping, and their placement relative to the Fermi energy (chemical potential), provides a range of Wilson ratio behavior in agreement with experiment on a substantial number of heavy fermion systems. (In treating thermodynamic behavior for heavy fermions, given the extreme narrowness of the magnetic-nonmagnetic subbands, it is essential to take into account the temperature dependence of the Fermi energy.) The behavior in this enhanced mass band regime is discussed in Section IV. In Section III, we first discuss the strong ferromagnetic ordering that occurs in the solid-solution-like RLFS phase.

III. MAGNETIC ORDERING OF URANIUM SYSTEMS IN THE WEAKLY HYBRIDIZING SUBREGIME: RECOGNITION AND QUANTIFICATION OF KEY PHYSICS CONCEPTS

A great simplification is possible when considering *magnetic ordering* for *light actinide* SCE systems. Magnetic ordering involves the electronic behavior *globally* (i.e. involves the spectral density in an integrated sense). This greatly simplifies treating the consequences of the two-electron dynamics with ab-initio-based methodology in chemical environments such that the behavior of the $5f$ electrons is neither closely atomiclike nor significantly influenced by direct overlap with $5f$ electrons from other lattice sites. In this *hybridizing regime*, as described in the preceding section, *in the relatively weakly hybridizing, but not too atomiclike, $5f$ subregime* that describes magnetic ordering in a number of uranium-based compounds and alloys thereof, the localized $5f$ spectral density, which provides any *ordered magnetic moment, is randomly distributed* among the uranium lattice sites. The coupling between these sites is then provided by the remaining, itinerant, component of the $5f$ spectral density.[12,16] (This is apt not so simply to be the situation in cerium-based systems where the f-electron behavior is more atomiclike.[27] The variation in the calculated correlation energy U serves as a guide to how atomiclike the f electron behavior is. For CeSb, U=6.7 eV; for the more hybridizing CeTe, U=6.0 eV; for fcc Pu, U=4.1 eV; for PuSb, U=3.97 eV; and for US, U=3.4 eV.) Being in a subregime of behavior that allows us to use this simplified picture for the $5f$ electronic behavior, and the consequent intersite magnetic interactions, provides a very important simplification. This is so because coulomb exchange is a true two-body interaction (involves two creation and two annihilation operators), while hybridization is a one-body interaction (involves one creation and one annihilation operator, see Eq (2c) above). Thus we have the situation of *randomly located localized-$5f$ sites locked together coherently (i.e., magnetically ordered) by coupling provided via itinerant $5f$ electrons* *originating from the other fluctuating-valence sites.* This has enabled us to make absolute material-specific predictions of alloying or high-pressure effects using LDA input into many-electron dynamics with only one *non-ab-initio* number (the baseline percentage separation between sites with localized and those with delocalized $5f$ electrons). This one number is chosen for the pure (unalloyed) material at ambient pressure. (Below in Section V, we will discuss the situation in the pure material, e.g. US, as T approaches zero.) Experimentally, the alloying effects can be dramatic, e.g. in $U_xLa_{1-x}S$, the strong magnetic ordering abruptly disappears[21-24] at about 55% uranium.

Thus magnetic ordering behavior provides a powerful tool for diagnosing the nature of the correlated (two-electron) electronic structure; and this leads to recognition of, and a measure of, the quantitative division in occupied composition between the two kinds of two-electron states and the far-reaching consequences of that bifurcated behavior. The theory is highly successful in its detailed absolute predictions, and this has important implications for the overall understanding of electronic behavior in SCE systems including heavy fermion systems. In the final section (Section V), we discuss the implications for systems beyond the particular class of extremely anisotropic cubic ferromagnets for which the results of our detailed calculations give extraordinary agreement with experiment for the Curie temperature and low-temperature ordered moment.

A. Development of the Magnetic Interaction Model
Incorporating the Dynamic Effects of Hybridization

As discussed above, through a combination of first principles model calculations[26] and ab-initio-based materially predictive calculations,[12-16,38] we have come to understand that there are two kinds of occupied many-electron states involving f-electrons. Since hybridization does not mix spin, the virtually-bound l-l (two electron p/d "ligand") excited singlet, embodying the "tails" of electrons of p/d atomic parentage centered on other sites, can only mix with the f-l singlet; and the hybridization gives configuration interaction, i.e. locally gives valence fluctuations and correspondingly enhanced Pauli susceptibility, only by acting through the singlet two-electron states. Thus, the f spectral density in magnetic (triplet f-l) states remains highly localized, hence sharp in energy, while the f spectral density in nonmagnetic (singlet f-l) states is of mixed local and itinerant character. (The generalization of the treatment in Ref. 4, which involves only one f electron, i.e. the cerium situation, to that involving more than one f electron, i.e. the actinide situation, is given in Ref. 20. It should be noted that, given the permutation symmetry of all electrons, the magnetic f-l states cannot remain totally isolated from the nonmagnetic states, and thus cannot be perfectly sharp in energy; nevertheless we expect this strong distinction in character of the spectral distribution to be essentially preserved, at least so long as the f electron role in the overall dynamics is neither too atomistic nor significantly affected by direct overlap.) It should be emphasized that the hybridization is strongly reduced by the strong onsite f-f coulomb repulsion energy (the f-f correlation energy, U, for uranium is somewhat above 3 eV e.g., 3.4 eV in US).

Since higher order hybridization effects give intersite singlet-triplet mixing, there is hybridization-mediated intersite f-f coupling[31,33,39,40] between the sites having a stable localized (for uranium, f^3) configuration that is transmitted through the itinerant $5f$ electrons associated with the fluctuating (for uranium, f^3/f^2) configuration sites. This intersite f-f coupling[39] gives highly anisotropic[40] magnetic ordering. Highly anisotropic ferromagnetic ordering, as discussed in Refs. 40, is a characteristic feature of the Coqblin-Schrieffer interaction both in its original form[25] or in its generalized form.[31,33] In its pure manifestation (without the presence of any significant antiferromagnetic kinetic superchange interaction of the type discussed in the Appendix of Ref. 33), this gives very strongly anisotropic ferromagnetism. This is a consequence of the situation that hybridization acts to give orbitally-driven magnetism, while exchange acts to give spin-driven magnetism. (The two types of magnetic ordering, orbitally-driven and spin-driven, are coupled together by the strong spin-orbit coupling for f electrons.) Both the generalized Coqblin-Schrieffer (C-S) and the kinetic superexchange interaction arise from a combination of hybridization and exchange effects. As one approaches the phase-transition boundary discussed above within the hybridizing regime from the more weakly hybridizing side, the generalized C-S interaction becomes relatively stronger. When the generalized Coqblin-Schrieffer interaction is overwhelmingly dominant, one expects extremely anisotropic ferromagnetism. Thus the extraordinarily anisotropic strong ferromagnetic ordering behavior of the uranium monochalcogenides offers a benchmark series of compounds for testing the theory and predictions presented here. (For a discussion of the magnetic ordering of these compounds presented in the overall context of actinide and cerium-based magnetically ordered systems, see References 41 and 42.) When the generalized Coqblin-Schrieffer interaction is dominant, but there is a relatively weak antiferromagnetic kinetic superexchange, the dominant C-S interaction manifests itself in the form of strongly ferromagnetic planes weakly antiferromagnetically coupled[39] as is characteristic of colossal magnetoresistance [CMR] manganese perovskites as well as cerium and many light actinide correlated-electron magnetically-ordered systems.

Thus, the key physics is the recognition and quantification of the concept that there are two kinds of occupied $5f$-p/d (band) two-electron (correlated) states, and correspondingly there is a random distribution of two kinds of, e.g., uranium, sites. The composition of one of these includes (one-electron) localized uranium $5f$ states and is magnetic in character. (The $5f$ spectral weight resides stably around a single site.) The composition of the other kind of $5f$-p/d two-electron states includes a mixture of (one-electron) uranium $5f$ states of localized and delocalized nature and is nonmagnetic in character. (For these sites, the $5f$ spectral weight does not remain stably around a single site. Viewed from a single site, such a state fluctuates rapidly between valences.) Both pressure and dilution alloying (substituting a lanthanum for a uranium) increase the $5f$-with-p/d (band) hybridization driving $5f$ electron occupation from the well-localized (stable f configuration) to the itinerant (rapidly fluctuating f configuration) two-electron states.

Lanthanum and yttrium used as the dilutent for the uranium monochalcogenides,[21-23,43] are chemically very similar (and close in size) to uranium, but do not have partially filled f shells. Thus their substitution for uranium increases the hybridization of each of the remaining uraniums in a system such as $U_xLa_{1-x}S$. *The analogy to dissolving a solute [the uranium 5f electrons] in a solvent [the band electrons of non-f atomic parentage] to form a solution may be useful in picturing this situation. In effect, one is decreasing the amount of solute while keeping the amount of solvent the same, where the mixing is provided by the hybridization.* This increase in hybridization gives a *decrease in the low-temperature ordered moment*, and interestingly we expect a *correspondingly enhanced Pauli paramagnetism.*

Thus both the pressure and dilution alloying effects act in the same qualitative way to increase the average number of "holes" in the magnetic lattice; however, increasing dilution alloying has a more strongly highly nonlinear effect on increasing the number of holes, in part presumably because the alloying (e.g., of lanthanum onto uranium sites) itself introduces the key element of having a *random distribution* of different types of occupied uranium sites. In addition (as discussed and illustrated below) there is an increased localization of the remaining "undissolved" localized $5f$ electrons. This has the consequence that, opposite to the effect of hydrostatic pressure, the hybridization-mediated magnetic coupling between the remaining localized $5f$ sites *decreases* with increasing dilution of uranium by the substitution of lanthanum. Thus for dilution alloying, at a certain point the well-localized $5f$ component falls abruptly below a critical value necessary for magnetic ordering, and a catastrophic collapse in magnetic ordering occurs as[21-24] in $U_xLa_{1-x}S$ at 55% uranium.

It is important to recognize that this physical picture of random "holes" in the magnetic lattice, and its implementation[12-16] for calculating the Curie temperature as

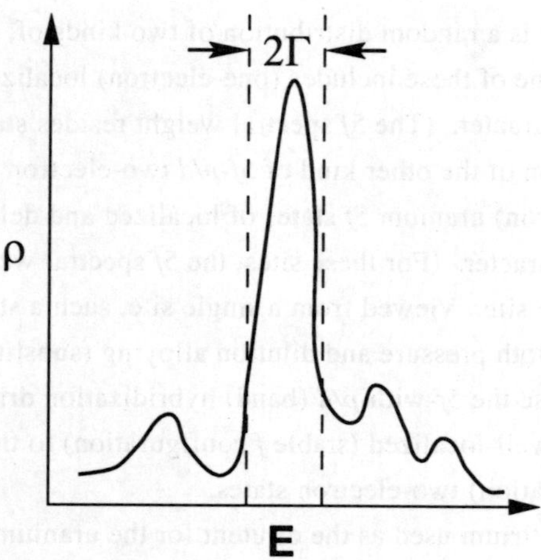

Fig. 1. Schematic drawing of typical behavior for the *f* spectral distribution in energy as provided by an LDA (one-electron dynamics) calculation.

discussed below using random holes in an Ising lattice, violates Bloch's Theorem, which itself is a consequence of lattice translational symmetry. However, Bloch's Theorem is not required for a random alloy nor is it required at finite temperature for an unalloyed lattice. (Nor is there a requirement even at zero temperature except for an element. However, the bonding energy advantage in lowering the free energy would tend to restore lattice periodicity on approaching zero temperature.)

The essential recognition is that *we are dealing with a randomly disordered system.* For that reason, the physics treated here and in Refs. 12-16 supersedes that of References 27 and 38 (developed for systems with lattice periodicity) for the subregime being discussed here where the hybridization is sufficiently weak to be out of the $5f$ correlated-narrow-band regime but sufficiently strong so that the dynamics are not altered by exchange effects. (This is true at least in the limit discussed here where exchange effects are adequately captured by including them in the LDA exchange-correlation potential for properly exchange-symmetrized two-electron dynamics.) The success of the theory of Sheng and Cooper[27] for the cerium NaCl-structure monopnictides and monochalcogenides may point to a *role for exchange, on going toward the atomic limit for the f electron behavior, in acting to homogenize the random ortho/para distribution on the lattice* in that limit.

A schematic drawing of typical behavior for the f spectral distribution in energy as provided by an LDA (one-electron dynamics) calculation is shown in Fig. 1. The true many-electron states (well approximated for this purpose by two-electron states, as discussed above) are expected to give spectral densities consisting of (1) a very sharp peak corresponding to the localized magnetic states, and (2) a distribution that is similar in appearance to that provided by the LDA calculation, with a broadened peak and diffuse tails, corresponding to the nonmagnetic locally fluctuating plus itinerant states. The broadened peak corresponds to the period of circulation around a given nucleus, and the diffuse tail corresponds to the period of itinerancy between lattice sites. Thus when projected onto the LDA one-electron states, we would expect the magnetic component to be wholly contained within the LDA broadened peak and the nonmagnetic component to be contained in the remainder of that peak plus the diffuse tail. This leads to the *crucial ansatz* of our LDA *ab-initio-based* materially-specific prediction of dilution alloying and pressure effects on magnetic ordering. This ansatz[13-16] is that the *LDA calculation of the shift in f-electron spectral weight from the broadened peak to the diffuse tail with dilution alloying or pressure accurately approximates the true shift of spectral density from localized (magnetic) to locally-fluctuating/itinerant (nonmagnetic) states.*

With this ansatz, our *ab initio* full-potential LDA calculations allow us to predict the pressure and uranium-dilution dependence of transfer of f spectral weight from the localized magnetic f states to the hybridizing nonmagnetic f states. When used as input into our phenomenological theory of magnetic ordering, including correlation effects,[12] this allows us to predict concentration and pressure dependence of ordered moments and Curie temperatures with remarkable absolute accuracy. The approximation that in practice probably limits this accuracy is the relatively simple method used to evaluate the intersite coupling giving magnetic ordering. This is chosen for the ideal hybridization-mediated coupling limit. Thus this approximation will fail both at the limit where direct overlap effects are significant for the f-electrons, i.e. where the f-electrons become too itinerant, and at the limit where the f-electrons become too localized, so that[27] the role of the f-ligand coulomb exchange in the intersite magnetic coupling becomes more complicated.

B. Predictive Calculation of Low-Temperature Ordered Moment and of Curie Temperature

The remarkable successful predictive methodology we have used to calculate the ferromagnetic ordering (Curie) temperatures, and ordered moments (as temperature approaches 0K), of the NaCl-structure uranium monochalcogenides developed in the context of the successful calculations reported in References 27 and 38. In Reference 27, using the hamiltonian of Equation (2), a technique incorporating ab initio LDA LMTO information was used to calculate *ab initio* absolute antiferromagnetic ordering (Neél) temperatures (T_N) and zero-temperature ordered moments (m_o) for several weakly hybridizing unalloyed cerium compounds at ambient pressure. A key point is that a resonance-width scheme developed by Wills and Cooper, first for cerium[31] systems and then generalized to light actinide[32] systems, was used to calculate the hybridization matrix elements, in effect giving the *f*-ion intersite magnetic coupling. The calculations of Reference 27 provided very close absolute agreement with T_N and m_o for CeBi, CeSb, and CeTe including the change of almost an order of magnitude in these quantities between the values for the first two compounds and those for CeTe.

The generalization presented in Reference 38 to unalloyed uranium compounds at ambient pressure, to calculate the Curie temperatures and ordered moments, was almost as successful. The somewhat poorer absolute agreement with experiment was attributed to the need for incorporation of an atomiclike treatment of coulomb exchange, more appropriate for cerium, into the treatment of hybridization allowing for the distinctly stronger hybridization of light actinides. However, when the experimental results for the pressure variation of T_c in UTe up to about GPa became available,[44] showing dramatic nonmonotonic behavior, the straightforward extension of the work in Reference 38 to the effect of pressure did not yield sufficiently strong change with pressure. This led[12] us to decide, on an ad hoc, basis, to investigate the consequences of making the most important of a series of ansatzes. This ansatz was to consider having two types of uranium present occupying different sites, one of stable $5f^3$ configuration, the other rapidly fluctuating between $5f^3$ and $5f^2$ thereby providing magnetically ineffective sites, "holes" in the magnetically-coupled lattice. By having a baseline (ambient pressure) $5f^3$ occupation less than 100%, one could expect to have accentuated effects of pressure.

Thus, instead of a totally unadjustable ab initio calculation using the technique of Reference 38, we now had a technique with one adjustable parameter, the baseline fraction of $5f$ uranium[12] at ambient pressure. Thus we developed[12] a model based on the physics of Reference 38 but adapted to incorporate this one change. Since we regarded the anisotropic ferromagnetism provided by the generalized Coqblin-Schrieffer interaction of Reference 27 as the key physics that explained the remarkable anisotropic ferromagnetism of the uranium monochalcogenides, we decided to map the physics onto an exact magnetic ordering behavior of a model system incorporating both strong magnetic anisotropy and the presence of a number of magnetic holes on a lattice, which increased with pressure. (A detailed discussion of the way in which an extremely anisotropic ferromagnetic interaction, as occurs in the uranium chalcogenides, results from the hybridization-mediated correlated electronic behavior is given in Ref. 31. A very complete derivation and discussion of this physics is provided in the appendix of Ref. 33. In the final summary paragraph on page 12,518 of Ref. 33, it is pointed out that the magnetic interaction mediated by the interchange of a particle-hole excitation of the conduction electrons [generalized Coqblin-Schrieffer interaction] is ferromagnetic at short ranges [this interaction is also very anisotropic];[31] while the interaction mediated by the particle-particle and hole-hole interchange is antiferromagnetic [kinetic superexchange]. We expect this antiferromagnetic superexchange to be enhanced as the band-f coulomb exchange strengthens.) Thus we decided[12] to map the behavior onto an Ising lattice with holes, and with a nearest-neighbor ferromagnetic interaction between sites not occupied by a hole. *Since it seemed the only reasonable thing to do, we located the holes randomly on the lattice.* The free energy variation with temperature was calculated by computer simulation to determine the Curie temperature, T_C in units of J (nearest-neighbor ferromagnetic interaction) for a specified fraction of holes. T_C was identified as the temperature where the double minimum in free energy at finite positive and negative magnetization disappears, with the minimum of the free energy then occurring at zero magnetization. The calculations were done for randomly placed holes on an fcc lattice with $2^9 = 512$ sites.

To complete the "recipe", *which is then deterministic (unadjustable),* for T_C and m_o at a given pressure, we need to specify J and the number of holes at that pressure. To specify both these quantities, we used an LDA density of states calculation. At each pressure the $5f$-projected density of states typically (see Figure 1) consists of a "localized" peak and itinerant "tail". We decided on a technical rule to define the boundary between peak and tail (dashed lines in Figure 1), and tested for sensitivity of our results to a few reasonable choices of that rule. Then we specified the fraction of nonhole (stable $5f^3$) sites) at each pressure by simply decreasing the starting (ambient pressure) baseline fraction of sites occupied by stable $5f^3$ uranium by the decrease in occupied $5f$ states per uranium within the peak.[13-15] At each pressure J is determined[13-15] from the width, G, of the $5f$ density-of-states peak. We were motivated to do this by the f-resonance width point-of-view of References 31 and 32, as generalized in Reference 38. To quantify the correspondence between G and J, we used[12] a nearest-neighbor Hubbard model

$$H_H = \left(-t \sum_{\substack{in \\ mm'}} c_{im}^+ c_{i+n,\,m'} + h.c. \right) + U \sum_{i,\,m \neq m'} n_{im} n_{im'} \tag{3}$$

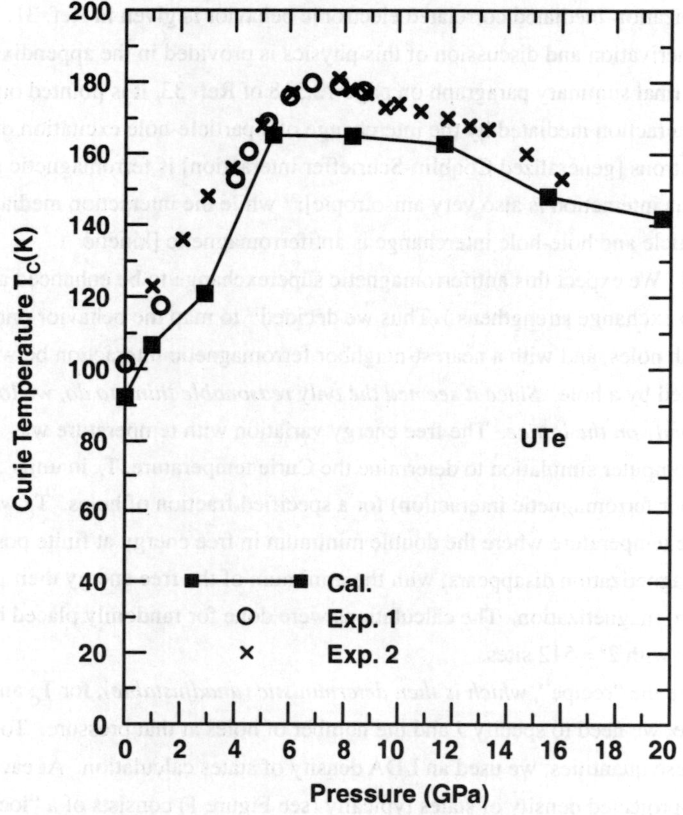

Exp. 1: Cornelius et. al., J. Magn. Magn. Mater. **161** (1996) 169.
Exp. 2: Link et. al., J. Phys: Condens. Matter **4** (1992) 5585.

Fig. 2.   Predicted variation of Curie temperature with pressure for UTe compared to experiment[44,45].

(where t is the interatomic interaction and U is the on-site coulomb repulsion as in Equation (2b)) to generate a tight binding band which has a width $\Gamma=1.04\ t$ for an fcc lattice. For $U \gg t$, this can be mapped into a hamiltonian with a nearest neighbor magnetic coupling $J = t^2/U$. Thus by evaluating Γ at each pressure and calculating U (which varies little with pressure), we obtain J at each pressure.

Thus we obtained a totally deterministic recipe for T$_c$ and m_o at every pressure.

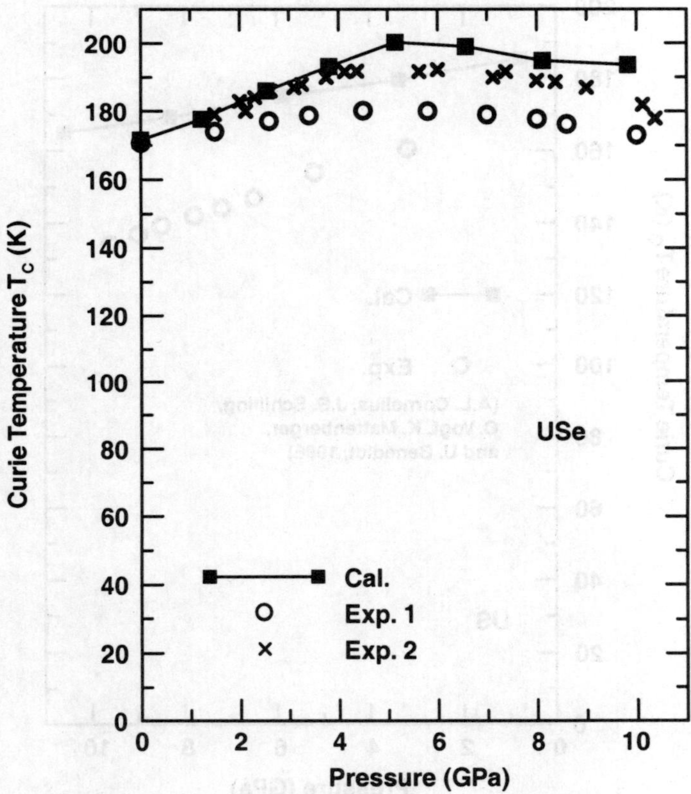

Exp. 1: Cornelius et. al., J. Magn. Magn. Mater. **161** (1996) 169.
Exp. 2: Link et. al., Physica B **190** (1993) 68.

Fig. 3.   Predicted variation of Curie temperature with pressure for USe compared to experiment[45,46].

Figure 2 shows[14,15] the impressive absolute agreement obtained[12,14,15] with experiment for UTe. Agreement of similarly impressive quality was obtained[12,15] for USe and US (Figures 3 and 4, respectively). This quality of agreement led us to adopt exactly the same deterministic model to calculate T_C and m_o for the uranium-dilution alloying, e.g., in $U_xLa_{1-x}S$. We simply added the fraction of lanthanum to the hole fraction per occupied uranium calculated from the uranium $5f$ density-of-states peak in the same way as in the pressure calculation (using a supercell with the appropriate occupation of uranium sites

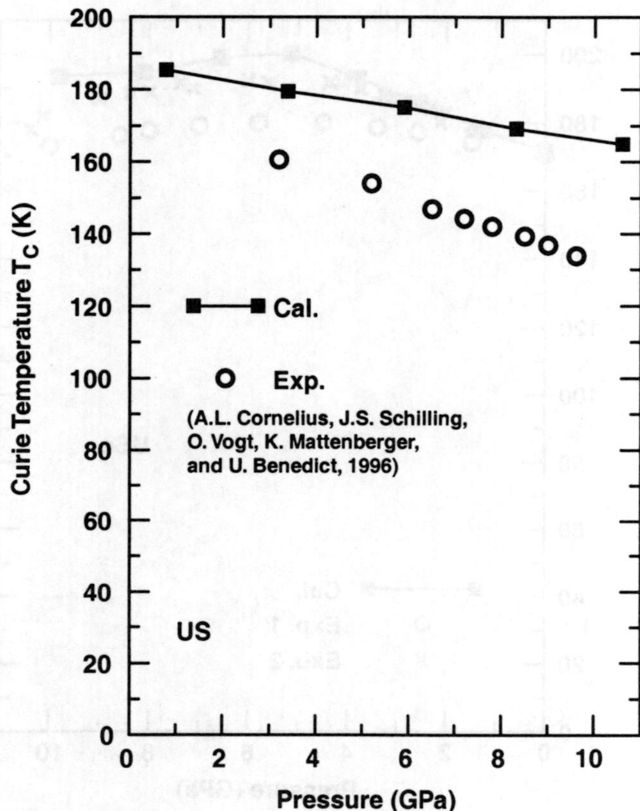

Fig. 4.   Predicted variation of Curie temperature with pressure for US compared to experiment[45].

by lanthanum). The predicted variation of Curie temperature with f-dilution alloying for US compared to experiment[22,23] is shown in Fig. 5. It is very important to recognize the essential role of the randomness of the holes in the Ising lattice in the behavior found. A uniform distribution of the same average moment would not yield the observed behavior. Neither will an ad hoc modification of LDA to improve localization effects, such as a spin-and-orbitally polarized LDA calculation, yield results as shown in Figure 5.

The calculation of Figure 5 was done using LDA LMTO $5f$-electron density of states information from a supercell calculation containing four uranium and four sulfur sites, substituting lanthanum for uranium at one, two, and then three of these

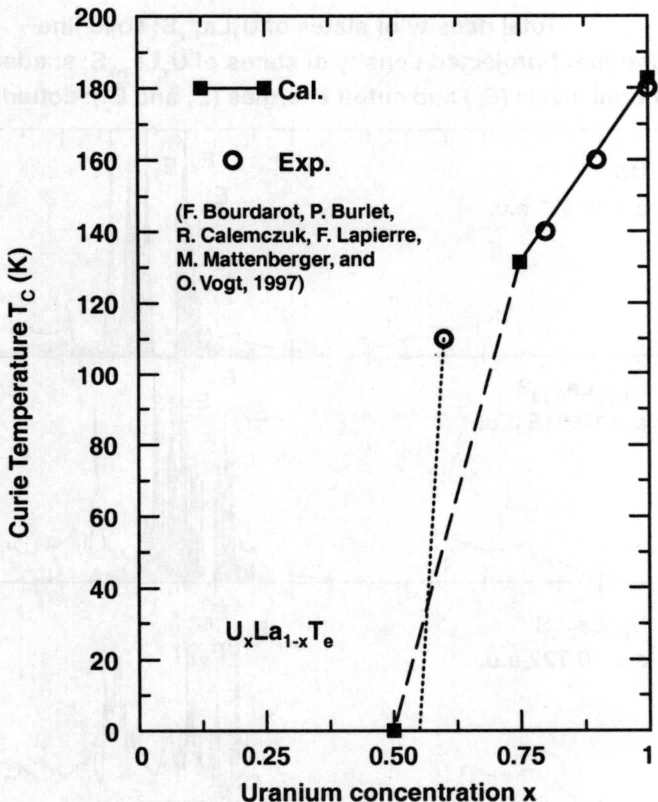

Fig. 5.   Predicted variation of Curie temperature with dilution alloying for US compared to experiment[22,23].

sites as shown[47] in Fig. 6.

Our *ab-initio-based* model, as described above, gives absolute material-specific predictions using input from the LDA paramagnetic uranium f-electron-projected density of states plus the ab-initio calculated value of the correlation energy U (3.4 eV for uranium in US). Figure 6 shows the density of states behavior[47] for a superlattice LDA calculation for $U_xLa_{1-x}S$ (4 uranium-or-lanthanum, 4 sulfur sites per unit cell) using the full-potential linear combination of muffin-tin-orbitals technique.[31,48,49] The shaded area shows the uranium f-projected density of states. To calculate the LDA shift in f-electron spectral weight from the broadened peak to the diffuse tails with dilution alloying or pressure, we have to specify an objective (non-adjustable) quantitative criterion to

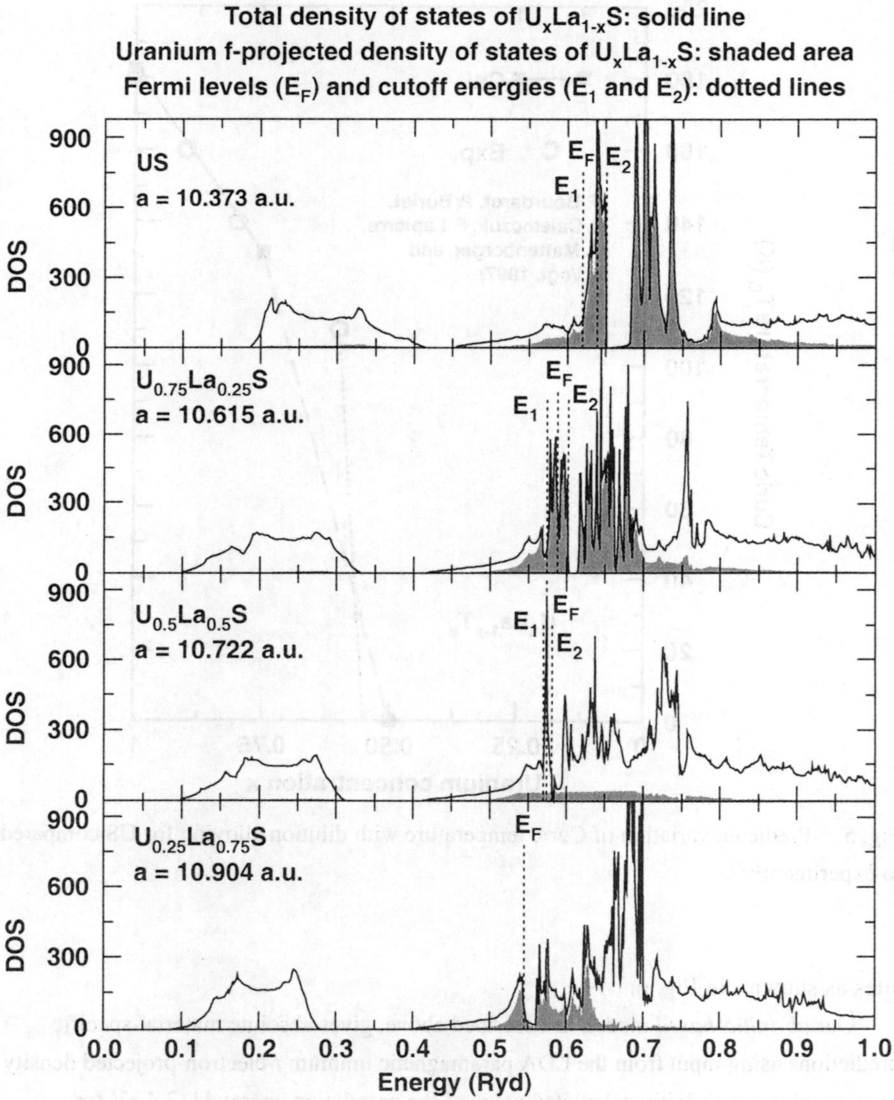

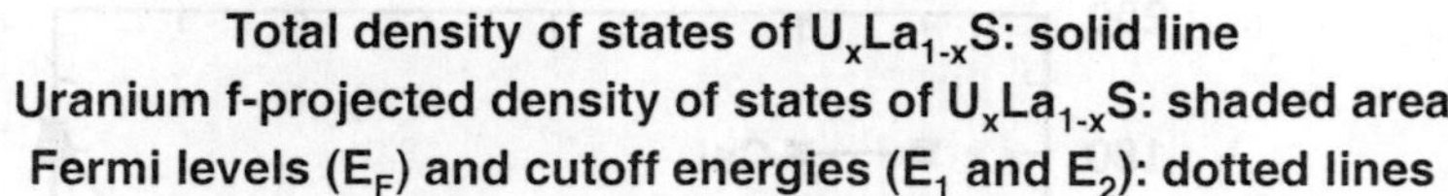

Fig. 6.   Density of states for a superlattice LDA Calculation for $U_xLa_{1-x}S$ (4 uranium-or-lanthanum, 4 sulfur sites per unit cell) from a full-potential LMTO calculation. The shaded area shows the uranium f-projected density of states.

set the energy separating the localized and delocalized f spectral weight, i.e. defining the "peak" by the selection of energies such as E_1 and E_2 shown in Fig. 6. We adopt a boundary defining the energy where, upon decreasing energy from the $j = 5/2$ peak in the density of states, there is a distinct difference between the total and the f-projected density of states. (In effect, this is the energy where significant hybridization begins to occur between the f electrons and the non-f band electrons. Operationally, we define this as the energy where the density of states first stops dropping monotonically from the peak, i.e. the first local minimum in the density of states.) This gives E_1; and E_2 is chosen as the energy on the high-energy side of the peak where the density of states has the same value as at E_1. Once this is done, there is only one *non-ab-initio* number that must be provided[12] for a given material, e.g. US. This number sets the starting (baseline) division between uranium sites where an f^3 configuration exists for long enough for those sites to be able to magnetically order and those sites which fluctuate between an f^3 configuration and an f^2 configuration sufficiently rapidly to act as magnetic holes in much the same way as a La site. We choose that single number (i.e., the baseline fraction of uranium sites with stable f^3 configuration) for the pure (unalloyed) material (e.g. US) at ambient pressure.

The catastrophic ordered magnetic moment collapse that occurs at 55% per cent uranium for $U_xLa_{1-x}S$ is then a consequence of three effects, as reflected in the LDA density of states such as that of Fig. 6, providing an extremely nonlinear magnetic response to the dilution: most importantly, the increased delocalization acts wholly to reduce the stable f^3 component; secondarily that f^3 *component becomes more localized* (because of exchange symmetry and the consequent intersite coupling through higher order hybridization effects, the f^3 component can never be strictly localized), hence *giving less coupling for a given f^3-f^3 separation*; and finally in addition the f^3 sites are spread further apart with increasing lanthanum dilution.

While probably secondary in triggering the ordered moment collapse, the extreme localization, as it disappears, of the remaining f^3 component does contribution to the collapse of magnetic ordering as uranium concentration decreases. This is because the magnetic coupling between the remaining localized $5f$ sites decreases with decreasing uranium concentration. This is quite different than the situation for hydrostatic pressure where there are two competing effects because the coupling between the remaining localized $5f$ sites increases as pressure increases. In addition, this increased localization at the localized $5f$ sites as their number decreases— may be important for the observation of correlated narrow-band behavior (with the possibility of heavy fermion effects) in the reconstructed state of the system after moment collapse as discussed in Section IV. This concentrating of spectral density may be significant for specific heat and susceptibility enhancement in that reconstructed ground state.

The way in which f electrons are counted has to be done with considerable thought and care in implementing the LDA part of the physics. Given the physical context, we do this counting by integrating the $5f$ charge within the uranium muffin-tin spheres (where in the full-potential calculation, space is divided between muffin-tin spheres centered about nuclei and interstitial). There is a difference between the way the uranium muffin-tin radius is chosen in calculating pressure effects and is calculating dilution alloying effects. For the pressure effects, the muffin-tin radius is chosen as part of the self-consistent procedure that determines the lattice constant through total-energy minimization, and thus corresponds to the lattice constant used at a specified pressure. (The muffin-tin radius is chosen to give continuity of the self-consistent potential at the radius, i.e. between muffin-tin sphere and interstitial. In the pressure calculation, this sphere radius decreases with increasing pressure.) This is not what is done for the dilution alloying. There the lattice constant is chosen from the experimental behavior for US, or for USe and UTe using Vegard's law (linear dependence on concentration), for the lattice constant, i.e. substituting lanthanum gives an expansion of the lattice. (Experimentally for US there is evidence for a departure from Vegard's law behavior near the ordered moment collapse [see Bourdarot *et al*, Refs. 22 and 23].) This means that one is faced with using some physical criterion to specify the way the muffin-tin radius is chosen. If one uses the same, now inconsistent, choice as in the pressure calculations (continuity of the self-consistent potential), this will artificially increase the uranium muffin-tin sphere, thereby altering the LDA selection of f-electrons. This changes the nature of the LDA spectral distribution. A better choice, consistent with Vegard's law, is to use fixed uranium and lanthanum muffin-tin sphere radii corresponding to the end points of the concentration variation. This is what we do. (For comparison, we have also used the first, inconsistent choice and compared the behavior found. In fact, there is very little change in the predicted magnetic ordering behavior.) A self-consistent choice of lattice constant and muffin-tin radius could be done since the supercell calculation is, in effect, the calculation for an ordered intermetallic compound, e.g. U_3LaTe_4. This would give a departure from Vegard's law. It may be that the experimental departure from Vegard's law close to the moment collapse is, wholly or in part, associated with some degree of ordering of the lattice; and it could therefore be interesting to also compare results using this way of choosing the muffin-tin radius.

Using the Hubbard model, following the procedure described in Reference 12, the coupling *(J)* between localized moments on an *fcc* lattice is obtained from the width of the localized peak in the uranium *f*-electron-projected density of states and the calculated correlation energy *(U)*. If *t* is the inter-atomic interaction, then for an *fcc* lattice, the localized peak width Γ is related to *t* through $\Gamma = 1.04$ and $J = t^2/U$. The magnetic ordering is then calculated exactly, as in Reference 12, *for an Ising lattice with randomly-located holes*. (As discussed earlier in this subsection, an Ising lattice is used to mimic the very large magnetic anisotropy of these ferromagnets; and it is very important to recognize the *essential role of the randomness* of the holes in the Ising lattice in the behavior found. A uniform distribution of the same average moment would not yield the observed behavior.)

The key point is to identify the number of holes, i.e. the number of sites not occupied by stable f^3 uranium. As stated above, we do this[12] by first establishing a baseline fraction of stable f^3 sites for the pure unalloyed material at atmospheric pressure. This is the one point at which we either insert a physically-motivated guess, or use one piece of experimental data, or use a value chosen to achieve some balance between fitting several pieces of data. The obvious choices for this purpose are the low-temperature ordered moment, the Curie temperature, or the LDA calculated *f*-electron count (which will vary slightly depending on the choice of technical method used in determining it). For US, in fact, all three of these indicate very closely the same baseline value for the praction of stable f^3 sites. The experimental low-temperature ordered moment per uranium of US from the recent neutron measurements of Bourdarot *et al*[22,23] is $1.60\mu_B$, and from the magnetization measurements of Schoenes *et al*[21] is $1.55\mu_B$ per uranium. This is slightly less than half the ordered moment expected for U^{3+} (f^3 uranium). (For atomic U^{3+}, this is $3.27\mu_B$; in the solid state, because of the breaking of spherical symmetry, one expects a slightly different value.) This would give a baseline fraction of stable f^3 of 0.49. There are several ways to technically define the *f*-count from a band calculation which give slightly different values; but these all give a value of n_f close to 2.5 for US, indicating a baseline f^3 fraction of 0.5. In addition, for US as discussed below with that choice of 0.5, the calculated T_c is 183K compared to the experimental value[41,42] of 180K. So we take the baseline value of holes as 50% for pure US at atmospheric pressure. (We note that this is a different baseline than was used in Reference 12, and there are also some changes in the technical details of the LDA calculation and of extracting the desired input information from them. Thus the predicted variation of Curie temperature with pressure shown in Figs. 2, 3 and 4 differs somewhat from that of Reference 6.)

For USe and UTe, the choice of the starting baseline fraction of stable f^3 uranium is not so clearcut. The reason for this is that our way of determining the magnetic coupling between localized moments tends to underestimate that coupling with increasing degree of localization.[27,38] Matching the low-temperature experimental ordered moment[41,42] from neutron measurements ($2.0\mu_B$ for USe and $2.25\mu_B$ for UTe) would give 0.61 and 0.69, respectively, for the baseline f^3 fractions, which are close to those expected from the f-count from band calculations. On the other hand, matching the Curie temperatures[41,42] (160K and 104K, respectively) **requires** 0.72 and 0.91, respectively. Guided by these values, we have chosen 0.68 and 0.79 as the baseline fractions of f^3 for USe and UTe, respectively. It is this greater degree of localization, of having more stable f^3, on going from US to USe to UTe that gives rise to the dramatically different variation of T_C with pressure.

Then the key input information from the LDA calculations is how this baseline number of holes increases with pressure and dilution alloying as the hybridization with the p/d electrons increases. We do this by calculating the decrease in integrated f-spectral weight per uranium contained in the localized peak of the LDA density of states. Thus for US, for pressure effects the increase in holes comes wholly from the size of that spectral transfer, which decreases the 0.5 electron per uranium that on average defines the baseline fraction of well ordered f^3 uranium.

For dilution alloying, there are two ways in which the number of holes increases. One is by the shift of spectral weight per uranium from the localized peak to the itinerant tail, i.e. the same effect as caused by pressure. In addition, substitution of a lanthanum by itself gives a hole. Thus dilution alloying has a much enhanced effect on magnetic ordering. The actual LDA calculations for dilution alloying have been carried out using supercells containing four uranium/lanthanum sites and four anion sites. Thus, at present, we can predict alloying effects only at intervals of 25% dilution.

C. Results for Low-Temperature Ordered Moment and Curie Temperature

The predicted variation of Curie Temperature with pressure and dilution alloying is compared to experiment for US in Fig. 4 and Fig. 5, respectively; while the nonmonotonic variation with pressure predicted in close agreement with experiment for USe and UTe is shown in Figs. 3 and 2. We emphasize that the predictions are absolute. All input information is ab initio except for the baseline specification of the fraction of stable f^3 uranium for unalloyed US at atmospheric pressure. In Fig. 7, we show the comparison between the predicted and experimental [Bourdarot *et al*, Ref. 22 and 23] low temperature ordered moment for $U_xLa_{1-x}S$. The remarkable agreement of prediction with experiment for $U_xLa_{1-x}S$ shown in Figures 4, 5 and 7, including the collapse of magnetic

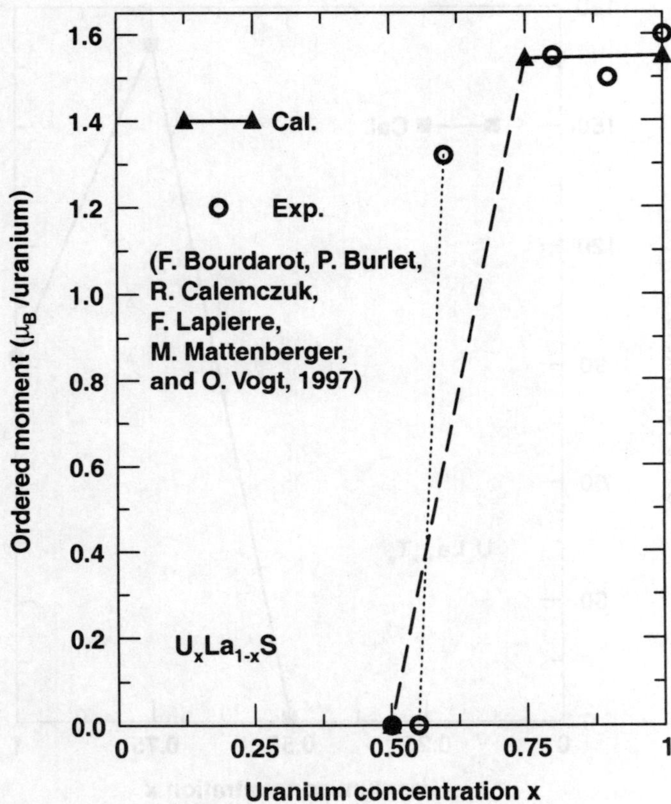

Fig. 7. Predicted variation of low-temperature ordered moment with dilution alloying for US compared to experiment[22,23].

ordering somewhat above 50% uranium, provides strong confirmation of the correctness of the physical picture used. Curie temperature and ordered moment calculations of the sort shown in Figs. 5 and 7 for the diluted US system have also been performed for the diluted USe and UTe systems, and our predictions for the Curie temperature and ordered moment in diluted UTe are shown in Figures 8 and 9, respectively. While the present calculations at intervals of 25% dilution are consistent with the experimental[43] observation that the moment collapse in diluted USe and UTe is similar to that in US in occurring at uranium concentrations not too far above 50%, calculations at 62.5% uranium (8 uranium sites in the supercell) are necessary to confirm that the detailed trend in behavior predicted between materials for the ordered moment collapse is

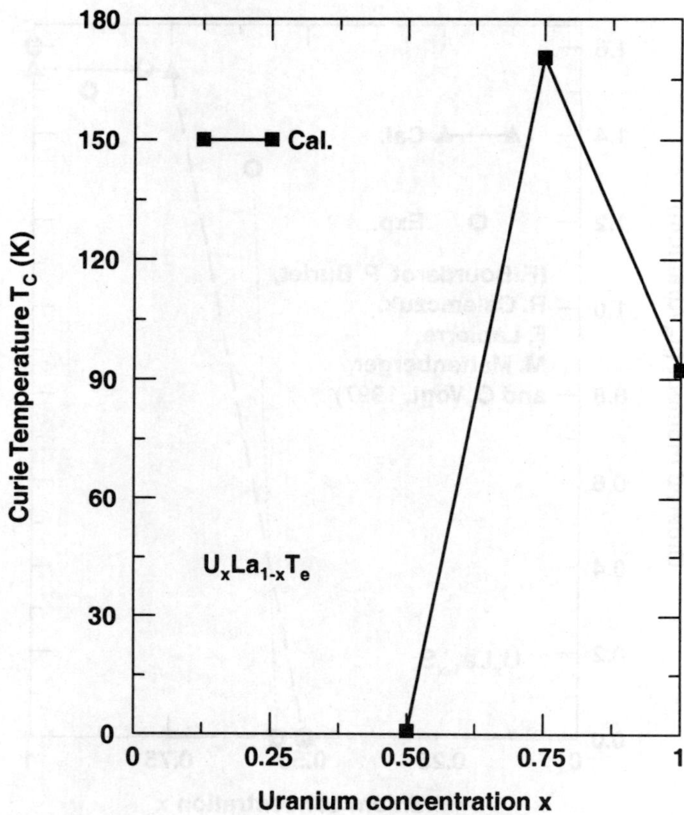

Fig. 8.   Predicted variation of Curie temperature with dilution alloying for UTe.

consistent with experiment.[21-24,43] For the baseline stable f^3 fractions used (0.50, 0.68 and 0.79 for US, USe and UTe, respectively), the predicted Curie temperature and ordered moment are 183K and $1.64\mu_B$ (experimentally 180K and $1.6\mu_B$) for US, 149K and $2.19\mu_B$ (experimentally 160K and $2.0\mu_B$) for USe, and 91K and $2.6\mu_B$ (experimentally 104K and $2.25\mu_B$) for UTe. It will be interesting to have neutron diffraction results for the UTe system of the sort already available[22,23] and shown for the US system in Figs. 5 and 7. The very substantial increase in T_c predicted (see Fig. 8) at 75% uranium in the diluted UTe is a consequence of the more localized behavior for the pure material (baseline f^3 fraction of 0.79 for pure UTe at ambient pressure) as compared to US (baseline f^3 fraction of 0.50). If not substantiated by experiment, this rise could be

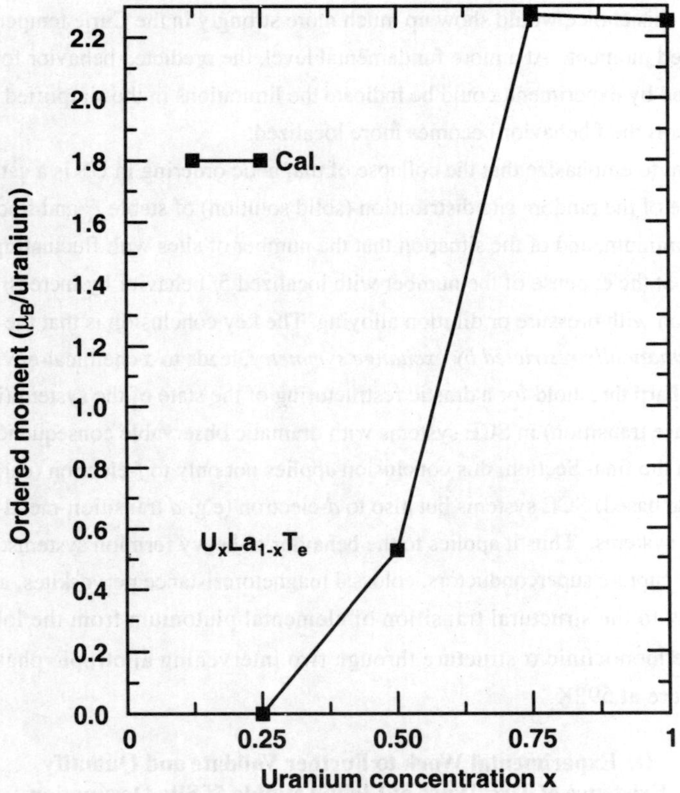

Fig. 9.   Predicted variation of low-temperature ordered moment with dilution alloying for UTe.

indicative of the detailed quantitative limitations in the simplified treatment of the intersite coupling. In fact, the predicted ordered moment behavior (see Fig. 9) for UTe is almost flat between 100% and 75% uranium, with only a very slight rise at 75% uranium. Since the Curie temperature varies quadratically with the ordered moment and linearly with the magnetical coupling, it is a much more sensitive quantity to calculate than the ordered moment. Thus the likeliest explanation for the substantial increase in the predicted Curie temperature and the very slight rise in the predicted ordered moment at 75% uranium is that the boundary of the LDA peak has been chosen at somewhat too low an energy. Since this determines both the counting of states, and hence the change in ordered moment, and the LDA f-peak width, and hence the intersite magnetic coupling, any error in this choice would show up much more strongly in the Curie temperature than in the ordered moment. At a more fundamental level, the predicted behavior for UTe, if not supported by experiment, could be indicate the limitations in the simplified treatment of exchange as the f behavior becomes more localized.

We want to emphasize that the collapse of magnetic ordering in US is a catastrophic consequence of the random site distribution (solid solution) of stable f^3 and fluctuating f^3/f^2 valence uranium, and of the situation that the number of sites with fluctuating valence is increased at the expense of the number with localized $5f$ behavior by increasing the f delocalization with pressure or dilution alloying. The key conclusion is that the hybridization, as *kinematically restricted by exchange symmetry*, leads to a chemical-environment-dependent sharp threshold for a drastic restructuring of the state of the system (i.e., a possible phase transition) in SCE systems with dramatic observable consequences. As discussed in the final Section, this conclusion applies not only to f-electron (cerium and light actinide based) SCE systems but also to d-electron (e.g. d transition-metal-oxide based) SCE systems. Thus it applies to the behavior of heavy fermion systems, high-temperature cuprate superconductors, colossal magnetoresistance pervoskites, and very interestingly to the structural transition of elemental plutonium from the low-temperature monoclinic α structure through two intervening allotropic phases to the fcc δ structure at 592K.

D. Experimental Work to Further Validate and Quantify Existence of Two Types of Light Actinide 5f Site Occupation

There are several experimental methods that can be used to quantify the existence of two types of $5f$ sites. The first method is a relatively straightforward use of magnetization and susceptibility data. The uranium monochalcogenides, and their alloys as diluted by lanthanum or yttrium,[21] as well as fcc (δ)-stabilized plutonium,[50a] have a high-temperature susceptibility (χ) that combines Curie-Weiss and enhanced Pauli contributions.

$$\chi = \chi_{CW} + \chi_{Pauli} \tag{4a}$$

$$\chi_{CW} = \frac{C}{T - \theta_p} \tag{4b}$$

$$\chi_{Pauli} = \chi_0 \left\{ 1 - \frac{\pi^2}{12} \left(\frac{k_B T}{E_F} \right)^2 \right\} \tag{4c}$$

This method is then based on the feature of the physics predicting that the increase of hybridization gives a decrease in the low temperature ordered moment that corresponds to an increase in the Pauli paramagnetic component of the paramagnetic susceptibility above T_C. This correspondence (that the increase in Pauli susceptibility quantitatively corresponds to the decrease in ordered moment) is presently being experimentally examined[50] for a number of magnetically ordered uranium compounds.

A definitive experimental technique[51] that can be used in the magnetically-ordered diluted uranium systems is neutron diffuse scattering. This requires the use of single crystals, say for 60% uranium in $U_x La_{1-x}S$, but single crystals of the necessary size should be available.

A third type of experiment involves sophisticated use[52] of x-ray absorption spectroscopy. There are two possibilities. The first of these is the more uncertain of success, but could be extended to plutonium (i.e., non-magnetically-ordered) systems. This is x-ray absorption near edge spectroscopy (XANES). This could quantitatively detect the ortho/para mixture. The second possible experiment involving x-ray absorption is magnetic XAFS (x-ray absorption fine structure) and involves comparison of measurements on single crystals of a ferromagnetic material (e.g., US) and its nonmagnetic analogue (e.g., LaS).

IV. CORRELATED NARROW-BAND BEHAVIOR IN THE STRONGLY HYBRIDIZING SUBREGIME: HEAVY FERMION BEHAVIOR

A central question is that of the relationship between the picture and results presented in the preceding section and that provided by the competing nonmagnetic and magnetic correlated ultra-narrow band picture of References 26 and 20. In Reference 26, the relationship of an ultra-narrow correlated band picture to the characteristic specific heat, susceptibility and Wilson-Sommerfeld ratio behavior for heavy fermion systems is discussed. The existence of these ultra-narrow correlated bands follows from treating two-body electronic correlations, i.e. means again as in the discussion in Section III dealing with singlet (nonmagnetic) and triplet (magnetic) two-electron correlations. In Reference 20, a recipe is presented in outline form for obtaining these ultra-narrow bands through ab initio calculations. Furthermore, in Reference 20 it is noted that the resonance widths and hybridization dressings in Tables II and III of Ref. 31 and Tables I and II of Ref. 32 correspond to widths of the 1-10 mRy (1 mRy = 13.6 meV = 158K) range needed for understanding observed heavy fermion behavior. Our overall picture then is that this ultra-narrow band picture describes the behavior *after the moment collapse phase transition* (treated in the preceding Section) and the *restructuring of the state of the system.*

It is highly desirable to develop an *ab initio* method for including two-electron correlations (i.e., for performing *ab initio* two-electron band calculations) in the $4f$ and $5f$ electronic structure for two related reasons. First, it is extremely desirable to calculate ab initio electronic and consequent properties dependent on effective enhanced mass behavior. Second, this capability opens up the possibility of direct verification of the Anderson localization mechanism we have conjectured for the phase transition between the correlated-narrow-band (enhanced mass) and the homogeneously disordered (solid-state-solution-like) phase having localized and fluctuating-valence $5f$ sites. If the domain size for Anderson localization is no larger than 20 to 30 atoms cubed (i.e. less than 1000 atoms) in size, one could possibly directly verify the stabilization of the Anderson localized state with a supercell calculation. Presumably this would be most straightforward for fcc plutonium where we hypothesize that a metastable state should exist at $T = 0$ with a 50/50 para/ortho composition.

Here we first present the main results of Reference 4 as they pertain to heavy fermion phenomenology. An important feature of these results is that the anticipated fine structuring of the $5f$ density of states into para and ortho components leads quite naturally to providing the quantitative separation between the Wilson-Sommerfeld ratio behavior for nonmagnetic and magnetic heavy fermion materials. Reference 20 contains a summary of a practical methodology for performing *ab initio* two-electron banding calculations; however, at present the only quantitative results available for the correlated-narrow band phase is that from Reference 26 based on our expectation of the results for the very narrow para/ortho fine structure imposed on the density of states by the two-electron correlations. Therefore after presenting the first-principles-based phenomenology as developed in Reference 26, in subsection B we only briefly present the way of proceeding with an *ab initio* two-electron banding calculation as discussed in Reference 20.

A. Two-Electron Banding and Heavy Fermion Phenomenology

To demonstrate the two-body approach, we start from a "crystal" of 2 atoms with one f electron each, as would be the case for cerium. The generalization to the light actinides, with more than one f electron is given in Ref. 20. The physics is dominated by the low lying states Φ_0 (nonmagnetic singlet) and Φ_1 (magnetic triplet). For simplicity, we neglect the degeneracy of Φ_1. Take $\Phi_0(n)$ and $\Phi_1(n)$ as basis functions, with $n = a,b$, indicating atom a and b. Move atom a and b close to each other so their wave functions overlap:

$$\langle \Phi_\eta(a) | \Delta H | \Phi_\eta(b) \rangle \equiv w_\eta; \tag{5a}$$

$$\langle \Phi_0(a) | \Delta H | \Phi_1(b) \rangle \equiv t, \tag{5b}$$

where $\eta = 0, 1$. (here ΔH contains the normal one-electron banding effects. Thus $t = 0$ in the spin-only case.) Let Φ_0 be at $-d$ and Φ_1 at $+d$. then, in the two-body basis $\Phi_\eta(n)$, we have the Hamiltonian

$$\begin{pmatrix} +d & 0 & w_1 & t \\ 0 & -d & t & w_0 \\ w_1 & t & +d & 0 \\ t & w_0 & 0 & -d \end{pmatrix} \tag{5c}$$

Hence, a four electron system is described by a two-body Hamiltonian. This two-body approximation is justifiable, because breaking the two-body states Φ_η will put 3 or 4 electrons on one atom, incurring high energy cost.

We set $t = 0$ first, in order to see the consequences purely of w_η. The resulting energies and eigenstates are:

$$\lambda = -d \pm w_0: \quad \Psi_0^\pm = \frac{1}{\sqrt{2}}(\Phi_0(a) \pm \Phi_0(b)); \tag{6a}$$

$$\lambda = +d \pm w_1: \quad \Psi_1^\pm = \frac{1}{\sqrt{2}}(\Phi_1(a) \pm \Phi_1(b)). \tag{6b}$$

$\psi_0^\pm$ are two nonmagnetic singlet states bonded (anti-bonded) together; and $\psi_1^\pm$ are two magnetic states bonded (anti-bonded) together. When the degeneracy of Φ_1 is included, w_1 is a matrix and t a vector. This can lead to a complicated magnetic self-splitting of the Φ_1^k band.

If we place the two atoms at $R_a = 0$ and $R_b = a$, the eigenstates can be written as

$$\Psi_\eta^k = \frac{1}{\sqrt{N}} \sum_n \Phi_\eta(n) e^{-ik \cdot R_n}, \quad k = \tag{7}$$

where $N = 2$; $\eta = 0,1$; and $k = 0 \left(k = \frac{\pi}{a} \right)$ corresponds to bonding (anti-bonding). This logic for two atoms also applies to $N \to \infty$ atoms. The system of $2N$ electrons is described by a two-body Hamiltonian and the gaps between $\pm w_\eta$ are filled with N levels. This results in a two-body nonmagnetic "singlet like" band Ψ_0^k with a width $\sim 2|w_0|$, and a two-body magnetic band Ψ_1^k with a width $\sim 2|w_1|$. $\Psi_0^\pm$ and Ψ_1^k belong to the entire crystal while preserving the integrity of the two-body atomic states, i.e. two-electron correlations. Hence, electrons in such states are itinerant, but also characterized by local correlations-being nonmagnetic or magnetic. The separation in energy between the centers of the two bands is $2|d|$.

The possible situations are best envisioned by thinking in terms of two narrow bands of fixed widths w shifting relative to each other as given by d. The ratio d/w embodies the competition between the dominance of the nonmagnetic "singlet-like" and the magnetic extended states. Since $2d$ is the band-center-to-band-center separation, for $d/w > +1$, the two bands are clearly separated, and the singlet-like nonmagnetic correlations dominate; while for $d/w < -1$, magnetic correlations dominate. For $|d/w| < 1$, there is a complicated correlated state mixing magnetic and nonmagnetic character.

We now turn on t. For simplicity, continue to let $w_0 = w_1 = w$. For the case of two atoms, the resulting energies and eigenstates are,

$$\lambda = -\beta d \pm w: \quad \Upsilon_0^\pm = \tau \, \Psi_0^\pm \pm (1 - \beta) \Psi_1^\pm; \tag{8a}$$

$$\lambda = +\beta d \pm w: \quad \Upsilon_1^\pm = \tau \Psi_1^\pm \pm (1 - \beta) \Psi_0^\pm, \tag{8b}$$

where $\beta = \sqrt{1 + \tau^2}$, and $\tau = \frac{t}{d}$. For a crystal, we have

$$\Upsilon_\eta^k = \Psi_\eta^k + (-1)^\eta \gamma^k(\tau) \Psi_{1-\eta}^k, \tag{9}$$

where $\eta = 0,1$, and $\gamma^k(\tau)$ depends on crystal structure. We then see that t mixes the Ψ_0^k and Ψ_1^k bands, and the extent of mixing increases with $|t/d|$. Accordingly, the competition between magnetic and nonmagnetic singlet-like correlations is also conditioned by $|t/d|$ in a different way, in addition to being conditioned by $|w/d|$. The following situation induced by t is interesting. When a system is totally in the nonmagnetic band which is well below the magnetic band, t still can mix a small magnetic part into the nonmagnetic band. Many "Kondo-like" materials are not totally non-magnetic, but possess very small moments with long-range ordering; and for those materials that are totally non-magnetic, there are short-range magnetic correlations. Thus weak magnetism may be attributed to the t-induced mixing described here.

In Fig. 10, we schematically show the various possibilities when we consider a nonmagnetic (para) and a magnetic (ortho) f subband and the non-f band with which they hybridize all lying in the vicinity of the Fermi energy. The various situations can explain the wide diversity of f magnetism. Figure 10 illustrates three possibilities, of how Ψ_0^k, Ψ_1^k, and Ψ_{ex}^k bands can be placed in energy space: (a) Ψ_0^k (nonmagnetic f subband) is occupied but Ψ_1^k (magnetic f subband) is mostly empty, describing the "Kondo-like" systems; (b) Ψ_0^k and Ψ_1^k are both occupied, describing the systems of reduced moments; (c) Ψ_1^k is occupied but Ψ_0^k is mostly empty, describing the strongly magnetized systems. Also, a small population in Ψ_{ex}^k describes the non-integer f population. As a tangential point, we also want to mention that if the local-environment coupling is sufficiently strong relative to extended interactions, one can have quasiparticle states which contain components having bose statistics. One then opens up the question of phenomenon such as bose condensation with possible ramifications for heavy fermion superconductivity. This situation would be analogous to that for BCS superconductivity where the statistics for higher temperatures would be fermion statistics, but the low temperature (superconducting) behavior would be caused by a relatively small number of low lying excitations of the Fermi sea having boson characteristics. (The situation for BCS theory is described at an elementary level with clear schematic figures in Refs. 53 and 54.)

Heavy Fermion Phenomenology

We now use the model situation of Figure 10 to perform numerical calculations of heavy fermion phenomena, namely the specific heat C and susceptibility χ. The essential features can be seen by neglecting the wide band Ψ_{ex}^k, and focusing on the narrow bands Ψ_0^k and Ψ_1^k, which respectively embody the singlet (nonmagnetic, para) and triplet

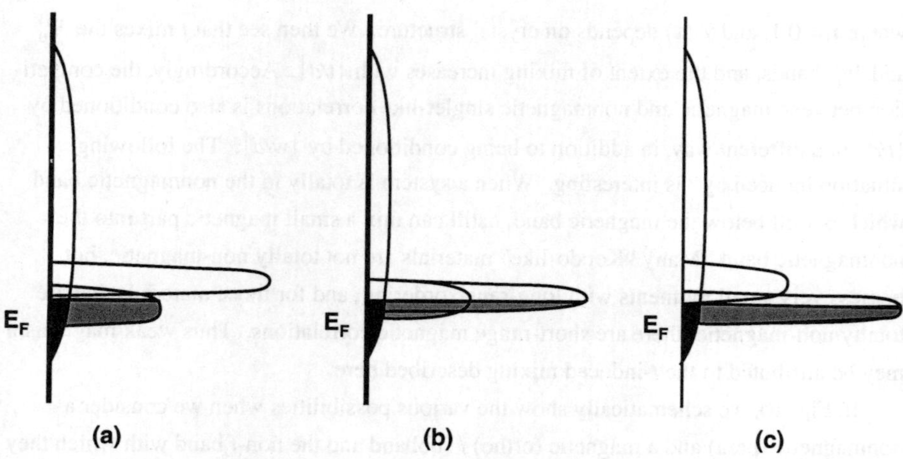

(a)   (b)   (c)

Fig. 10. Schematic display of nonmagnetic (singlet) Ψ_0^k and magnetic Ψ_1^k narrow bands, and wide band Ψ_{ex}^k. The Ψ_1^k band is indicated by a higher peak that the Ψ_0^k band. See explanations in text.

(magnetic, ortho) f-ligand correlations. We let each band have a density of states in the peak-like form $\rho(\varepsilon) = \frac{2}{\pi}\frac{w}{\varepsilon^2 + w^2}$, $2w$ being the width of the peak, and choose the origin of ε so that Ψ_0^k is centered at $-d$ and Ψ_1^k at $+d$.

Because Ψ_0^k and Ψ_1^k are narrow, care must be taken since when the temperature T is comparable to or larger than w, the chemical potential (Fermi energy) μ of the system depends on T sensitively. In the following, we take $k_B = \mu_B = 1$. then, μ is determined by

$$\int_{-\infty}^{\infty} \frac{\rho(\varepsilon+d) + \sum_{m=-1}^{1}\rho(\varepsilon+gmH-d)}{\exp\left[(\varepsilon-\mu)/T\right]+1}\,d\varepsilon = 2, \tag{10}$$

where H is an external field and $g = 2$. This equation comes from the fact that the system has two electrons per atom, with the first density of states term being for the singlets and the second for the triplets. Equation (10), which determines μ has to be solved numerically. Taking the derivative with regard to T on both sides, one obtains the equation to determine $\frac{d\mu}{dT}$, which cannot be ignored in the calculation. However, since $\frac{d\mu}{dH} = 0$

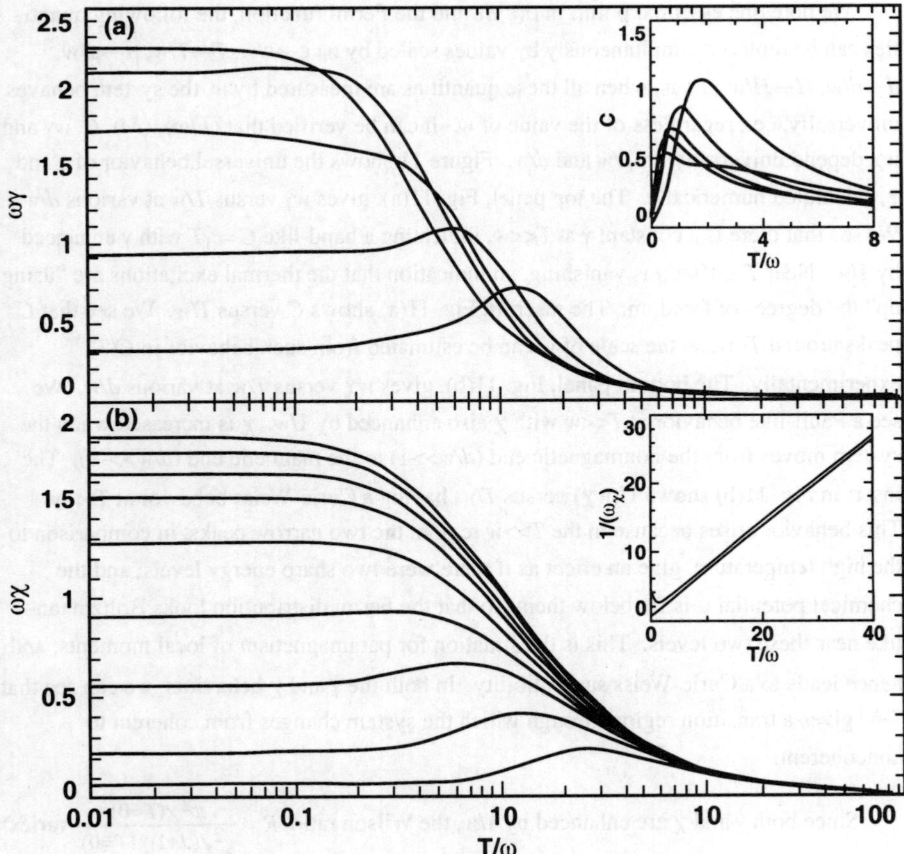

Fig. 11. (a): $\omega\gamma$ versus T/ω. From the bottom, for the five curves shown, $d/\omega = 4, 2, 1, 0, -4$, respectively. Insert: C versus T/ω. From the top, for the five curves shown, $[g^2 j(J + 1)]$, $= 4, 2, 1, 0, -4$, respectively. (b): $\omega\chi$ versus T/ω. From the bottom, for the nine curves shown, $d/\omega = 4, 2, 1, 0.5, 0, -0.5, -1, -2, -4$, respectively. Insert: $1/(\omega\chi)$ versus T/ω for $d/\omega = 2$(top), -2(bottom).

when H approaches zero, $\frac{d\mu}{dH}$ can be ignored in a zero field calculation. Once $\mu\,(T)$ is

determined, it is straight forward to calculate $C = \frac{d\langle E \rangle}{dT}$, $\gamma = \frac{C}{T}$, and $\chi = \frac{d\langle M \rangle}{dH}$ through

statistical integrals.

We note one essential point: in $\rho(\varepsilon)d\varepsilon$ and the Fermi function, the following quantities can be replaced simultaneously by values scaled by w: $\varepsilon \to \varepsilon/w$, $T \to T/w$, $\mu \to \mu/w$, $d \to d/w$, $H \to H/w$. Thus, when all these quantities are measured by w, the system behaves universally, i.e., regardless of the value of w. It can be verified that $\langle E/\omega \rangle$, $\langle M \rangle$, C, $w\gamma$ and $w\chi$ depend universally on T/w and d/w. Figure 11 shows the universal behavior of γ and χ, calculated numerically. The top panel, Fig. 11(a), gives $w\gamma$ versus T/w at various d/w. We see that there is a constant γ at $T<<w$, indicating a band-like $C = \gamma T$ with γ enhanced by $1/w$. Near $T \geq 10w$, γ is vanishing, an indication that the thermal excitations are "using up" the degrees of freedom. The insert in Fig. 11(a) shows C versus T/w. We see that C peaks around $T \sim w$, so the scale of w can be estimated from such behavior in $C(T)$ experimentally. The bottom panel, Fig. 11(b), gives $w\chi$ versus T/w at various d/w. We see a Pauli-like behavior at $T<<w$ with χ also enhanced by $1/w$. χ is increased when the system moves from the nonmagnetic end ($d/w>>1$) to the magnetic end ($d/w>>-1$). The insert in Fig. 11(b) shows $1/(w\chi)$ versus T/w, having a Curie-Weiss behavior at $T>>w$. This behavior arises because in the $T>>w$ region, the two narrow peaks, in comparison to the high temperature, give an effect as if there were two sharp energy levels; and the chemical potential μ is far below them, so that the Fermi distribution looks Boltzmann-like near these two levels. This is the situation for paramagnetism of local moments, and hence leads to a Curie-Weiss susceptibility. In both the γ and χ behaviors, we can see that $T \sim w$ gives a transition region through which the system changes from coherent to noncoherent.

Since both γ and χ are enhanced by $1/w$, the Wilson ratio $R = \dfrac{\pi^2 \chi(T=0)}{g^2 J(J+1)\gamma(T=0)}$ varies with d/w only, and by multiplying R by $g^2J(J+1)$ we can give a prediction for this quantity in absolute form as shown in Fig. 12. The lowest curve in Fig. 12 is for the case where the magnetic band has the same width as the nonmagnetic singlet band, i.e., $w \equiv w_0 = w_1$. For this case, $[g^2J(J+1)]R$ rises from 2 at the nonmagnetic end and approaches 8 at the magnetic end. As might be expected, for $|d/w|>1$, where the two narrow bands separate, the curve flattens. This $w_0 = w_1$ calculation involves the simplification that the two narrow bands ψ_0 and ψ_1 have the same width. This need not be the case. For example, since the hybridization is driven through the nonmagnetic subband, ψ_0 could be wider than ψ_1. When $w_0 \neq w_1$, the qualitative behaviors of γ and χ do not change, and $[g^2J(J+1)]R$ varies between $8\dfrac{w_0}{w_0+3w_1}$, at the nonmagnetic end, and 8, at the magnetic end. This variation is shown by the three curves in Fig. 12, corresponding to $w_0=w_1$, $2w_1$, and $3w_1$ respectively on going from bottom to top. We see

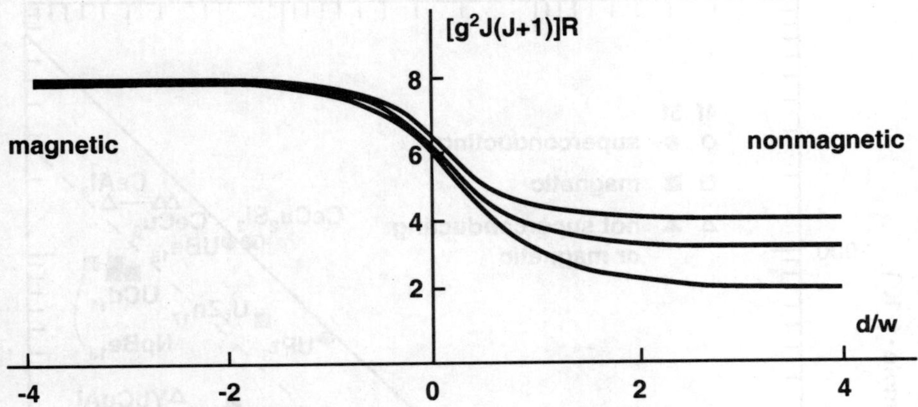

Fig. 12. The Wilson ratio put into absolute form by factoring out $[g^2j(j + 1)]$, i.e., $[g^2j(j + 1)]R$, versus d/ω, calculated at $T/\omega = 0.01$. From the bottom, the three curves shown are for $\omega \equiv \omega_0 = \omega_1$, $2\omega_1$, and $3\omega_1$ respectively.

that R maintains the same trend of increasing on going from nonmagnetic to magnetic regime when $w_0 \neq w_1$. This variation of R between the nonmagnetic and magnetic heavy fermion systems is in agreement with experiment. As summarized in Reference[55] for a group of typical heavy fermion materials, $[g^2J(J+1)]R{\sim}3.6\text{-}4.8$ for nonmagnetic systems, and $[g^2J(J+1)]R{\sim}5.1\text{-}13.5$ for magnetic systems. The calculated $[g^2J(J+1)]R$ shown in Fig. 12 going from 2-4 (nonmagnetic end) to 8 (magnetic end) reproduces the trend and the scale of the R given by these experiments.

This trend is also demonstrated in Fig. 13, (from P.A. Lee et.al.)[56] which shows the positions of about twenty heavy fermion materials in the parameter space of their γ and χ values. In addition to the original straight line given in Ref. 56 corresponding to $R=1$ calculated with $g = 2$ and $J = \frac{1}{2}$, we have added a dashed line corresponding to $R=2.5$ to this figure. We can see that most of the magnetic materials (except U_2Zn_{17}) fall in the $R>2.5$ region; and most of the nonmagnetic materials (except YbCuAl) fall in the $R<2.5$ region. Using $g=2$, $J=\frac{1}{2}$, and $w_0=w_1$, our calculated R goes from $\frac{2}{3}$ at the nonmagnetic end to $\frac{8}{3}$ at the magnetic end, reproducing the trend and the scale of the R values shown by Fig. 13.

In conclusion, we note that this model also has the potential to explain another important heavy fermion phenomenon-the anomalous resistivity behavior. The model

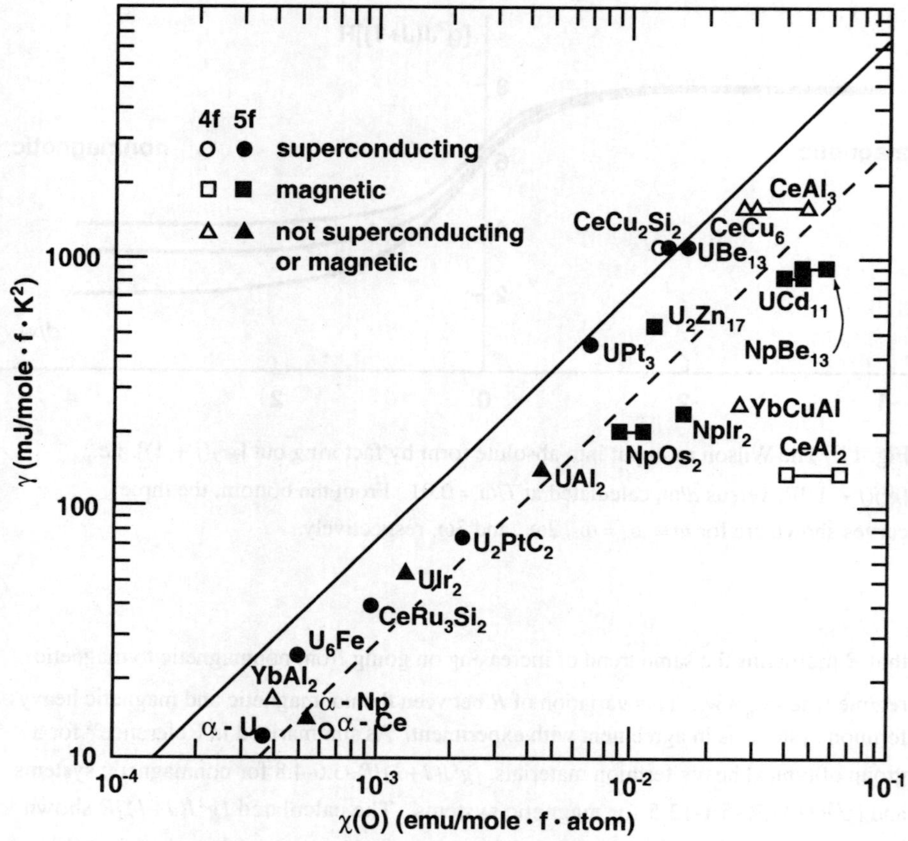

Fig. 13. γ versus χ for some heavy fermion materials. The figure is from P.A. Lee *et al.*[56] The solid line corresponds to $R = 1$ calculated with $g = 2$ and $J = $,. We have added a dashed line corresponding to $R = 2.5$. The magnetic materials, represented by squares, mostly fall in the R>2.5 region. The nonmagnetic and superconducting materials, represented by triangles and circles, mostly fall in the R<2.5 region.

suggests a mechanism where electrons transfer between the magnetic and nonmagnetic bands; and this is equivalent to emitting or absorbing magnetic excitations. Since magnetic excitations increase in population with T from $T = 0$, electronic scattering on them leads to a resistivity also increasing with T from $T = 0$. The resistivity would arrive at a maximum when magnetic excitations are saturated, i.e., when all the magnetic degrees of freedom are thermally occupied.

B. Electronic Structure Calculations Including
Locally-Based Two-Electron Correlations

We briefly describe a practical electronic structure scheme for *ab initio* calculations for real materials based on a sequence in which a one-electron linearized combination of muffin-tin orbitals (LMTO) LDA+U[57] calculation is followed by a calculation for the lattice with a helium-like two-electron hamiltonian at the f-atom sites (i.e., two-electron atoms, where initially for the core two electrons worth of charge are removed from the LMTO f-atom site). This procedure will reconstruct the LMTO bands to include two-electron texturing.

We should note that we anticipate having a deal with dispersion in bands with widths as low as the range between one to ten mRy. This is based on the simple consideration that to describe bands with effective electron masses 100 to 1000 times a free electron mass, we expect a narrowing of two to three orders of magnitude from the 1 Ry range characteristic of free-electron-like bands. This expectation is supported by the experimental observation[11] of such narrow dispersion. Furthermore, existing *ab initio* calculations of the $4f$ and $5f$ resonance widths and corresponding hybridization-induced crystal-field dressings in cerium and plutonium monopnictides, respectively, provide hybridization-induced f spectral widths ranging between less than 2 mRy (for cerium compounds and PuTe) to about 8 mRy (for PuAs). (See Tables II and III of Ref. 31 and Tables I and II of Ref. 32.) Thus a modification of the existing full potential methodology as used by us in both LDA and LDA+U calculations[13-15,20,48,49,57] is promising since it is capable of the requisite accuracy.

One would first do an LDA+U calculation[57] (i.e., a one-electron calculation). Then using the local spherical (muffin-tin) component of that potential, one can obtain fl and ll two-electron para (spatially symmetric, singlet) and ortho (spatially antisymmetric, triplet) states (where we denote p/d states by l), by constructing para (and ortho) basis states centered around the f nuclear site using the muffin-tin l and f one-electron states resulting from the LDA+U calculation. Next one would diagonalize that basis for the two-electron hamiltonian for a cerium-like (or actinide-like) atom where one has subtracted off two "appropriately averaged" electrons worth of the LDA+U charge to give the core. This then will give the two-electron correlated para and ortho states centered around the f nucleus. The basis states centered around the non-f sites can be obtained from simple Hartree products of the one-electron muffin-tin functions resulting from the LDA+U calculation. In the iterative banding process, the hamiltonian would have the helium-like two-electron hamiltonian in the f-site spheres, a spherically averaged one-electron potential at non-f sites, and would have a spatially averaged one-electron potential in the interstitial. A great advantage of this hamiltonian is that the two-electron texture would be preserved in the iterative process through both the direct and the exchange coulomb terms. The helium-like pseudohamiltonian is used only to reconstruct the LDA+U bands to include two-electron texturing, so one would populate the resulting bands with the same number of electrons as in the LDA+U calculation. (As noted above, the resonance widths and hybridization dressings in Tables II and III and Ref. 31 and Tables I and II of Ref. 32 correspond to widths of the desired 1-10 mRy range.)

We note that the resulting band structure and density of states contains all of the normal one-electron bonding, width an overlay (fine structure) of two-electron singlet (para) and triplet (ortho) structure. Thus whether one sees heavy fermion characteristics or not will depend on how well separated some of this two-electron structure is in the vicinity of the Fermi energy. On the basis of our previous calculations of reduced moment behavior,[27] our expectation is that this will depend on having a substantial p/d one-electron density of states near the Fermi energy. For uranium systems, the overall moment reduction may involve typical itinerant magnetism band broadening effects as well as correlated singlet admixture effects.

We briefly remark on the practical steps to calculate the two-body matrix elements. The nature of the on-site two-electron matrix elements to be evaluated is such that they can be "organized" from products of one-electron integrals (matrix elements). Following construction of the on-site matrix elements, the crystal hamiltonian and overlap matrices must be constructed and diagonalized to obtain the eigenstates. The intersite two-electron matrix elements giving the banding can also be constructed from one-electron integrals, but the construction is more involved than in the on-site case. The main practical difficulty of going from one-electron to two-electron calculations is that the consequent squaring of linear dimensions might result in matrices to be diagonalized of linear dimension as large as a few thousand. At this order of matrix, rather than use standard diagonalization techniques, improved times are possible with iterative techniques, but this still calls for substantial calculating.

V. BROADER IMPLICATIONS

What then is the unifying theme of this overall physics? *Why does the drastic restructuring of the ground state occur as the hybridization decreases through a critical value?* We recapitulate our reasoning that leads to the conjecture of self-induced Anderson localization that can be enhanced by magnetic ordering and modified by alloying.

We believe that the thermal stabilization of the fcc (δ) phase of elemental plutonium at 592K is an example of purely self-induced Anderson localization. By self-induced Anderson localization we mean the following situation. In the narrow band phase all $5f$ electron sites locally have fluctuating $5f$ configuration (e.g., between f^2 and f^3 for uranium) and thereby serve as the sources of the itinerant electrons which form the narrow bands. Stable (localized) configuration at some sites then has the effect of creating impurities that act as scattering centers for these itinerant electrons. When the localized sites provide sufficiently strong scattering of the itinerant $5f$ electrons originating from the remaining (fluctuating-configuration) sites, self-induced Anderson localization occurs. This scattering *must* reach critical strength at the point when the minimization of free energy requires the self-induced localization.

The discussion and results in Sections 2 and 3 provide several important clues leading to this hypothesis of Anderson localization. First, we can say that the collapse of magnetic ordering with dilution of uranium by lanthanum or yttrium in the uranium monochalcogenides clearly is not a percolation-controlled phenomenon. The concentration of uranium at the magnetic ordering collapse[21-24,43] is 55% in $U_xLa_{1-x}S$ and $U_xLa_{1-x}Se$ and 65% in $U_xY_{1-x}Te$. Furthermore, in the Cu_3Au-structure cubic systems[17,35,36] $U(In_{1-x}Sn_x)_3$, $U(Ga_{1-x}Ge_x)_3$, and $U(Ga_{1-x}Al_x)_3$ the magnetic ordering collapse occurs without dilution of the uranium. Interestingly in $U(In_{1-x}Sn_x)_3$, where the magnetic ordering collapse occurs as the Sn concentration increases to about 43%, the specific heat γ is enhanced to a distinctly heavy fermion-like value of about 500 mj/moleK^2 at a concentration of about 60% Sn. (The increase of hybridization and destruction of magnetic ordering in the $U(In_{1-x}Sn_x)_3$ and $U(Ga_{1-x}Ge_x)_3$ systems on substituting a quadravalent IVB anion for a trivalent IIIB anion is analogous to the effects on going from CeSb to CeTe with the attendant decrease in ordering temperature and ordered moment by an order of magnitudes as discussed in Ref. 27; while the change in hybridization in the $U(Ga_{1-x}Al_x)_3$ is analogous to that in the cerium monopnictides on going from the heavier strongly ordered antiferromagnets to the valence fluctuating CeN, i.e., an increase of hybridization on going to a small anion in the same column.) The next point to note is that the destruction of magnetic ordering is very sharply defined in alloying concentration and occurs with a precipitous drop of the ordered moment and Curie temperature from very substantial values to vanishing (see Figs. 5 and 7). A provocative aspect of the detailed calculations of Section III is that the concentration of stable f^3 for the pure uranium monochalcogenides at ambient pressure arrived at empirically by analyzing the experimental ordered moment, Curie temperature, and (to a lesser extend) LDA f count, are greater than 50% for all three compounds; and in pure US at ambient pressure all three indicators used provide a stable f^3 concentration remarkably close to 50%. With regard to the theory and calculations, we have already emphasized the central role of randomness in the stable f^3 versus valence fluctuating f^2/f^3 site distribution.

These clues all point to a phase transition associated with $5f$ localization, and the *centrality of randomness* points to a centrality of *configurational entropy, i.e., the entropy associated with a solid solution (homogeneously disordered alloy)* in driving the phase transition. Therefore, we arrive at the conclusion that we are looking at an instance of[58] Anderson localization. (For sufficiently strong scattering by randomly-distributed imperfections the integrated intensity of a wave initiated at some starting point becomes localized within some finite distance.) Alternatively, one could describe this as an instance of Mott localization. As Mott has pointed out,[30a] equivalent results are obtained from the Anderson and Mott points of view for noncrystalline systems, i.e. if there is no lattice periodicity. Since the *random distribution* of localized and delocalized transition-shell atoms is the *central point*, the *Anderson point of view* seems to be the *more natural one*. Furthermore, it ties in with the use of our exchange-symmetry-constrained generalization of the Coqblin-Schrieffer resonant scattering point of view to quantify the transition, for the magnetic ordering collapse in the dilution-alloyed cubic uranium monochalcogenides and Cu_3Au compounds.[21-24,43,17,35,36] However, the Mott point of view may be relevant to the departure from Vegard's law (linear dependence on alloy concentration) for the lattice parameter in the vicinity of the magnetic ordering collapse for[23] $U_xLa_{1-x}S$ and[36] $U(Ga_{1-x}Al_x)_3$.

Thus, we recognize that prior to the ordered magnetic moment collapse, even without any alloying, the relatively weakly hybridizing SCE system behaves like a random alloy. The stable f^3 sites break up the ability of the rapidly fluctuating f^2/f^3 sites to establish a coherent lattice state. Once the increase of hybridization destroys all of the stable f^3 sites, the correlated ultra-narrow band lattice state, with properties as discussed in Section IV, abruptly sets in. A significantly enhanced electronic specific heat and Pauli susceptibility will occur, which in a specific case may be sufficient to be labeled heavy fermion behavior. As hybridization increases further these bands will broaden, perhaps quite nonlinearly.

To examine the *role of configurational entropy (the entropy of mixing para and ortho sites to form a homogeneous solution)*, we are lead to consider the possibility of a phase transition in the absence of magnetic ordering, i.e., a *self-induced Anderson localization*. Given the context of SCE light actinide and cerium systems and the experimental evidence for enhanced specific heat and Pauli susceptibility in the nonmagnetically-ordered regime not far from the critical concentration for magnetic ordering, we then consider a *phase transition* via the trade-off between narrow-band f-bonding on the one hand and configurational entropy for the solid solution of stable f valence and fluctuating f valence actinide (or cerium) lattice sites. For a lattice with N f-atom sites subdivided into ortho (stable valence) (N_o) and para (fluctuating valence) (N_p) sites the configurational entropy per f-atom (site) is given by

$$S = kln\frac{N!}{N_o!N_p!}$$

(11a)

The maximum entropy is obtained when $N_o = N_p$, i.e., exact bifurcation of f-atom sites, so that

$$S_{max} = kln2$$

(11b)

Then for an f-electron contribution to the bonding energy of E_{fbond} in mRy ($1mRy=158K$ in temperature equivalent), we would expect a self-induced Anderson localization temperature given by,

$$kT_{Aloc}ln2 \approx 158kE_{fbond}$$

(11c)

$$T_{Aloc} \approx 228E_{fbond}$$

(11d)

So that a self-induced Anderson transition would occur at 228K for an f-bonding energy of 1 mRy if we consider only the ortho/para configurational entropy. This self-induced Anderson localization temperature is decreased by the excess of entropy per site associated with the degeneracy of the occupied ortho (stable valence) levels relative to that of para sites. However, unless magnetic ordering, which removes this degeneracy and thus removes the associated entropy, occurs at or above T_{Aloc} as given by Eqs., (11c) and (11d), this entropy does not aid the f-localization transition. If magnetic ordering would occur above this temperature then the lowering of the internal energy (U) by magnetic ordering outweighs the loss of magnetic-disorder entropy in contributing to the overall lowering of the free energy (F = U − TS). Therefore it will be favorable for minimization of free energy for any magnetic ordering temperature to coincide with the localization temperature.

For this magnetic-ordering enhancement of the f localization to occur, the f bonding energy to be overcome cannot be too high, otherwise the temperature will be too high for magnetic ordering. In the uranium monochalcogenides and uranium Cu_3Au compounds, UGa_3 and UIn_3, this condition is clearly satisfied. When hybridization increases sufficiently to have the lowest free energy in an f-bonding situation, the transition to the correlated narrow-band regime discussed in Section IV occurs. Clearly, further study of the $U(In_{1-x}Sn_x)$ system, where[17] the specific heat enhancement reaches values characterized as heavy fermion-like, would be interesting.

Thus in the f-dilution-alloyed uranium monochalcogenides ($U_xLa_{1-x}S$, $U_xLa_{1-x}Se$, and $U_xY_{1-x}Te$), both the ortho/para configurational entropy and the magnetic ordering energy of the ortho uranium sites favor $5f$ localization. Hence magnetic ordering and localization occur simultaneously as hybridization decreases, i.e., uranium concentration increases *(remember that the hybridization increases as there is more p/d band "solvent" per uranium f electron to "dissolve" in)*. Entropy favors at least half of the uranium localizing and thus magnetically ordering. That is why the ordered magnetic moment and Curie temperature go up so precipitously at the critical uranium concentration, and it is energetically favorable for the magnetic ordering to approach its pure uranium behavior rapidly with increasing uranium concentration.

An interesting question presents itself on approaching $T = 0$ for the pure (unalloyed) compounds where lattice periodicity (and hence Bloch's theorem) is favored by the gain in bonding energy in the coherent (periodic) state and the absence of an entropy contribution to the free energy. The solid-solution-like behavior may gradually disappear and be replaced by a situation with lattice periodicity such as uniformly-reduced-moment behavior on approaching 100% uranium at $T = 0$, or the system may settle into a metastable state. Another possibility for a state with lattice periodicity is one where the randomness disappears by the ortho and para uranium sites being located on ordered sublattices of the overall fcc uranium lattice, i.e., an ortho-para analogue to an ordered intermetallic, (a line compound). For US, with its division into half ortho and half para sites, this would then provide ordered moment behavior per uranium like that provided by[38] a spin-and-orbitally-polarized LDA calculation. The most interesting possibility, however, is that uniform magentic ordering and lattice periodicity would not be recovered gradually, but could involve an abrupt low temperature reconstruction (condensation) into a strongly magnetic state with lattice periodicity, perhaps an ultra heavy fermion state. This would give a ordered moment per uranium consistent with the LDA result[38]. (The LDA calculation cannot yield heavy fermion behavior.)

Once we recognize this "random alloy" versus "correlated lattice banding (bonding) state" competition, the winner of which is decided by the magnitude and nature of the f-band hybridization mechanism as treated via our generalization of the original Coqblin-Schrieffer[25] picture, we can ask whether there is evidence for the wide applicability of this picture to SCE phenomenology. Three important instances come to mind. One is found in the behavior of the 1·2·3 copper-oxide-based high-temperature superconductors when praseodymium or curium is substituted for yttrium.[59,60] The second is found in the behavior of the colossal magnetoresistance (CMR) perovskite manganese oxides.[61] The third instance is found in the highly unusual thermo-structural behavior of elemental plutonium[29] and the stabilization of its nominally face-centered-cubic δ phase at low temperatures, in preference to the monoclinic α phase, by the addition of small amounts of a trivalent III B additive such as gallium. In discussing the behavior of plutonium, we focus on the thermally-induced transition to the low-density fcc δ phase (The anomalies of the δ phase include negative thermal expansion and elastic properties falling at the extreme in behavior of cubic metals.) at 592K. This transition occurs after two intervening allotropic transitions (to the monoclinic β structure at 395K and the orthorhombic γ phase at 479K). As stated above and discussed further below, we attribute this transition to self-induced Anderson Localization driven by the configurational entropy of a random distribution (solid solution) of (ortho) stable f^6 and (para) fluctuating f^5/f^4 Pu sites. In this case there is no ordered magnetism. We attribute this absence of magnetic ordering to an increase in f^5 bonding energy to about 2-3 mRy for α plutonium relative to δ plutonium as compared to 1mRy or less for the uranium chalcogenides where the "spacing out" of the actinide sites by the intervening anion decreases the hybridization; and thus the temperature required for the necessary configurational entropy contribution to the free energy is too high to have magnetic ordering.

We now give further discussion of each of these three instances of the wider applicability of the physics of the hybridizing regime of behavior. For the 1·2·3 superconductors, as discussed in Reference 59, the disappearance of superconductivity and the appearance of Pr magnetic ordering point to competing hybridization mechanisms determining the choice between magnetic and superconducting behavior. Thus the orbitally driven praseodymium-praseodymium magnetic coupling, resulting from the hybridization of Pr f electrons with oxygen-p-derived band electrons, can explain the distinctly high magnetic ordering temperature of the 100% Pr system compared to other rare earth subsitutents and the low ordered moment. Depriving the copper d electrons from hybridizing with these same oxygen p electrons then kills the superconductivity. Interestingly, this picture lends some support to one class of the proposed superconductivity mechanisms. This is the class of mechanism, such as the interlayer tunneling (ILT) model[62,63] of Anderson, depending on interlayer interactions. This is because the ability to control the process via the yttrium site implies the centrality of interlayer coupling.

A central feature of ideas and theories for understanding CMR behavior in doped manganese oxides is the ferromagnetic double exchange interaction between Mn^{3+} and Mn^{4+} ions. In fact, the motivation for the original ideas of Zener[64] and the development of these ideas into the theory of double exchange[65] by de Gennes was provided by the behavior of systems such as $(La_{1-x}Ca_x)(Mn_{1-x}{}^{3+}Mn^4{}_x)O_3$. At both ends of the composition range, the manganite series $(La_{1-x}Ca_x)(Mn_{1-x}{}^{3+}Mn^4{}_x)O_3$ is an antiferromagnetic insulator. However, at intermediate compositions, both a dramatic increase in conductivity and the appearance of ferromagnetism occur. This occurrence is associated with the presence of mixed valence behavior, i.e., Mn^{3+} and Mn^{4+} occurring at crystallographically equivalent sites to compensate for the presence of both trivalent lanthanum and divalent calcium. This double exchange mechanism[65], associated with mixed valence situations in the doped manganese oxides and other such systems, is a kinetic ferromagnetic interaction mediated through charge transfer, e.g. between Mn^{3+} and Mn^{4+} via a $Mn^{3+}-O^2-Mn^{4+}$ coupling. Double exchange provides a ferromagnetic interaction between two different valances within a static random distribution. The generalized Coqblin-Schrieffer interaction[31,33] may add a dynamic enhancement of this coupling.

We attribute the transition in elemental plutonium to the low density δ phase at 592K to a self-induced Anderson localization of the $5f$ electrons as driven by the configurational entropy of a random distribution (solid solution) of (ortho) stable f^5 and (para) fluctuating f^5/f^4 Pu sites. From Equations (11c) and (11d), the transition at 592K would correspond to an f^5 bonding energy per atom of 2.5mRy which falls at the expected value. (This is about 1% of the[65a] cohesive energy per atom of 255mRy for plutonium.)

For the stabilization of fcc δ Pu (rather than monoclinic α Pu) at low temperatures, presumably the few percent of randomly located trivalent additive that stabilizes the δ structure nucleates the breakup of the coherent f bonding that stabilizes the monoclinic α

structure. This breakup of coherence would then result from the Pu-to-Pu f-bonding possibilities via hybridization being broken up by a geometrically random pattern of hybridization of Pu $5f$ electrons from one Pu atom with non-f electrons from the additive species, e.g., Ga, rather than with non-f electrons from other Pu atoms. Given the highly nonlinear nature of the effects of alloying in driving the restructuring of the ground state, this nucleation is likely to give rise to an avalanche effect, with the formation of randomly located stable f^5 Pu sites further breaking up the f bonding coherency between the fluctuating f^5/f^4 Pu sites. This is another way of saying that the randomly located additive species atoms provide sufficiently strong scattering to lower the temperature for the self-induced Anderson localization transition into the RLFS phase to be below room temperature. (For many years there has been a fruitless attempt to find ordered magnetism in the stabilized δ Pu alloys. Presumably, by randomly breaking up the hybridization-mediated coupling, the bonding to the trivalent impurities destroys the coherent magnetic coupling between the stable f^5 sites.)

The effect of gallium substituting for plutonium in elemental plutonium is quite different from that of lanthanum or yttrium substituting for uranium in the uranium monochalcogenides. When lanthanum or yttrium substitute for uranium, the dominant effect is of the valence d electrons from the lanthanum ($5d$) or yttrium ($4d$) simply blending into the hybridizing d band formed from the uranium $6d$ valence electrons. Thus one has an essentially unchanged band solvent containing less uranium $5f$ solute; and the effect is to *increase* the d-band hybridization per Pu $5f$ electron, giving $5f$ delocalization as the dominant effect. On the other hand, for gallium substituted in plutonium the valence $4p$ electrons from the gallium compete with the plutonium $5f$ electrons to hybridize with the band electrons originating from the plutonium $6d$ electrons. This not only effectively *decreases* the $6d$ hybridization per Pu $5f$ electron, it provides a severe disordered disruption of the $6d$-mediated $5f$ banding; and the effect is to localize the $5f$ electrons. Indeed the disruption of plutonium-to-plutonium $6d$-mediated $5f$-bonding is sufficiently great so that even the vestigial $5f$-bonding preserved by magnetic ordering is not present.

The avalanche effect triggered by the gallium may provide an explanation of the behavior found in the experimental extended x-ray absorption fine structure (EXAFS) and x-ray diffraction work of Faure *et al.*[66] These experiments provide both Ga-Pu and Pu-Pu interatomic distances as Ga content varies over a range between 1.89 and 10.43 at %. Each substitutional gallium atom creates a Ga-Pu cluster where the 12 nearest neighbor Ga-Pu bonds are shortened. The observed bond shortening has a minimum near or at the 7.7% Ga corresponding to one Ga for every 12 Pu's. The x-ray diffraction data and the Pu-Pu EXAFS data provide an identical negative deviation from Vegard's law (defined by the straight line between the end points defined by the extrapolation to pure Pu at one end and to Pu_3Ga at the other end) over the composition range to 10.43 at % Ga. Faure *et al*[66] state that increasing correlation between Ga-Pu clusters offers a qualitative explanation for the decrease of the Ga-Pu bond shortening up to 7.7 at % Ga, but state that they cannot understand why there is then a substantial increase in bond shortening on going to 10.4 at % Ga.

It is possible that the avalanche effect provides an explanation for the increase in bond shortening beyond 7.7 at % by providing a reinforcement of the effect of the Ga nucleation centers in breaking up the Pu-Pu bonding. The function of gallium is not only that it decreases the Pu-Pu f-f bonding by interfering with the coherent Pu-f-to-Pu-non$f$ hybridization that provides that bonding, but that is does so by random scattering of the delocalized f electron. Thereby the randomly located additive atoms break up the coherence of the overall Pu-Pu bonding (banding) for the lattice. This being the case, the gallium is less effective in stabilizing the δ structure as its own arrangement becomes more coherent over the lattice. As indicated by Faure *et al*,[66] for any physically reasonable mechanism for the additive-induced stabilization of δ Pu one would expect the stabilizing influence per Ga to decrease until saturation is reached, i.e., until every Pu site has the same opportunity to be affected by a Ga site. However, the present mechanism does have the advantage of providing an explanation for the increase in Pu-Ga bond-length shortening beyond 7.7 at % Ga. The stabilizing effect per Ga depends both on deminishing and on breaking up the coherency of Pu-Pu bonding. Thus, in our picture, as coherent hybridization per Pu is decreased by the increase of Ga, at some point the threshold for the spontaneous generation of stable localized f^5 Pu sites is reached, i.e. the point where if the magnetic coupling between these sites were not interfered with by the Ga, spontaneous strong magnetic ordering would occur. This gives a strengthening of stabilization with added Ga beyond 7.7 at % Ga.

A further related interesting question is, how does the physics of the thermal stabilization of δ plutonium relate to the anomalous melting behavior of plutonium and neptunium? The melting temperature of plutonium is 913K and that of neptunium, which immediately precedes plutonium in the periodic table, is 912K. This is quite low compared to those of thorium (2028K), protactinium (1845K), and uranium (1408K) which precede neptunium, and compared to those of americium (1449K) and curium (1618K) which follow plutonium in the $5f$ row of the periodic table. Thus the melting temperatures of Pu and Np are depressed by 500-600K with respect to the immediately neighboring U and Am, and by a significantly larger amount on going further away in the actinide row of the periodic table. *Sublimation corresponds to a full debonding*, i.e., total destruction of coherent delocalization of the valence and transition-shell electrons in the solid. Thus one could think of *sublimation as the phase transition associated with Anderson localization* as the heating of a solid generates a sufficient number of imperfections to provide the critical strength of scattering. The Random-Localized-Fluctuating-Site (RLFS solid-solution-like) phase transition corresponds to the part of the *sublimation* provided by the loss of $5f$ bonding. Since the δ-Pu transition temperature corresponds to the configurational entropy gain necessary to counterbalance the $5f$ bonding, that temperature (592K) should also correspond rather closely to the depression of the *sublimation* temperature since correspondingly less additional thermally-induced lattice defects are required for the overall-*sublimation* Anderson localization. We then suggest that the reduced melting temperature reflects this expected reduced sublimation temperature. *In a sense, the melting behavior of plutonium is analagous to having the eutectic of ortho and para plutonium since the ratio of para to ortho plutonium has adjusted itself, and hence the melting temperature, to minimize the free energy in the solid state.* (The eutectic temperature is the temperature, and hence alloy composition, where the minima in the upper temperature of the solid phase and lower temperature of the liquid phase coincide.)

The transitions to the monoclinic β phase at 395K and to the orthorhombic γ phase at 479K may be thought of as partial meltings of the $5f$ bonding. Presumably the immediately neighboring elements uranium and americium, have a much decreased melting-temperature depression because they are sufficiently closer to full-$5f$ delocalization and coherent-$5f$ localization, respectively. Uranium presumably would undergo a self-induced RLFS transition sufficiently close to the overall melting temperature so that one lowered phase-transition temperature, rather than two distinctly different ones, occurs; and americium probably is in an RLFS phase over much of the temperature range below melting.

The behavior of elemental cerium may provide another example of a structural phase transition being engendered by a self-induced Anderson localization transition (with possible enhancement by magnetic ordering). The low temperature structure of cerium[34,67] is the fcc α phase. At room temperature, the stable structure is the expanded fcc γ phase. At atmospheric pressure, cycling between room temperature and approximately 78K (liquid nitrogen temperature) gives a β double hexagonal close packed[34] form which by room temperature is restored to the γ phase. So that there is a "sluggish" phase transition from α to β to γ as temperature increases. By applying hydrostatic pressure, a transition directly from the α fcc phase to the expanded γ fcc phase is obtained. The α phase has somewhat enhanced Pauli susceptibility;[34] while the β

phase has[34] antiferromagnetic ordering at 13K with Cure-Weiss susceptibility above T_N; and the γ phase has Curie-Weiss susceptibility.

For cerium, for which the $4f$ behavior is more atomiclike than the $5f$ behavior in the light actinides, the role of exchange is more complicated than that presented in Section II and applied to light actinide materials. Specifically, for cerium exchange may complicate the behavior in the randomly localized RLFS phase by to some extent "homogenizing" the behavior at the two types of cerium site as may be the case in CeTe. (This conjecture for CeTe is based on the situation that in CeTe, our treatment[27] for CeTe and CeSb based on Eq. (2) *for a periodic lattice gives very close ab-initio-based absolute agreement with the experimental Néel temperature and ordered moment.* For CeTe both these quantities are decreased by about an order of magnitude with respect to those CeSb. In CeSb, the $4f$ behavior is sufficiently atomiclike so that the predicted and observed ordered moment is very close to free Ce^{3+} behavior, but the weak hybridization does give an enhanced Néel temperature.) However, for elemental cerium a magnetic enhancement of the Anderson localization that is decreased compared to a case like US could explain the sequence of transitions as temperature decreases: from expanded fcc γ, to antiferromagnetic β *dhcp* contracted by less than 1% in volume, to nonmagnetic fcc α contracted by more than 10% in atomic volume. The "sluggishness" in the $\gamma - \beta$ transition for elemental cerium and the antiferromagnetism of the β phase would both be expected for the increasingly important effects of the coulomb exchange on the $5f$ electron dynamics and the magnetic ordering as the f behavior becomes more atomiclike. For cerium at the pressures of a few kilobars used to go directly between the γ and α phases on cooling, the $4f$ bonding is presumably not sufficient to cause a departure from the close-packed lattice preferred by non-f bonding, but does cause an increase in density.

CONCLUDING REMARKS

In this Section I have discussed a few examples of the broader implications of the physics treated in this chapter for the electronic and magnetic properties of solids composed in part or fully from d and f transition-shell atoms. In the broadest sense, the physical picture developed here shows that hybridization treated via the Coqblin-Schrieffer resonant-scattering point of view[25] as generalized herein provides the connection (under certain conditions described herein, an actual phase transition) between localized (i.e., coupled magnetic ions) magnetism and weakly-magnetic or nonmagnetic heavy fermion behavior (which can be treated by implementing the methodology described in Section IV B) and which itself merges into itinerant-electron magnetism as hybridization strengthens.

ACKNOWLEDGMENTS

This work was supported by National Science Foundation Grant No. DMR-91-20333. I have benefited from discussion of their data with F. Bourdarot, P. Burlet, G.M. Kalvius, and O. Vogt, and I appreciate their permission to quote their results. Valuable discussion was provided by S. Beiden, G.H. Lander, Y.L. Lin, P.A. Montano, L. Muratov, and S. Mukherjee. Invaluable support was provided by my association with Los Alamos National Laboratory, and especially by L.E. Cox, A.C. Lawson, J.L. Smith, and M.F. Stevens of LANL.

REFERENCES

1.	Z. Fisk, D. W. Hess, C. J. Pethick, D. Pines, J. L. Smith, J. D. Thompson and J. O. Willis, Science **239**, 33 (1988).

2.	D. W. Hess, P. S. Riseborough and J. L. Smith, "Heavy Fermion Phenomena", pp 435-463, Encyclopedia of Applied Physics, Vol. **7** (ed, G. Trigg, VCH Publishers Inc., New York, 1993).

3.	J. M. Lawrence and D. L. Mills, Comments Cond. Mat. Phys. **15**, 163, (1991).

4.	N. Grewe and F. Steglich, chapter 97, pp 343-474 in vol. 14, Handbook on the Physics and Chemistry of the Rare Earths (eds K.A. Gschneidner, Jr. and L. Eyring, Elsevier Science Publishers, Amsterdam, 1991).

5.	J. W. Allen, S. -J. Oh, O. Gunnarsson, K. Schönhammer, M. B. Maple, M. S. Torikachvili and I. Lindau, Adv. In Physics **35**, 275, (1986).

6.	An extensive discussion of the Kondo effect including a review of the development of the theory with key references can be found in N. E. Bickers, D. L. Cox and J. W. Wilkins, Phys. Rev. B **36**, 2036, (1987).

6a.	A. C. Hewson, "The Kondo Problem to Heavy Fermions", (Cambridge University Press, 1993).

7.	A succinct discussion of the Kondo resonance with some convenient diagrams is provided by Section IV and Figs. 1 and 46 in P. Thalmeier and B. Lüthi, chapter 96 in vol. 14, "Handbook on the Physics and Chemistry of the Rare Earths" (eds K. A. Gschneidner, Jr. and L. Eyring, Elsevier Science Publishers, Amsterdam, 1991).

8.	F. Patthey, J. M. Imer, W. D. Schneider, H. Beck, Y. Baer, and B. Delley, Phys. Rev. B **42**, 8864, (1990).

9. J. Lawrence, A. J. Arko, J. J. Joyce, P. C. Canfield, Z. Fisk, J. D. Thompson and R. J. Bartlett, J. Magn. Magn. Mater. **108**, 215, (1991).

10. J. J. Joyce, A. J. Arko, J. Lawrence, P. C. Canfield, Z. Fisk, R. J. Bartlett and J. D. Thompson, Phys. Rev. Lett., **68**, 236, (1992).

11. A. J. Arko, J. J. Joyce, A. B. Andres, J. D. Thompson, J. L. Smith, E. Moshopoulou, Z Fisk, A. A. Menovsky, P. C. Canfield and C. G. Olsen, Physica B **230-232**, 16,(1997).

12. Q. G. Sheng and B. R. Cooper, J. Magn. Magn. Mater. **164**, 335, (1996).

13. B. R. Cooper and Y. -L. Lin, J. Appl. Phys. **83**, 6432, (1998).

14. B. R. Cooper, Y. -L. Lin and Q. G. Sheng, Physica B **259-261**, 176, (1999).

15. B. R. Cooper, Y. -L. Lin and Q. G. Sheng, J. Appl. Phys. **85**, 5338, (1999).

16. B. R. Cooper, O. Vogt, Q.-G. Sheng and Y. -L. Lin, Phil. Mag. B **79**, 683, (1999).

17. L. W. Zhou, C. L. Lin, J. E. Crow, S. Bloom, R. P. Guertin and S. Foner, Phys. Rev. B **34**, 483, (1986).

18. D. Forster, "Hydrodynamics, Fluctuations, Broken Symmetry, and Correlation Functions", (W. A. Benjamin, Inc., Reading, MA.,1975).

19. G.-J. Hu and B. R. Cooper, Physica B **163**, 483, (1990); G.-J. Hu and B. R. Cooper, Phys. Rev. B **48**, 12743, (1993).

20. B. R. Cooper, Y. -L. Lin and Q. G. Sheng, J. Appl. Phys. **81**, 3856, (1997).

21. J. Schoenes, O. Vogt, J. Löhle, F. Hulliger and K. Mattenberger, Phys. Rev. B **53**, 14, 987, (1996).

22. F. Bourdarot, P. Burlet, R. Calemczuk, F. LaPierre, K. Mattenberger and O. Vogt, J. de Actinides extended abstract 08.2 pp 79-80. Dijon, France, April (1997).

23. F. Bouradarot, A. Bombardi, P. Burlet, R. Calemczuk, G. H. Lander, F. Lapierre, J. P. Sanchez, K. Mattenberger and O. Vogt, European Physical Journal B **9**, 605, (1999) and extended abstract P1 (pp 74-75) 29th Journees des Actinides, Luso, Portugal (1999).

24. G. Grosse, G. M. Kalvius, A. Kratzer, E. Schreier, F. J. Burghart, K. Mattenberter and O. Vogt, J. Magn. Magn. Mater. **205**, 79, (1999).

25. B. Coqblin and J. R. Schrieffer, Phys. Rev. **185**, 847, (1969).

26. Q. G. Sheng and B.R. Cooper, Phil. Mag. Lett. **72**, 123, (1995).

27. Q.G. Sheng and B. R. Cooper, Phys. Rev. B **50**, 965, (1994).

28. J. W. Allen and R. M. Martin, Phys. Rev. Lett. **49**, 1106, (1982).

29. For a discussion of the six allotropic forms of plutonium see pp 158-169 in "The Structures of the Elements" by Jerry Donohue, Robert E. Krieger Publishing Co., Malabar, FL (1982).

30. A. Georges, G. Kotliar, W. Krauth and M. J. Rozenberg, Rev. Mod. Phys. **68**, 13, (1996).

30a. See N. F. Mott, pp 588-591 (specifically discussion in the latter part of column 2 on page 590), *Encyclopedia of Physics* (eds R.G. Lerner and G.L. Trigg, Addison-Wesley Publishing Co., Reading, MA, 1980). Also see remarks at end of column 2 on page 718 of the article by N.F. Mott in *Encyclopedia of Physics* Second Edition (R. G. Lerner and G. L. Trigg eds, VCH Publishers, New York, 1990).

31. J. M. Wills and B. R. Cooper, Phys. Rev. B **36**, 3809, (1987).

32. J. M. Wills and B. R. Cooper, Phys. Rev. B **42**, 4682, (1990).

33. C. Sanchez-Castro, B. R. Cooper and K. S. Bedell, Phys. Rev. B **51**, 12506, (1995).

34. D. C. Koskenmaki and K. A. Gschneider Jr, Chapter 4, pp 337-377 in Volume 1, "Handbook on the Physics and Chemistry of the Rare Earths" (eds K.A. Gschneidner Jr, and J. Eyring, North-Holland Publishing Co., 1978).

35. L. W. Zhou, C. S. Jee, C. L. Lin, J. E. Crow, S. Bloom and R. P. Guertin, J. Appl. Phys. **61**, 3377, (1987).

36. C. Jee, T. Yuen, C. L. Lin, J. E. Crow, S. Bloom and R. P. Guertin, Abstract KK6, Bull. Am. Phys. Soc. **32**, 720, (1987).

37. J. H. Van Vleck, "The Theory of Electric and Magnetic Susceptibilities", pp 287-297, Oxford University Press, London, 1932; for a simple physical picture see page 426 in C. Kittel *Introduction to Solid State Physics*, Seventh Edition, John Wiley & Sons, New York, 1996.

38. Q. G. Sheng, B. R. Cooper and S. P. Lim, Phys. Rev. B **50**, 9215, (1994).

39. See B. R. Cooper, R. Siemann, D. Yang, P. Thayamballi and A. Banerjea, Chapter 6 in *Handbook on the Physics and Chemistry of the Actinides*, edited by A. J. Freeman and G. H. Lander, Elsevier Science Publishers, Amsterdam, (1985).

40. B. R. Cooper, G. -J. Hu, N. Kioussis and J. M. Wills, J. Magn. Magn. Mater. **63-64**, 121 (1987); B.R. Cooper, J. Less-Common Metals **133**, 31 (1987); B. R. Cooper, J. M. Wills, N. Kioussis, and Q. G. Sheng, J. de Phys. **49**, C8-463, (1988).

41. J. -M. Fournier and R. Troc, Chapter 2 "Bulk Properties of the Actinides" in *Handbook on the Physics and Chemistry of the Actinides*, Volume 2 (eds A. J. Freeman and G. H. Lander, Elsevier Science Publishers, Amsterdam, 1985).

42. O. Vogt and K. Mattenburger, Chapter 114 "Magnetic Measurements on Rare Earth and Actinide Monopnictides and Monochalcogenides" in volume 17 of the *Handbook on the Physics and Chemistry of Rare Earths* (eds K. A. Gschneidner, Jr., L. Eyring, G. H. Lander, and G. R. Chopin, Elsevier Science Publishers, Amsterdam, 1993).

43. O. Vogt, private communication, (1998).

44. P. Link, U. Benedict, J. Wittig, and H. Wühl, J. Phys.: Condens. Matter **4**, 5585, (1992).

45. A. L. Cornelius, J. S. Schilling, O. Vogt, K. Mattenberger and U. Benedict, J. Magn. Magn. Mater. **161**, 169, (1996).

46. P. Link, U. Benedict, J. Wittig and H. Wühl, Physica B **190**, 68, (1993).

47. B. R. Cooper and Y -L. Lin, pp 251-265 in *Electron Correlations and Materials Properties* (eds., A. Gonis, N. Kioussis and M. Cliftan, Klewer Academic/Plenum Publishers, 1999).

48. D. L. Price and B. R. Cooper, Phys. Rev. B **39**, 4945, (1989).

49. D. L. Price, J. M. Wills and B. R. Cooper, Phys. Rev. B **48**, 15301, (1993).

50. O. Vogt, private communication, October, 1998; O. Vogt, K. Mattenberger, and J. Löhle, extended abstract, pp 58-59, 29th Journees des Actinides, May, 1999, Luso, Portugal.

50a. S. Méot-Reymond and J. M. Fournier, J. of Alloys and Compounds **232**, 119,(1996).

51. G. H. Lander, private communication, April, 1999.

52. P. A. Montano, private communication, December, 1988.

53. See discussion of pp 355-357 of C. Kittel, *Introduction to Solid State Physics* (7th edition, Wiley, New York, 1996).

54. See Fig. 3.1, on page 56, and discussion thereon in M. Tinkham, *Introduction to Superconductivity* (Second Edition, McGraw-Hill, New York, 1996.)

55. G. R. Stewart, Rev. Mod. Phys. **56**, 755, (1984).

56. P. A. Lee *et. al.*, Comm. Cond. Mat. Phys., **12**, No. 3, pp 99-161, (1986).

57. The variation of the LDA+U method to be used is described in D. L. Price, B. R. Cooper, S. -P. Lim, and I. Avgin, to be published in Phys. Rev. B, (1999). It is functionally similar to the self-interaction-corrected pseudopotential method, and total angular **mementum** eigenstates are chosen for the occupied state.

58. P. W. Anderson, Phys. Rev. **109**, 1492, (1958).

59. B. R. Cooper, "Magnetism Versus High Temperature Superconductivity-The Role of Competing Hybridization" in *Proceedings of High-T$_c$ Superconductors: Magnetic Interactions*, pp 7-16 (eds. L. H. Bennett, Y. Flom, G. C. Vezzoli, World Publishing, Singapore, 1989).

60. L. Soderholm, K. Zhang, D. G. Hinks, M. A. Beno, J. D. Jorgenson, C. U. Segre, and I. K. Schuller, Nature (London) **328**, 604, (1987); H. B. Radousky, K. F. McCarthy, J. L. Peng and R. N. Shelton, Phys. Rev. B **39**, 12383, (1989); L. Soderholm, C. K. Loong, G. L. Goodman, U. Welp, J. Bolender and C. W. Williams, Physica B **163**, 655, (1990).

61. S. Jin, T. H. Tiefel, M. McCormack, R. A. Fastnacht, R. Ramesh and L. H. Chen, Science **264**, 413, (1994); K. von Helmolt, J. Wecker, B. Holzapfel, L. Schultz, and K. Samwer, Phys. Rev. Lett. **71**, 2331, (1993); K. Chahara, T. Ohno, M. Kasai, and Y. Kozono, Appl. Phys. Lett. **63**, 1990, (1993); R. M. Kusters, J. Singleton, D. A. Keen, R. McGreevy, and W. Hayes, Physica B **155**, 362, (1989).

62. P. W. Anderson, *The Theory of Superconductivity in the High T_c Cuprate Superconductors* (Princeton U. Press, Princeton, 1997); Science **279**, 1196, (1998).

63. A. J. Leggett, Science **279**, 1157, (1998).

64. C. Zener, Phys. Rev. **82**, 403, (1951).

65. P. G. DeGennes Phys. Rev. **18**, 141, (1960).

65a. See Fig. 3 page 158 in M. S. S. Brooks, B. Johansson and H. L. Skriver, vol. 1 *Handbook on the Physics and Chemistry of the Actinides* (eds, A. J. Freeman and G. H. Lander, Elsevier Science Publishers, Amsterdam, 1984).

66. Ph. Faure, B. Deslanders, D. Bazen, C. Tailland, R. Doukhan, J. M. Fournier and A. Falanga, J. of Alloys and Compounds **244**, 131, (1996).

67. For a discussion of the allotropic forms of cerium see pp 88-95 in *The Structures of the Elements* by Jerry Donohue, Robert E. Krieger Publishing Co., Malabar, FL 1982)

6

MAGNETISM IN THE Pr
CONTAINING CUPRATES

H. B. Radousky

Lawrence Livermore National Laboratory, Livermore, CA 94550
and UC-Davis Department of Physics, Davis, CA 95616

I. INTRODUCTION

The purpose of this chapter is to discuss the magnetism which has been observed in some of the Pr containing cuprate systems. The rationale for including these materials in the heavy fermion class of compounds is primarily their high values of the specific heat. For example, in $PrBa_2Cu_3O_7$ the value of γ is 100 mJ/K^2-mole Pr [1-4], and for $(Pr_{1.5}Ce_{0.5})Sr_2Cu_2NbO_{10}$ the value is 99 mJ/ K^2-mole Pr [5], which puts them in the heavy fermion range. In addition, the $PrBa_2Cu_3O_7$ material has been shown to have a γT cutoff near 50K which is characteristic of heavy fermion materials.[2]

There are several related structures which fall into the category of having high values of the specific heat, correlated with unusually large values of the Pr Néel temperature compared to the other rare earth (RE) members of the same series. These structures are $PrBa_2Cu_3O_7$, $PrBa_2Cu_2NbO_8$, $(Pr_{1.5}Ce_{0.5})Sr_2Cu_2NbO_{10}$, $PrBa_2Cu_2TlO_7$, $PrSr_2Cu_2TlO_8$, $PrSr_2Cu_3Pb_2O_8$, $PrSr_2Cu_2HgO_6$, and $PrSr_2Cu_2TlO_8$. There are also Pr

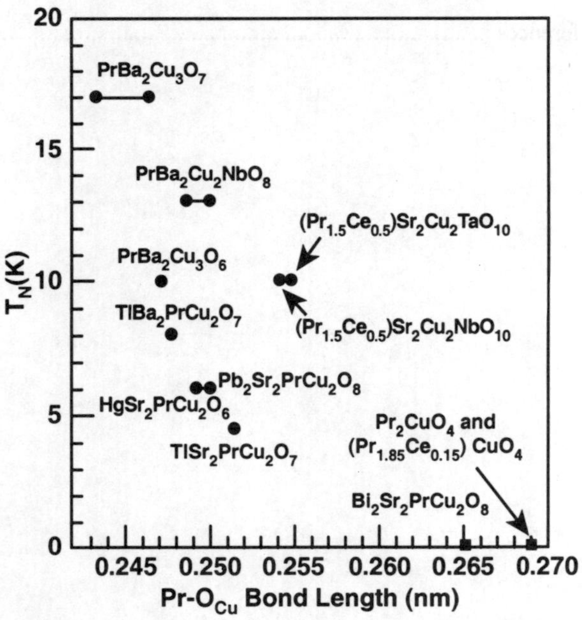

Fig. 1. T_N versus the Pr-CuO_2 planar oxygen bond length for a variety of Pr-containing cuprates. References for values are as follows: $PrBa_2Cu_3O_7$ and $PrBa_2Cu_3O_6$ (1, 7); $PrBa_2Cu_2NbO_8$ (8, 9); data for $TlBa_2PrCu_2O_7$, $TlSr_2PrCu_2O_7$, $HgSr_2PrCu_2O_6$, and $Pb_2Sr_2PrCu_3O_8$ (10); Pr_2CuO_4 (11), $Bi_2Sr_2PrCu_2O_8$ (5, 12), and $(Pr_{1.5}Ce_{0.5})Sr_2Cu_2NbO_{10-\delta}$ and $(Pr_{1.5}Ce_{0.5})Sr_2Cu_2TaO_{10-\delta}$ (5, 6, 13). After Goodwin *et al.*[6]

containing cuprate materials such as Pr_2CuO_4 and $PrSr_2Cu_2Bi_2O_8$ which do not show high specific heat values or anomalous Néel temperatures. The transition temperatures and Pr-CuO_2 bond distances for these materials are summarized[6] in Figure 1.

For the twelve materials shown in Fig. 1, those with Pr-CuO_2 bond lengths less than 2.6 Å show Pr Néel temperatures between 5 and 17K, while those with Pr-CuO_2 bond lengths above 2.6 Å have Néel temperatures below 1K, which are consistent with ordering temperatures in the other RE members of that series. This chapter will focus on the magnetism in these Pr containing cuprates, and not address any of the issues surrounding the loss of superconductivity as Pr replaces Y in $YBa_2Cu_3O_7$.[1] All of the materials discussed are Pr containing cuprates which are insulating, and have specific magnetic signatures which are absent from the other rare earth containing analogs of that structure.

In particular, the chapter will concentrate on the $PrBa_2Cu_3O_7$, $PrBa_2Cu_2NbO_8$ and $(Pr_{1.5}Ce_{0.5})Sr_2Cu_2NbO_{10}$ systems. For the latter two oxide materials, they have been well studied, but are less widely known than $PrBa_2Cu_3O_7$. They also have the

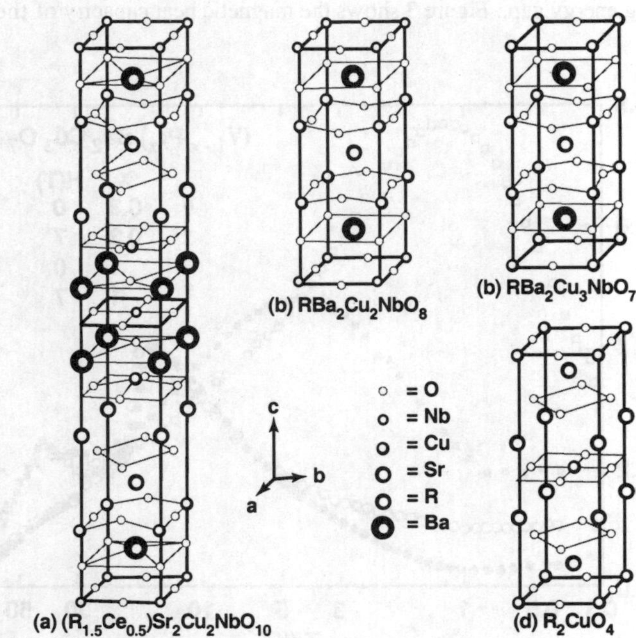

Fig. 2. Unit cells for the (a) $(R_{1.5}Ce_{0.5})Sr_2Cu_2NbO_{10-\delta}$, (b) $RBa_2Cu_2NbO_8$, (c) $RBa_2Cu_3O_7$, and (d) R_2CuO_4 T' high T_C cuprate structures. After Goodwin *et al.*[6]

advantage of removing the Cu-O chains from the picture.

Figure 2 shows the progression and similarities of these three structures. In going from $PrBa_2Cu_3O_7$ to $PrBa_2Cu_2NbO_8$, NbO_2 planes replace the CuO chains. In then going to the $(Pr_{1.5}Ce_{0.5})Sr_2Cu_2NbO_{10}$ structure, the Pr plane is replaced by a Pr_2O_2 fluorite structure, which is illustrated by Pr_2CuO_4. The introduction of the Pr_2O_2 structure has the effect of introducing a glide plane into the structure, which results in a doubling of the unit cell length to 30Å. The presence of the Pr_2O_2 structure also has the effect of greatly increasing the complexity of the magnetism which is observed relative to the $PrBa_2Cu_3O_7$ or $PrBa_2Cu_2NbO_8$ materials.

II. HEAT CAPACITY

The majority of the heat capacity work in the Pr containing cuprates has focused on the $PrBa_2Cu_3O_7$ structure. The heat capacity of the $Y_{1-x}Pr_xBa_2Cu_3O_7$ system has been presented in a number of papers[2-4,14-19] in an attempt to understand the magnetic ordering, role of hyperfine fields, value of the Sommerfeld γ, crystal fields, and the value of the superconducting energy gap. Figure 3 shows the magnetic heat capacity of the x = 0.3

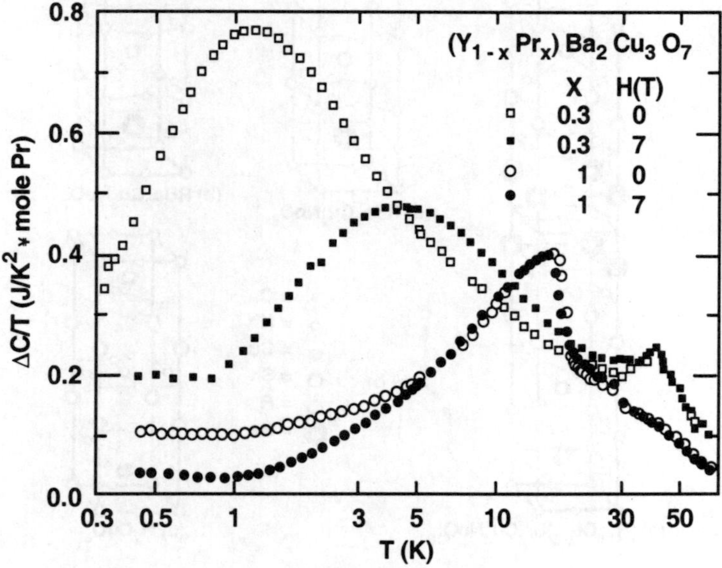

Fig. 3. $\Delta C/T$ for $Y_{1-x}Pr_xBa_2Cu_2O_7$ x = 0.3 and 1, where ΔC is the magnetic ordering plus "heavy-fermion-like," γT, contributions to C. After Fisher *et al.*[4]

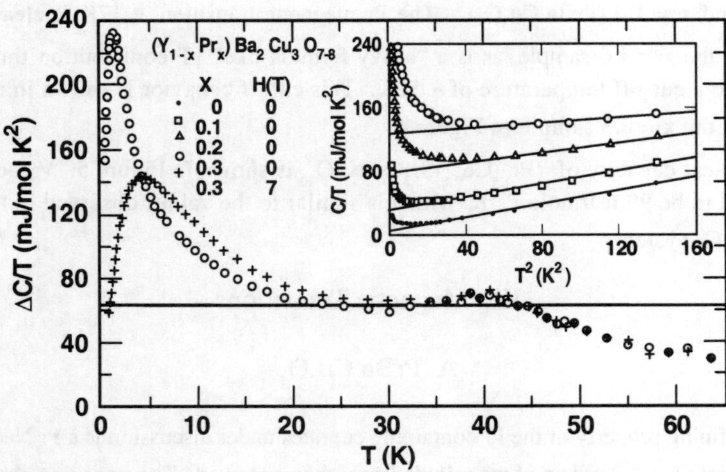

Fig. 4. ΔC/T vs T, where ΔC is the magnetic ordering plus "heavy-fermion-like," γT, contributions to C. The inset shows C/T vs T^2 for T < 13K. After Phillips *et al.*[2]

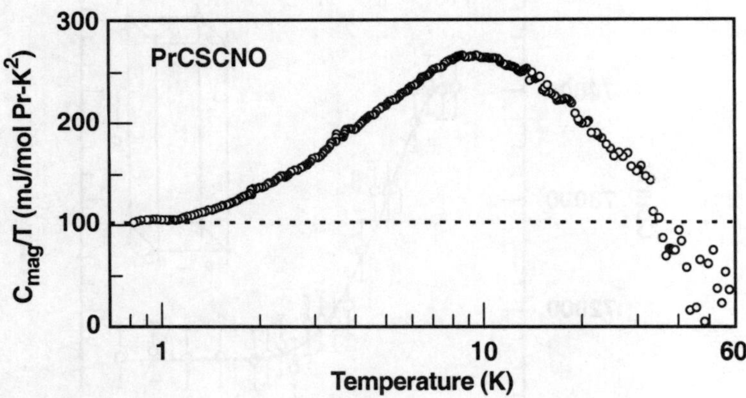

Fig. 5. The magnetic contribution to the specific heat of and (Pr$_{1.5}$Ce$_{0.5}$)Sr$_2$Cu$_2$TaO$_{10-\delta}$. After Goodwin *et al.*[5]

sample and x = 1.0 ($PrBa_2Cu_3O_7$).[4] The Pr magnetic transition at 17K is clearly visible in the x = 1.0 sample, as is a "heavy fermion-like" γT contribution that extends to a cut-off temperature of ~ 50K. This cutoff behavior is shown in more detail for the x = 0.3 sample in Figure 4.

The heat capacity of $(Pr_{1.5}Ce_{0.5})Sr_2Cu_2NbO_{10}$ is shown in Figure 5. Values of γ are found to be 99 mJ/mole Pr-K^2 which is similar to the values observed in the $PrBa_2Cu_3O_7$ system.

III. Magnetic Transitions

A. $PrBa_2Cu_3O_7$

A defining property of the Pr containing cuprates under discussion is a Pr Néel temperature (T_N) 1-2 orders of magnitude larger than expected. The transition shown in the heat capacity at 17K for $PrBa_2Cu_3O_7$ in Fig. 3, also shows up very clearly in the neutron scattering. This is illustrated in Fig. 6.

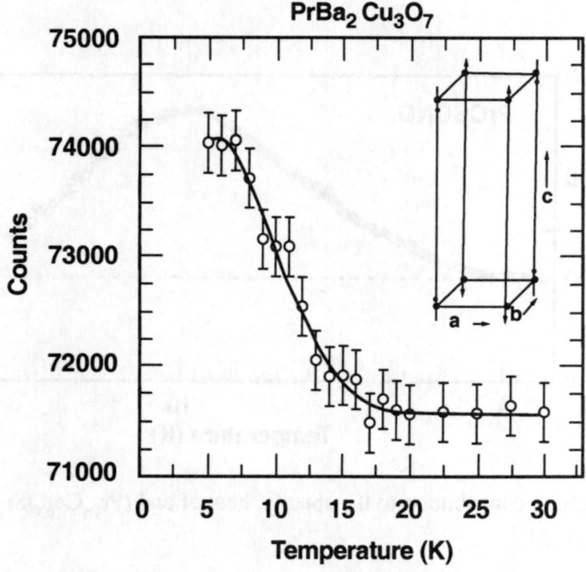

Fig. 6. Temperature dependence of the {1/2 1/2 1/2} peak intensity, showing the variation of the square of the staggered magnetization with temperature. The Néel temperature for this system is ~17 K, based on these data as well as specific heat data.[14] The solid curve is a guide to the eye. The insert shows the magnetic configuration of the Pr spins. After Li *et al.* [20]

From the magnetization data and neutron data, $PrBa_2Cu_3O_7$ has been shown by Soderholm *et al.*[21-22] to have a pseudo-triplet Pr 4f crystal field ground state marked by three broad and closely spaced energy levels which yield magnetic data that simulate a Curie-Weiss profile for a Pr ion with a reduced effective moment (μ_{eff}), and a large linear contribution to the low temperature specific heat . This level scheme is shown in Figure 7. Specific heat and magnetization measurements of $PrBa_2Cu_3O_7$ have shown the magnetic ordering at 17K to have an associated entropy[14] of approximately Rln2. Neutron diffraction line shapes and intensities assign this ordering at 17K to the Pr sublattice.[20] The Cu-O plane sites have already

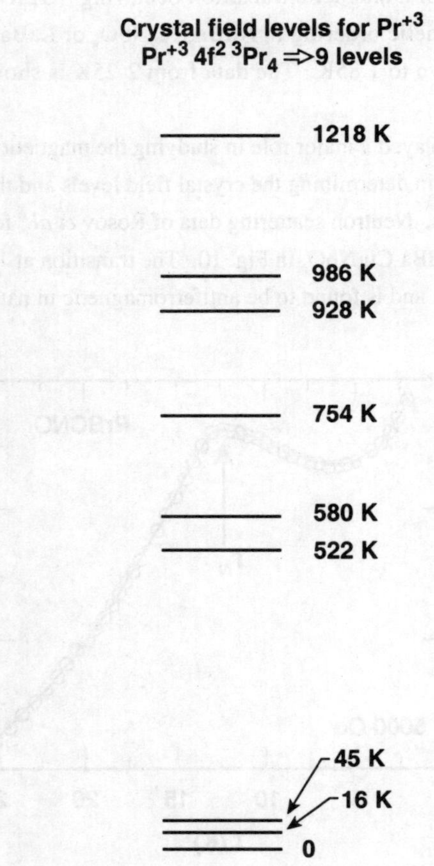

Fig. 7. Crystal field levels for Pr+3 4F2 as determined from neutron diffraction and calculation. $^3H_4 \Rightarrow 9$ levels. Adapted from Soderholm *et al.*[21]

352

ordered at 270K[23] for $PrBa_2Cu_3O_7$, and the Cu on the chain sites do not order. For $PrBa_2Cu_3O_6$ the T_N drops to 10K, and the chain Cu's order at 80 K.[24] For $PrBa_2Cu_3O_7$, the anti-ferromagnetic ordering temperature (T_N) for Pr is a factor of 30 higher than expected, based on a RKKY scaling of T_N from the other $RBa_2Cu_3O_7$ compounds.[25]

B. $PrBa_2Cu_2NbO_8$

The magnetic susceptibilities of the $PrBa_2Cu_2NbO_8$, $LaBa_2Cu_2NbO_8$ and $NdBa_2Cu_2NbO_8$ samples were examined by Bennahmias et al.[8] $PrBa_2Cu_2NbO_8$ shows a distinct signature of a magnetic transition occurring ~ 12K while there is no evidence of such a magnetic ordering in $NdBa_2Cu_2NbO_8$ or $LaBa_2Cu_2NbO_8$ from the magnetization data down to 1.85K. The data from 2-25K is shown for $PrBa_2Cu_2NbO_8$ in Fig. 8.

Neutron scattering has played a major role in studying the magnetic properties of the Pr containing cuprates, both in determining the crystal field levels and the nature of the magnetic order which occurs. Neutron scattering data of Rosov et al.[9] for $PrBa_2Cu_2NbO_8$ is shown in Fig. 9 and for $NdBa_2Cu_2NbO_8$ in Fig. 10. The transition at 12K is clearly observed in the neutron data, and is found to be antiferromagnetic in nature. The

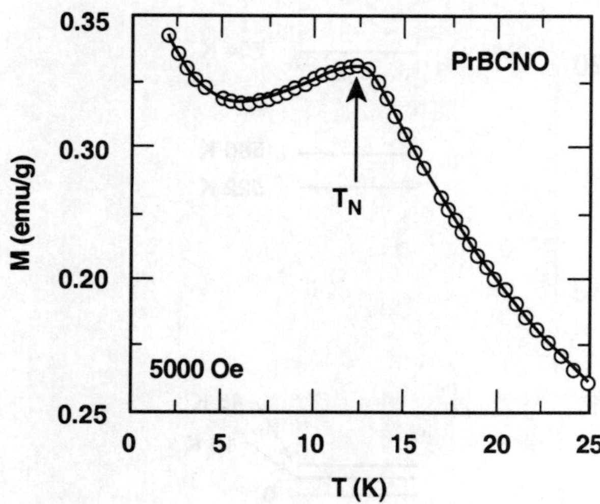

Fig. 8. Magnetization vs. temperature for $PrBa_2Cu_2NbO_8$ in the range near the magnetic phase transition at 12 K. After Radousky et al.[26]

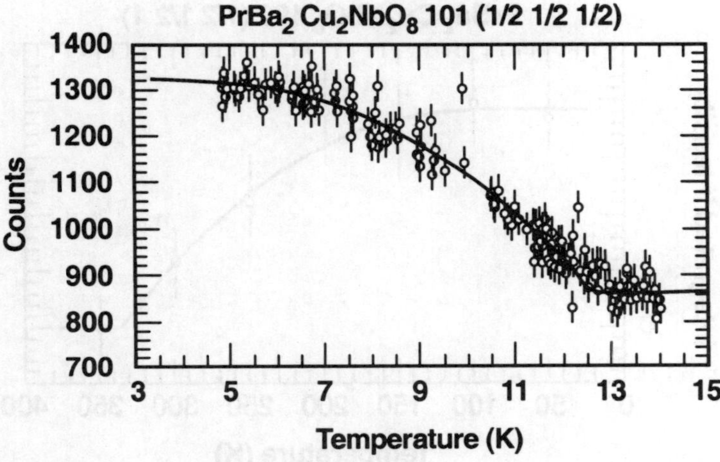

Fig. 9. Peak intensity of scattering for the 101 reflection for PrBa$_2$Cu$_2$NbO$_8$ vs T. The fitted ordering temperature for the Pr moments is 12.65 K. After Rosov *et al.*[9]

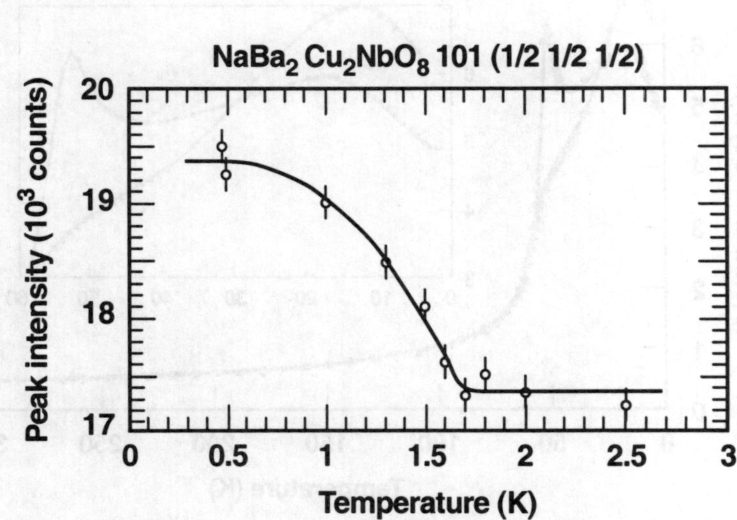

Fig. 10. Peak intensity of scattering for the 101 reflection for NdBa$_2$Cu$_2$NbO$_8$ vs. T. The fitted ordering temperature for the Nd moments is 1.69 K After Rosov *et al.*[9]

354

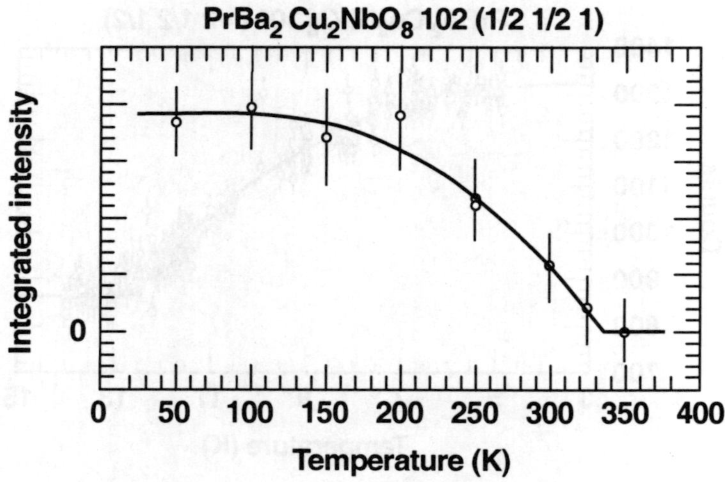

Fig. 11. Integrated intensity of scattering for the 102 reflection for PrBa$_2$Cu$_2$NbO$_8$ vs. T. The fitted ordering temperature for the Cu moments is 340 K After Rosov *et al.*[9]

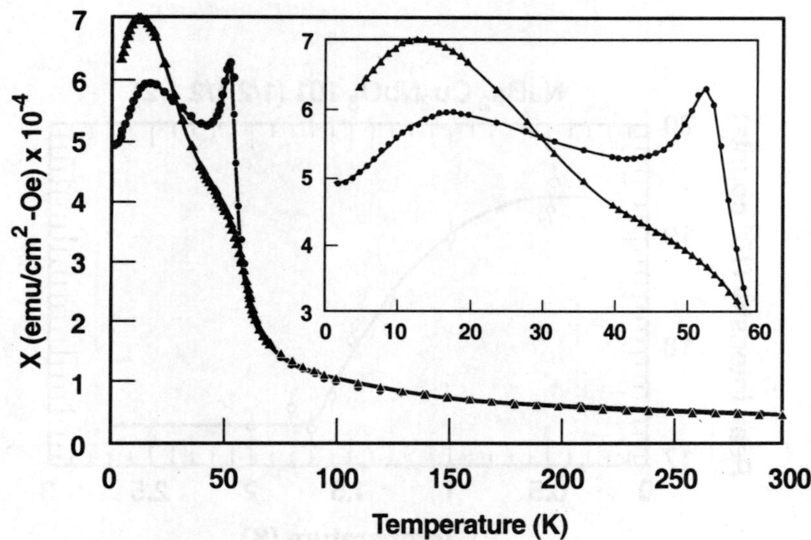

Fig. 12. Temperature dependent susceptibility for (Pr$_{1.5}$Ce$_{0.5}$)Sr$_2$Cu$_2$NbO$_{10}$ measured at 5kOe(▲) for T = 2 K to 300 K and 500 Oe(-) for T = 5 K to 300K. After Goodwin *et al.*[21]

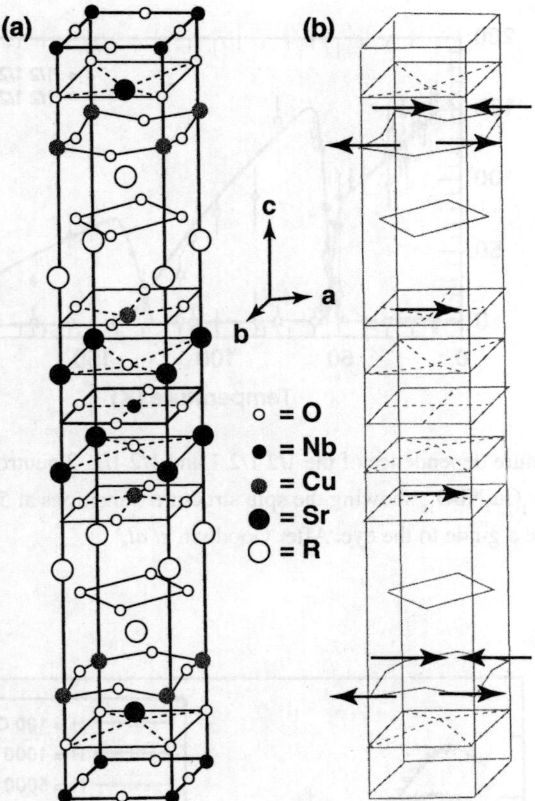

Fig. 13. (a) The crystal structure for the $(R_{1.5}Ce_{0.5})Sr_2Cu_2NbO_{10-\delta}$ compounds. (b) Illustration of the +—+ Cu spin structure proposed in the text with the Cu spins aligned parallel to the *a*-axis. After Goodwin *et al.*[5]

ordering is quasi two dimensional, with only short range order observed in the c - direction. This is contrasted to $PrBa_2Cu_3O_7$, where 3 dimensional order was observed for the 17K Pr ordering.[20] The $NdBa_2Cu_2NbO_8$ material is seen from the neutron data to order at 1.7K.[9] The ordering temperature for the Cu sublattice is near 350K, which is shown in Figure 11.

C. $(Pr_{1.5}Ce_{0.5})Sr_2Cu_2NbO_{10}$

The magnetic properties of non-superconducting $(Pr_{1.5}Ce_{0.5})Sr_2Cu_2MO_{10-\delta}$ with M=Nb, Ta have been characterized with magnetization, specific heat, and neutron diffraction experiments.[5-6,13,27-33] Data for $(Pr_{1.5}Ce_{0.5})Sr_2Cu_2NbO_{10-\delta}$ reveal complex Cu magnetism marked by antiferromagnetic order below 200K, spin-structure

356

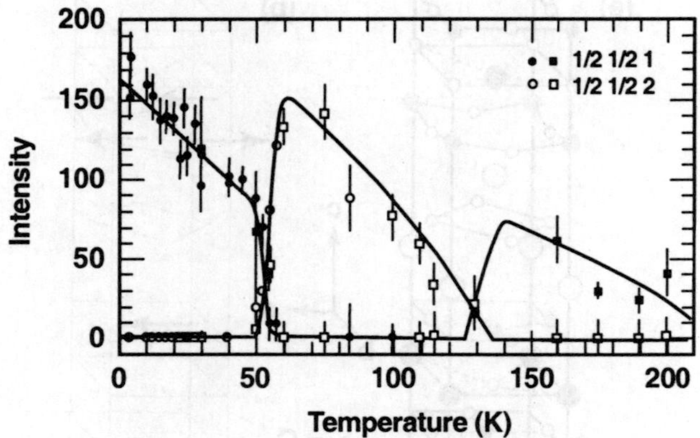

Fig. 14. Temperature dependence of the 1/2 1/2 1 and 1/2 1/2 2 neutron diffraction peaks from $(Pr_{1.5}Ce_{0.5})Sr_2Cu_2NbO_{10}$ showing the spin structure transitions at 55 K and 130 K. The solid lines are a guide to the eye. After Goodwin *et al.*[5]

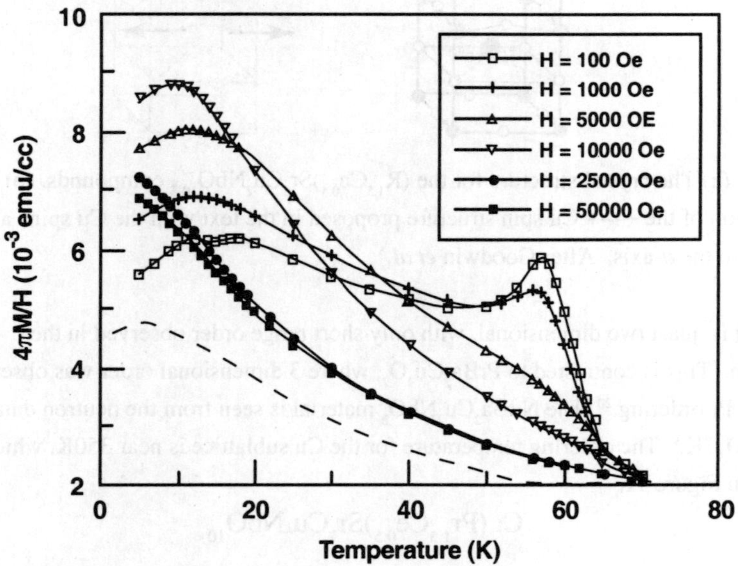

Fig. 15. Zero field cooled MDC(T)/H data for $(Pr_{1.5}Ce_{0.5})Sr_2Cu_2NbO_{10}$ at H = 100 Oe, 1 kOe, 5 kOe, 10 kOe, 25 kOe, and 50 kOe. [The dashed line represents the contribution of the Pr moments.] After Goodwin *et al.*[5]

transitions at 130K and 55K, both collinear and non-collinear antiferromagnetic spin structures, and weak ferromagnetic behavior below 130K. The data also indicate a possible ordering of the Pr spins near 10K, a large linear contribution to the low temperature specific heat, and a Pr $4f$ crystal field ground state similar to that found in $PrBa_2Cu_3O_7$. Furthermore, there is evidence that the weak ferromagnetic behavior couples to the Pr ordering near 10K. These points are now discussed in greater detail.

The susceptibility of $(Pr_{1.5}Ce_{0.5})Sr_2Cu_2NbO_{10}$ from Goodwin et al.[21] are shown in Figure 12 and reveal two distinct transitions at 17K and 55K (measured at 500 Oe). No such transitions were observed in other members of the $(RE_{1.5}Ce_{0.5})Sr_2Cu_2NbO_{10}$ series down to 2K. The transitions at 17K and 55K are very prominent at 500 Oe. However, in a 5 kOe field, the 17K transition shifts down to 13K and the 55K transition is nearly suppressed and appears also to be shifted. Measurements shown at 50 kOe down to 2K show both transitions to be almost completely suppressed.

Neutron scattering results from Goodwin et al.,[5] for $(Pr_{1.5}Ce_{0.5})Sr_2Cu_2NbO_{10}$ are summarized in Figure 13. The temperature dependence of the 1/2 1/2 1 and 1/2 1/2 2 peak intensities are shown in Figure 14. The data indicate an onset of the magnetic peaks above 200K and two sudden redistributions of the peak intensities near 130K and 55K. The onset of the magnetic peaks is difficult to determine due to the extremely weak scattering at higher temperatures. However, the magnetic peaks observed below 200K were not detected at 300K. The presence of 1/2 1/2 l reflections at high temperatures lead Goodwin et al.,[5] to conclude, that the Cu-spin configuration is antiferromagnetic in the CuO_2 planes as is observed in all other high T_C cuprate systems - where nearest-neighbor Cu spins within a CuO_2 planes are antiparallel and all the Cu spins are parallel to the a-b plane.

The sudden changes in peak intensities near 130K and 55K indicate reorientations of the spin structure at these temperatures. This yields three distinct spin structures for $(Pr_{1.5}Ce_{0.5})Sr_2Cu_2NbO_{10}$ with temperature regions from T_N (above 200K) down to 130K (phase I), 130K to 55K (phase II), and below 55K (phase III). However, the presence of the 1/2 1/2 l peaks for all three phases indicate that the proposed CuO_2-planar antiferromagnetic spin structure is present in all three spin structures. Finally, the 55K transition coincides with the cusp reported in the magnetization of $(Pr_{1.5}Ce_{0.5})Sr_2Cu_2NbO_{10}$.[6,9] Field cooled $M_{Cu}(T)$ data for $(Pr_{1.5}Ce_{0.5})Sr_2Cu_2NbO_{10}$ were collected at various fields by Goodwin et al,[5] and are shown in Figure 15.

The change in magnetization in these compounds in going from Nb to Ta was studied by Bennamias et al.[27] Several zero-field cooled and field cooled magnetic spectra for these compounds, under the influence of an external field of 500 Oe, are shown in Figure 16. The magnetic behavior of the $Pr_{1.5}Ce_{0.5}Sr_2Cu_2NbO_{10}$ sample (upper left) exhibits two

358

magnetic transitions, while only one ordering peak is seen in the data for the $Pr_{1.5}Ce_{0.5}Sr_2Cu_2TaO_{10}$ sample (lower right). The magnetic peak at the higher temperature is associated with a Cu spin reorientation[5] (upward pointing arrows), shifts downward in temperature from around 55K in the $Pr_{1.5}Ce_{0.5}Sr_2Cu_2NbO_{10}$ compound to about 19K in the

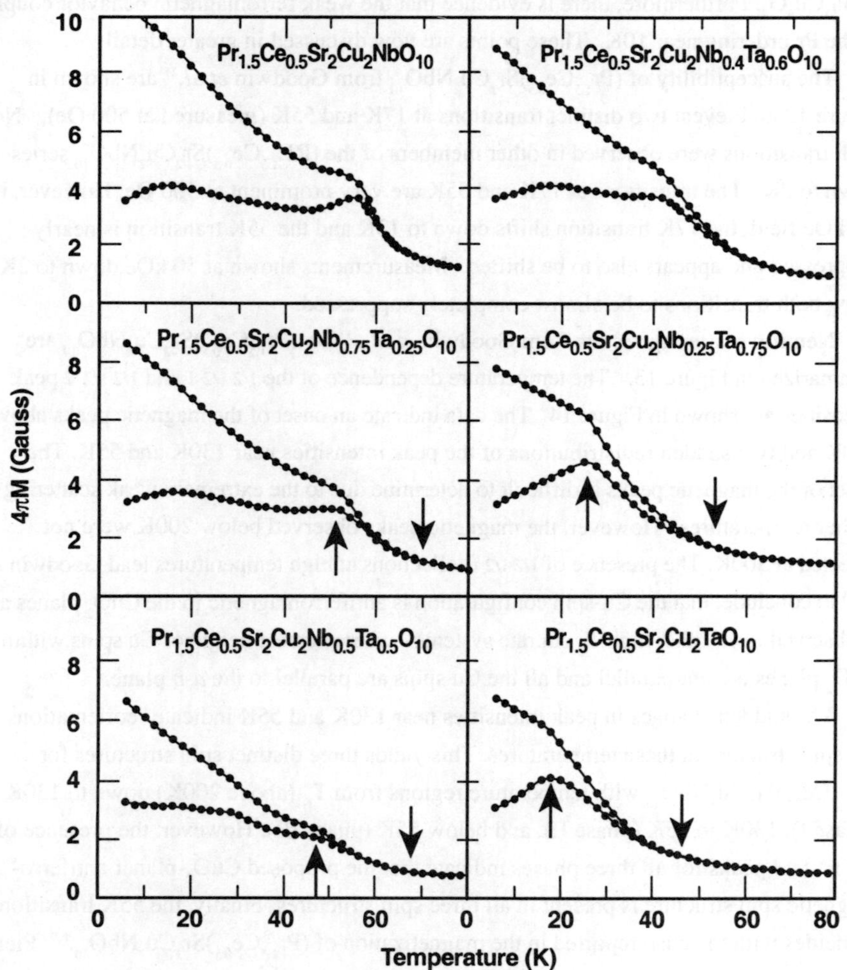

Fig. 16. Zero field cooled and field cooled magnetization spectra taken at 500 Oe for the compounds $Pr_{1.5}Ce_{0.5}Sr_2Cu_2Nb_{1-x}Ta_xO_{10}$ (x = 0.0, 0.25, 0.5, 0.6, 0.75, 1.0). The downward pointing arrow indicates the position of the irreversibility temperature and the upward pointing arrow identifies the temperature of the Cu moment reorientation peak. After Bennahmais *et al.*[27]

$Pr_{1.5}Ce_{0.5}Sr_2Cu_2TaO_{10}$ compound. Figure 16 indicates that all of the samples displayed irreversible behavior, which is associated with weak ferromagnetism in these compounds. The irreversibility temperature (downward pointing arrows) can also be seen to shift towards lower temperatures as one goes from the Nb side of the phase diagram to the Ta side.

The temperature at which the inter-layer Cu moment reorientation occurs in these materials under the application of a 500 Oe field, as a function of Nb content, is plotted in Figure 17. The data in Figure 17 illustrates that the nature of the magnetic interactions driving the Cu-spin reordering inherent to these materials is not a linear function of the amount of Ta contained in these compounds.

The effects on the intensity of the magnetic transition peaks due to varying the applied magnetic field is shown in Figure 18 over the temperature range 0K to 80K. Increasing the strength of the applied magnetic field leads to a subsequent reduction in the intensity of the Cu-spin transition and enhancement of the lower temperature magnetic transition. The exact nature of the 13K cusp is uncertain at this point. However, two likely candidates are a Cu-spin reorientation as suggested by Felner et al.[30] or a coupling of the Cu weak ferromagnetism to an antiferromagnetic Pr ordering, as

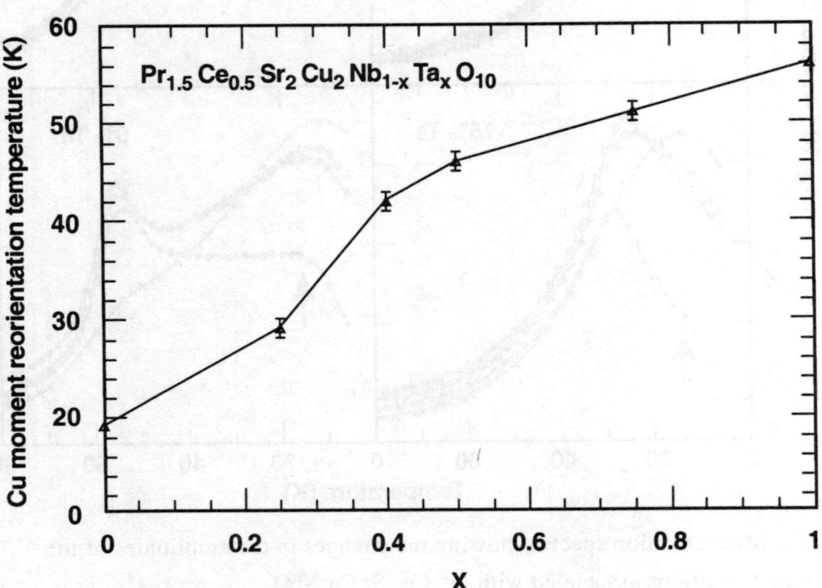

Fig. 17. The Cu-spin transition temperature as a function of Nb content in a field of 500 Oe. The vertical error bars in this figure are determined to be ± 1K from the slope bisection method used to obtain the data. After Bennahmais et al.[27]

suggested by Bennahmias *et al.*[27]

Figure 19 shows the phase diagram for the $(Eu_{1.5-x}Pr_xCe_{0.5})Sr_2Cu_2NbO_{10}$ system as Eu replaces Pr. Both the 55K and 13K transitions decrease with replacement of the Pr, and disappear near a concentration of x = 0.5 (1/3 Pr and 2/3 Eu). The phase diagram also shows the superconductivity in $(Eu_{1.5-x}Pr_xCe_{0.5})Sr_2Cu_2NbO_{10}$ disappearing at x = 0.3.

Recent results[34-41] indicate that $Pr_{1.5}Ce_{0.5}Sr_2Cu_2NbO_{10}$ can be made superconducting, but the accompanying magnetic properties of these samples, or of superconducting

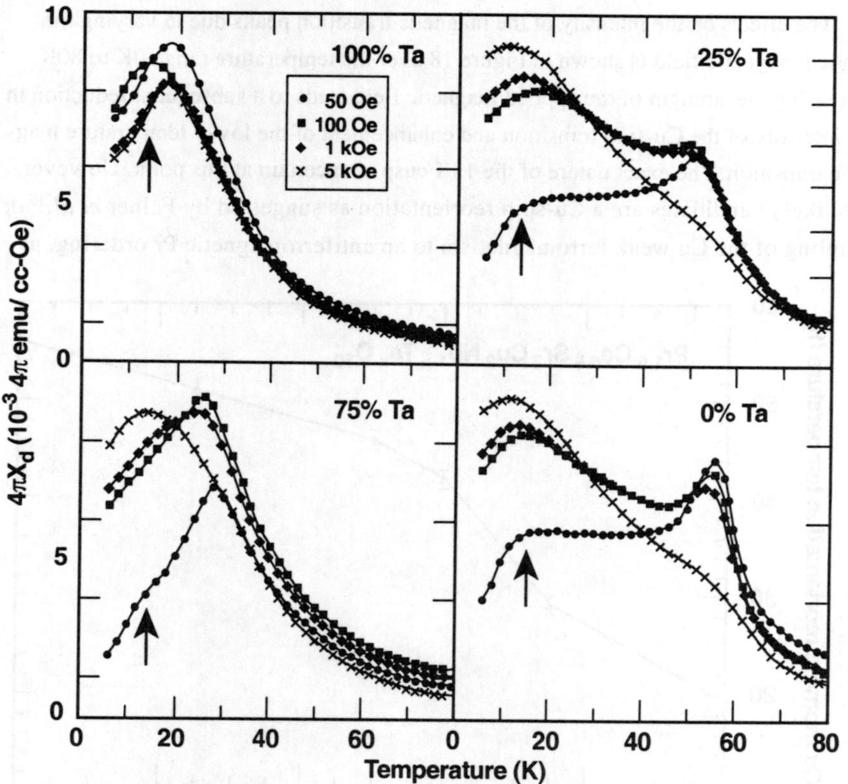

Fig. 18. Magnetization spectra showing the changes in the amplitudes of the magnetic transitions associated with $Pr_{1.5}Ce_{0.5}Sr_2Cu_2NbO_{10}$, $Pr_{1.5}Ce_{0.5}Sr_2Cu_2Nb_{0.75}Ta_{0.25}O_{10}$, $Pr_{1.5}Ce_{0.5}Sr_2Cu_2Nb_{0.25}Ta_{0.75}O_{10}$, and $Pr_{1.5}Ce_{0.5}Sr_2Cu_2TaO_{10}$ samples obtained in applied magnetic fields ranging from 50 Oe (circles), 500 Oe (squares), 1 kOe (diamonds), and 5 kOe (crosses). The position of the lower T peak is marked by the arrow. After Bennahmais *et al.*[27]

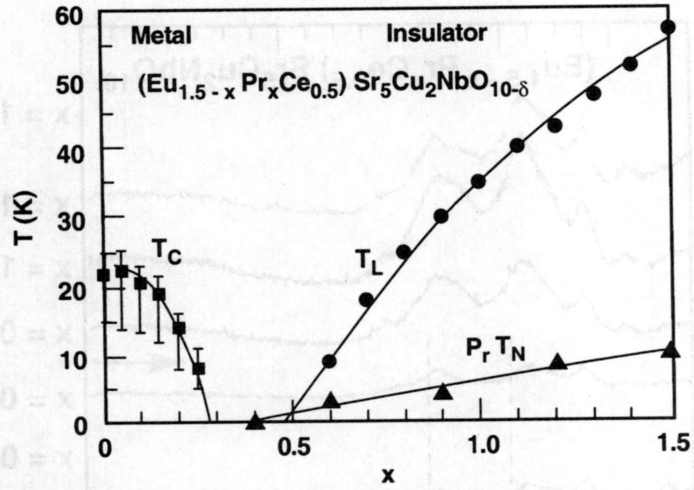

Fig. 19. Phase diagram for the $(Eu_{1.5-x}Pr_xCe_{0.5})Sr_2Cu_2NbO_{10-\delta}$ system taken from references 15 and 22. For x = 0, $(Eu_{1.5}Ce_{0.5})Sr_2Cu_2NbO_{10-\delta}$ is metallic and superconducting with $T_C \approx 24$ K. As Pr is doped in, T_C (■) is suppressed until $x \approx 0.28$ where the superconductivity is completely suppressed and the system undergoes a metal-to-insulator transition. For $0.3 \leq x < 1.5$, the system is electronically insulating. For $x > 0.6$, the system exhibits complex magnetism below T_N (▲) and a Cu spin reorientation at T_L (●). After Goodwin et al.[6]

$PrBa_2Cu_3O_7$, are not yet known. In particular, it is not known what is different between the non-superconducting and superconducting $PrBa_2Cu_3O_7$ and detailed structural information such as the $Pr\text{-}CuO_2$ bond lengths discussed in Figure 1 have not yet been reported. All the results summarized here for $Pr_{1.5}Ce_{0.5}Sr_2Cu_2NbO_{10}$ are for samples which are insulating and do not superconduct.

Finally, a series of Raman spectra were taken by Bennahmias et al.[21] for the following range of solid solutions: $(Eu_{1.5-x}Pr_xCe_{0.5})Sr_2Cu_2NbO_{10}$ ($0.0 \leq x \leq 1.5$). Spectra collected for various solutions are plotted in Fig. 20 from 250 to 1350 cm^{-1}. The corresponding values of the magnetic susceptibility at room temperature for these solid solutions is shown as a function of Pr content in Fig. 21. Both data sets indicate an abrupt change occurring at x = 1.0, which has been interpreted as evidence for a subtle phase transition which occurs at this concentration.

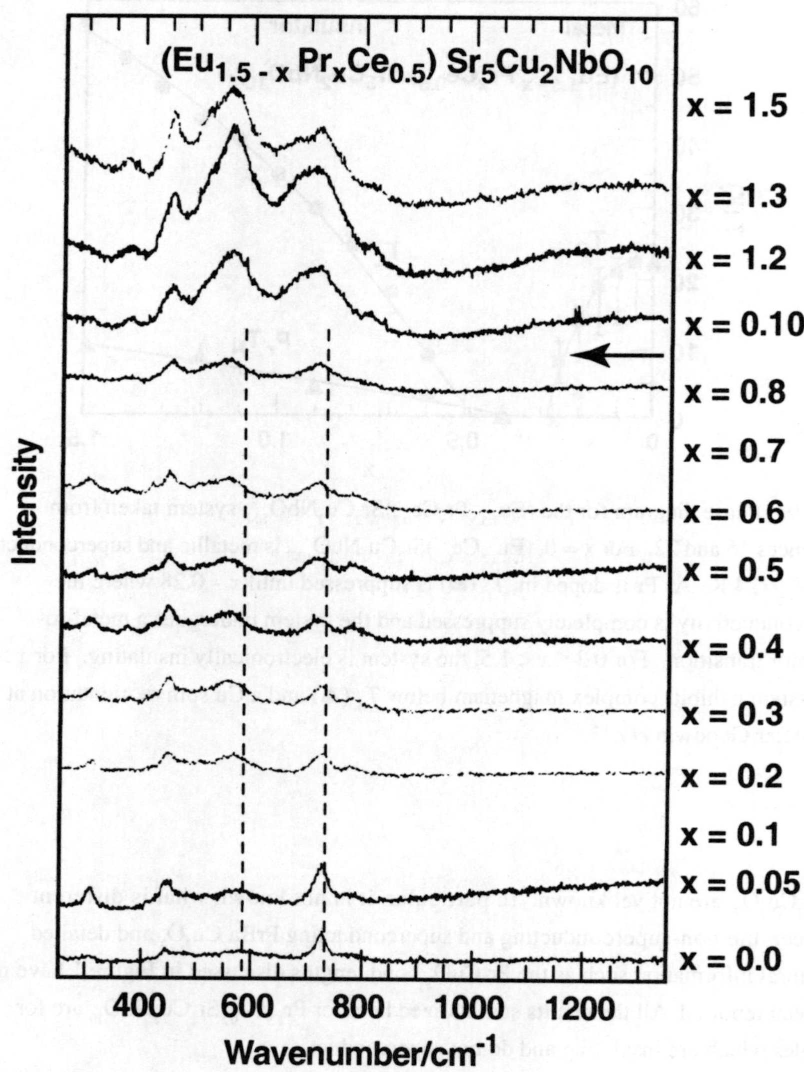

Fig. 20. Room temperature Raman spectra for the series of solid solutions (Eu$_{1.5-x}$Pr$_x$Ce$_{0.5}$)Sr$_2$Cu$_2$NbO$_{10}$ ($0.0 \leq x \leq 1.5$) in the energy range 250 cm^{-1} to 1350 cm^{-1}. The arrow in the figure marks the discontinuous character change at $x = 1$ that has been interpreted as the onset of a subtle structural transition. The dotted lines marking the position of two modes are provided as references to register the peak shifts with concentration. After Bennahmias *et al.*[29]

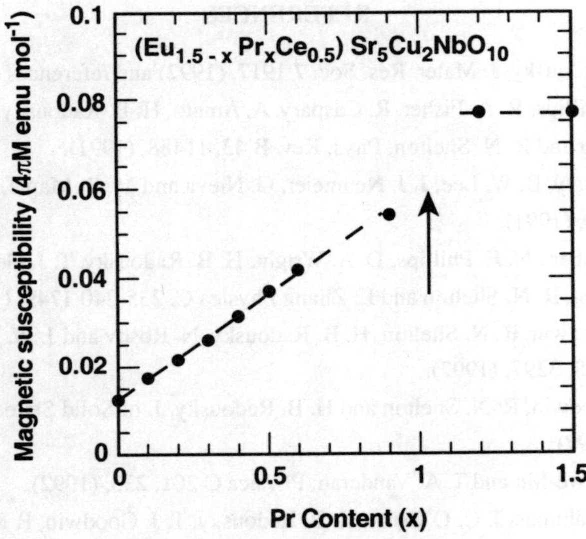

Fig. 21. Values of magnetic susceptibility at room temperature for the series of solid solutions (Eu$_{1.5-x}$Pr$_x$Ce$_{0.5}$)Sr$_2$Cu$_2$NbO$_{10}$ (0.0 ≤ x ≤ 1.5). The susceptibility exhibits a dramatic character change at a Pr content x = 1. After Bennahmias et al.[29]

IV. SUMMARY

Magnetism in the Pr containing cuprates is a very complex and still active field of research.[42-53] These materials offer a chance to observe heavy fermion behavior in a set of structures which are quite different from the heavy fermion systems described in the other chapters. I expect that work in the near future will continue to focus on characterizing the nature of the interactions mediating the high Pr ordering temperatures in these insulating materials, and to refine current theories[54-59] to the point where they can calculate the magnetic data which has been summarized in this chapter.

ACKNOWLEDGMENTS

This work was performed under the auspices of the U.S. Department of Energy by University of California Lawrence Livermore National Laboratory under contract No. W-7405-Eng-48, through the LLNL Materials Research Institute.

The author wishes to acknowledge many helpful discussions with Professor Carolus Boekema during his sabbatical in the LLNL Materials Research Institute.

REFERENCES

1. H. B. Radousky, J. Mater. Res. Soc. **7** 1917, (1992) and references therein.

2. N. E. Phillips, R. A. Fisher, R. Caspary, A. Amato, H. B. Radousky, J. L. Peng, L. Zhang and R. N. Shelton, Phys. Rev. B **43**, 11488, (1991).

3. S. Ghamaty, B. W. Lee, J. J. Neumeier, G. Nieva and M. B. Maple, Phys. Rev. B **43**, 5430, (1991).

4. R. A. Fisher , N. E. Phillips, D. A. Wright, H. B. Radousky, T. J. Goodwin, J. L. Peng, R. N. Shelton and L. Zhang Physica C. 235-240 1749, (1994).

5. T. J. Goodwin, R. N. Shelton, H. B. Radousky, N. Rosov and J. W. Lynn, Phys. Rev. B **55**, 3297, (1997).

6. T. J. Goodwin, R. N. Shelton and H. B. Radousky, J. of Solid State Chemistry, 133, 445, (1997).

7. C. K. Lowe-Ma and T. A. Vanderah, Physica C 201, 233, (1992).

8. M. Bennahmias, J. C. O'Brien, H. B. Radousky, T. J. Goodwin, P. Klavins, J. M. Link, C. A. Smith and R. N. Shelton, Phys. Rev. B **46** 11986, (1992).

9. N. Rosov, J. W. Lynn, H. B. Radousky, M. Bennahmias, T. J. Goodwin, P. Klavins and R. N. Shelton, Phys. Rev. B **47** 15256, (1993).

10. C. H. Chow, Y. Y. Hsu, J. H. Shieh, T. J. Lee, H. C. Ku, J. C. Ho and D. H. Chen, Phys. Rev. B **53**, 6729, (1996).

11. D. E. Cox, A. I. Goldman, M. A. Subramanian, J. Gopalakrishnan and A. W. Sleight, Phys. Rev. B **40**, 6998, (1989).

12. Y. Gao, P. Pernambuco-Wise, J. E. Crow, J. O'Reilly, N. Spencer, H. Chen and R. E. Saloman, Phys. Rev. B **45**, 7436, (1992).

13. T. J. Goodwin, H. B. Radousky and R. N. Shelton, Phys. Rev. B **56**, 5144, (1997).

14. A. Kebede, C. -S. Jee, J. Schwegler, J. E. Crow, T. Mihalisin, G. H. Myer, R. E. Solomon, P. Schlottmann, M. V. Kuric, S. H. Bloom and R. P. Guertin, Phys. Rev. B **40**, 4453 (1989); C. -S. Jee, A. Kebede, D. Nichols, J. E. Crow, T. Mihalisin, G. H. Myer, I. Perez, R. E. Solomon and P. Schlottmann, Solid State Comm. **69**, 379, (1989).

15. I. Felner, U. Yaron, I. Nowik, E. R. Bauminger, Y. Wolfus, E. R. Jacoby, G. Hilscher and N. Pillmayr, Phys. Rev. B **40**, 6739, (1989).

16. M. B. Maple, J. M. Ferreira, R. R. Hake, W. B. Lee, J. J. Neumeier, C. L. Seaman, K. N. Yang and H. Zhou, J. of the Less Common Metals **149**, 405, (1989).

17. N. Sankar, V. Sankaranarayanan, L. S. Vaidhyanathan, G. Rangarajan, R. Srinivasan, K. A. Thomas, U. V. Varadaraju and G. V. Subba Rao, Solid State. Comm. **67**, 391, (1988)

18. A. P. Ramirez, L. F. Schneemeyer and J. V. Waszczac, Phys. Rev. B **36**, 7145, (1987).

19. A. Amato, R. Caspary, R. A. Fisher, N. E. Phillips, H. B. Radousky, J. L. Peng, L. Zhang and R. N. Shelton, Physica B **165**, 1347, (1990).

20. W-H. Li, J. W. Lynn, S. Skanthakumar, T. W. Clinton, A. Kebede, C.-S. Jee, J. E. Crow and T. Mihalisin, Phys. Rev. B **40**, 5300, (1989).

21. L. Soderholm, C.-K. Loong, G. L. Goodman and B. D. Dabrowski, Phys. Rev. B **43**, 7923, (1991).

22. G. L. Goodman, C.-K. Loong and L. Soderholm, J. of Physics, Condensed Matter **3**, 49 (1991); G. L. Goodman and L. Soderholm, Physica C **171**, 528, (1990).

23. D. W. Cooke, R. S. Kwok, R. L. Lichti, T. R. Adams, C. Boekema, W. K. Dawson, A. Kebede, J. Schwegler, J. E. Crow and T. Mihalisin, Phys. Rev. B **41**, 4801 (1990).

24. N. Rosov, J. W. Lynn, G. Cao, J. W. O'Reilly, P. Pernambuco-Wise and J. E. Crow, Physica C **204**, 171, (1992)

25. J. Lynn, J. Alloy Compounds **181**, 419, (1992).

26. H. B. Radousky, J. C. O'Brien, M. Bennahmias, P. Klavins, C. A. Smith, J. M. Link, T. Goodwin and R. N. Shelton, Mat. Res. Soc. Symp. Proc. **275**, 113, (1992).

27. M. Bennahmias, H. B. Radousky, C. M. Buford, A. B. Kebede, M. McIntyre, T. J. Goodwin and R. N. Shelton, Phys. Rev. B **53** 2773, (1996).

28. T. J. Goodwin, H. B. Radousky and R. N. Shelton, Physica C **204** 212, (1992).

29. M. Bennahmias, H. B. Radousky, H. E. Lorenzana, T. J. Goodwin and R. N. Shelton, Journal of Raman Spectroscopy **30**, 543, (1999).

30. I. Felner, U. Yaron, U. Asaf, T. Kroner and V. Breit, Phys. Rev. B **49** 6903 (1994).

31. U. Asaf, I. Felner, D. Schmitt, D. B. Barbara, C. Goddart and E. Alleno, Phys. Rev. B **54**, 16160, (1996).

32. U. Asaf and I. Felner, Solid State Communications, **94**, 873, (1995).

33. U. Staub, L. Soderholm, R. Osborn, T. J. Goodwin, R. N. Shelton and H. B. Radousky. J. Of Physics-Condensed Matter, **10**, 4637, (1998).

34. H. A. Blackstead, J. D. Dow, International Journal Of Modern Physics B **29-31**, 3591, (1999).

35. V. N. Narozhnyi, D. Eckert, K. A. Nenkov, G. Fuchs, K. H. Muller, T. G. Uvarova, International Journal Of Modern Physics B **29-31**, 3712, (1999).

36. V. N. Narozhnyi, D. Eckert, K. A. Nenkov, G. Fuchs, T. G. Uvarova, K. H. Muller, Physica C **312**, 233, (1999).

37. T. Mizokawa, C. Kim, Z. X. Shen, A. Ino, A. Fujimori, M. Goto, H. Eisaki, S. Uchida, M. Tagami, K. Yoshida, A. I. Rykov, Y. Siohara, K. Tomimoto and S. Tajima, Phys. Rev. B **60** 12335, (1999).

38. Y. Nishihara, Z. Zou, J. Ye, K. Oka, T. Minawa, H. Kawanaka and H. Bando, Bulletin of Materials Science, **22**, 257, (1999)

39. J. H. Ye, Z. G. Zou, K. Oka, Y. Nishihara, T. Matsumoto, J. Of Alloys And Compounds, **288**, 319, (1999).

40. M. Muroi and R. Street, Physica C **314**, 172, (1999).

41. M. Luszczek, W. Sadowski, T. Klimczuk, J. Olchowik, B. Susla and R. Czajka, Physica C **322**, 57, (1999).

42. M. C. Qian, W. Y. Hu, Q. Q. Zheng and H. Q. Lin, J. of Appl. Physics, **85**, 4765, (1999).

43. T. Stein, G. A. Levin, C. C. Almasan, D. A. Gajewski and others, Phys. Rev. Lett., **82**, 2955, (1999).

44. L. Jansen and R. Block, Physica A **252**, 278, (1998).

45. B. Grevin, Y. Berthier, G. Collin and P. Mendels, Phys. Rev. Lett., **80**, 2405, (1998).

46. S. Uma, W. Schnelle, E. Gmelin, G. Rangarajan, S. Skanthakumar, J. W. Lynn, R. Walter, T. Lorenz, B. Buchner, E. Walker and A. Erb, J. of Physics-Condensed Matter, **10**, L33 (1998); E. Gmelin, G. Rangarajan and A. Erb, J. of Appl. Phys., **81**, 4227, (1997).

47. R. B. Thompson, M. Singh, M. Phys. Rev. B **57**, 1284 (1998).

48. S. Skanthakumar, J. W. Lynn, N. Rosov, G. Cao and J. E. Crow, Phys. Rev. B **55**, 3406, (1997).

49. M. Takata, T. Takayama, M. Sakata, S. Sasaki, K. Kodama and M. Sato, Physica C **263**, 340, (1996).

50. M. Rubhausen, N. Dieckmann, A. Bock, U. Merkt, W. Widder and H. F. Braun, Phys. Rev. B **53**, 8619, (1996).

51. K. Nehrke and M. W. Pieper, Phys. Rev. Lett., **76**, 1936, (1996).

52. L. Soderholm, and U. Staub, In Electron Correlations And Materials Properties, Ed. A.Gonis, N Kioussis and M. Ciftan, **115,** (1999). [Klewer Academic/Plenum Publishers]

53. S. S. Weng, I. P. Hong, C. F. Chang, H. L. Tsay, S. Chatterjee, H. D. Yang, J. Y. Lin, Phys. Rev. B **59**, 11205(1999); H. D. Yang, J. Y. Lin, S. S. Weng, C. W. Lin, H. L. Tsay, Y. C. Chen, T. H. Meen, T. I. Hsu and H. C. Ku, Phys. Rev. B **56**, 14180, (1997).

54. R. Fehrenbacher and T. M. Rice, Phys. Rev. Lett., **70**, 3471, (1993).

55. I. I. Mazin and A. I. Liechtenstein, Phys. Rev. B **57**, 150, (1998); A. I. Liechtenstein, I. I. Mazin and O. K. Andersen, Physica C **235**, 2121(1994); A. I. Liechtenstein and I. I. Mazin, Phys. Rev. Lett., **74**, 1000, (1995).

56. G. H. Cao, Y. B. Yu and Z. K. Jiao, Physica C **301**, 294, (1998).

57. H. D. Yang, H. I. Tsay, F. L. Juang, W. M. Wang, W. J. Huang, C. M. Chen, T. H. Meen, Chinese Journal of Physics, **34**, 252, (1996).

58. Z. Hu, R. Meier, C. Schussler-Langeheine, E. Weschke, G. Kaindl, I. Felner, M. Merz, N. Nucker, S. Schuppler, and A. Erb, Phys. Rev. B **60**, 1460, (1999).

59. I. I. Mazin, Phys. Rev B **60**, 92, (1999).

SUBJECT INDEX

COMPOUND INDEX

APPENDIX — SELECTED REVIEW ARTICLES

Included in this section are standard review articles, as well as some papers which have a broad perspective relevant to the magnetic behavior of heavy fermion systems.

1. A. Amato, "Magnetism in Heavy-Fermion Systems Probed by Mu(+)Sr Spectroscopy", Physica B, **199**, (1994).

2. A. Amato, "Heavy-Fermion Systems Studied by μSr Technique", Reviews of Modern Physics, **69**, 1119, (1997).

3. G. Boebinger, "Correlated Electrons in a Million Gauss", Physics Today, **49**, 36, (1996).

4. J. P. Brison, N. Keller, P. Lejay, J. L. Tholence and others,"Magnetism And Superconductivity in Heavy Fermion Systems", Journal of Low Temperature Physics, **95**, 145, (1994).

5. L. Degiorgi, "The Electrodynamic Response of Heavy-Electron Compounds", Reviews of Modern Physics, **71,** 687, (1999).

6. Z. Fisk, J. L. Sarrao and J. D. Thompson, "Heavy Fermions", Current Opinion in Solid State & Materials Science, **1**, 42, (1996).

7. Z. Fisk, H. Borges, M. McElfresh, J. L. Smith, J. D. Thompson, H. R. Ott, G. Aeppli, E. Bucher, S. E. Lambert, M. B. Maple, C. Broholm and J. K. Kjems, "Superconductivity of Heavy-Electron Uranium Compounds", Physica C **153-155**, 1728, (1988).

8. Z. Fisk, J. L. Sarrao, J. L. Smith and J. D. Thompson, "The Physics And Chemistry of Heavy Fermions", Proceedings of the National Academy of Sciences of the United States of America, **92**, 6663, (1995).

9. Z. Fisk, J. L. Sarrao, S. L. Cooper, P. Nyhus, G. S. Boebinger, A. Passner and P. C. Canfield, "Kondo Insulators", Physica B, **224**, 409, (1996).

10. L. Glemot J. P. Brison, J. Flouquet, A. I. Buzdin and D. Jaccard, "Strong Coupling Superconductivity in Heavy Fermion Systems", Physica C, **318**, 73, (1999).

11. T. J. Goodwin, H. B. Radousky, and R. N. Shelton, "Superconductivity and Magnetism in $(R_{1.5-x}Pr_xCe_{0.5})Sr_2Cu_2NbO_{10}$-Delta (R = Nd,Sm, Eu): Criteria for Modeling the Suppression of Superconductivity by Pr in High-T-C Cuprates", Physical Review B-Condensed Matter, **56**, 5144, (1997).

12. E. Guha, E. W. Scheidt and G. R. Stewart, "$U_2Pt_{15}S_{17}$ - Competition Between Magnetism and the Formation of a Heavy-Fermion Ground State", Physical Review B-Condensed Matter, **53**, 6477, (1996).

13. N. Grewe and F. Steglich, "Heavy Fermions, Handbook on the Physics and Chemistry of Rare Earths", edited by Gschneidner K.A. Jr., and L. Eyring (Elsevier Science), Vol **14**, p343, (1991).

14. K. Hanzawa, K. Ohara and K. Yosida, "Magnetization in Heavy-Fermion Compounds", Journal of the Physical Society of Japan, **66**, 3001, (1997).

15. R. H. Heffner and M. R. Norman, "Heavy Fermion Superconductivity," comments on Condensed Matter Physics, **17**, 361, (1996).

16. R. H. Heffner, "Muon Spin Relaxation Studies of the Interplay Between Magnetism and Superconductuvuty in Heavy Fermion Systems", Journal of Alloys and Compounds, **213**, 232, (1994).

17. D. W. Hess, P. S. Riseborough and J. L. Smith, "Heavy-Fermion Phenomena", Encyclopedia of Applied Physics **7**, 435, (1993).

18. K. Heuser, J. S. Kim, E. W. Scheidt, T. Schreiner, T. and G. R. Stewart, "Inducement of Non-Fermi-Liquidbehavior with Magnetic Field in Heavy-Fermion Antiferromagnets", Physica B, **261**, 392, (1999).

19. D. Jaccard, H. Wilhelm, K. Alami-Yadri and E. Vargoz, "Magnetism and Superconductivity in Heavy Fermion Compounds at High Pressure", Physica B, **261**, 1, (1999).

20. T. Jarlborg, "Temperature-Dependent Electronic Structure: from Heavy Fermion Behaviour to Phase Stability", reports on Progress In Physics, **60**, 1305, (1997).

21. E. Kim and D. L. Cox, "Knight-Shift Anomalies in Heavy-Electron Materials", Physical Review B-Condensed Matter, 58, 3313, (1998).

22. Y. Kitaoka, K. Ishida, G. Q. Zheng, H. Tou, and others, "Nmr Study in Novel Superconductors - Heavy-Fermion System and High-T-C Cuprate", Journal of Physics and Chemistry of Solids, **56**, 1931, (1995).

23. S. Koh, "Antiferromagnetic Moment in the Heavy-Fermion Superconductors". Physics Letters A, **253**, 98, (1999).

24. A. Kumar, P. K. Ahluwalia and K. C. Sharma, "Heavy Fermion Systems in the Presence of Cef: Thermal Properties", Solid State Communications, **102**, 887, (1997).

25. A. Kumar, P. K. Ahluwalia, S. Kumar and K. C. Sharma, "Competition Between Kondo Effect and RKKY Interactions in Heavy-Fermion Systems: Specific Heat", Physical Review B-Condensed Matter, **56**, 3145, (1997).

26. S. Kumar, P. K. Ahluwalia and K. C. Sharma, "Competition Between Kondo Effect and RKKY Interactions in Heavy-Fermion Systems: Transport Properties" Physical Review B-Condensed Matter, **56**, 3140, (1997).

27. H. Lin, Z. Z. Li, M. W. Xiao and X. H. Xu, "Effect of Pressure on Heavy-Fermion Metals", Communications in Theoretical Physics, **31**, 49, (1999).

28. S. Lipinski, "Magnetic Moments Contribution to the Specific Heat and Susceptibility of Heavy Fermion System", Journal of Magnetism and Magnetic Materials, **170**, L248, (1997).

29. S.Lipinski, "Dynamic Crystal Field in Heavy Fermion Systems", Journal of Magnetism and Magnetic Materials, **192**, 553, (1999).

30. B. Luthi, B. Wolf, D. Finsterbusch and G. Bruls, "Heavy Fermion Superconductivity", Physica B, **204**, 228, (1995).

31. J. W. N. Lynn, N. Rosov, S. N. Arilo, L. Kurnevitch and A. Zhoklov, "Pr Magnetic Order and Spin Dynamics in the Cuprates", Chinese Journal of Physics, **38**, (in press April, 2000).

32. M. B. Maple, M. C. Deandrade, J. Herrmann, Y. Dalichaouch, D. A. Gajewski, C. L. Seaman, R. Chau, R. Movshovich, M. C. Aronson and R. Osborn, "Non Fermi Liquid Ground States in Strongly Correlated F-Electron Materials".

33. M. B. Maple, Y. Dalichaouch, M. C. Deandrade, N. R. Dilley and others, "Recent Issues in the Physics of Heavy Fermion Materials", Journal of Physics and Chemistry of Solids, **56**, 1963, (1995).

34. N. D. Mathur, F. M. Grosche, S. R. Julian, I. R. Walker, D. M. Freye, R. K. W. Haselwimmer and G. G. Lonzarich, "Magnetically Mediated Superconductivity in Heavy Fermion Compounds", Nature, **394**, 39, (1998).

35. S. Misawa, "Fermi-Liquid Description for Susceptibility Maximum and Near-Metamagnetism in Heavy-Fermion Compounds", Journal of Magnetism and Magnetic Materials, **177**, 325, (1998).

36. A. C. Mota, K. Aupke, A. Amann, T. Teruzzi and others, "Quantum Tunneling of Vortices in Cuprate and Heavy Fermion Superconductors" Journal of Low Temperature Physics, **95**, 377, (1994).

37. E. Mullerhartmann, International Workshop on Cuprate and Heavy Fermion Superconductors - Preface, Journal of Low Temperature Physics, **95**, 1, (1994).

38. J. A. Mydosh, "Magnetism, Disorder and Frustration in Heavy-Fermion Compounds", Zeitschrift Fur Physik B-Condensed Matter, **103**, 251, (1997).

39. J. A. Mydosh, "Disorder and Frustration in Heavy-Fermion Compounds", Physica B, **261**, 882, (1999).

40. G. Oomi and T. Kagayama, "Effect of Pressure on the Magnetoresistance and Magnetostriction of Heavy Fermion Materials", Physica B, **201**, 235, (1994).

41. Y. G. Naidyuk and I. K. Yanson, "Point-Contact Spectroscopy of Heavy-Fermion Systems", Journal of Physics-Condensed Matter, **10**, 8905, (1998).

42. H. R. Ott, "Heavy Electron Phenomena", in Electron Correlations and Materials Properties, Ed. A.Gonis, N Kioussis And M. Ciftan, **115,** (1999).

43. H. B. Radousky, "A Review of the Superconducting and Normal State Properties $Y_{1-x}Pr_xBa_2Cu_3O_7$", Journal of Materials Research **7**, 1917, (1992).

44. S. Ramakrishnan, and G. Chandra, "Studies in Heavy-Fermion Systems", Current Science, **67**, 865, (1994).

45. U. Rauchschwalbe, "Magnetic Properties of Heavy-Fermion Superconductors", Physica B&C, **147**, 1-80, (1987).

46. M. Sigrist and U. Kaxuo, "Phenomenological Theory of Unconventional Superconductivity", Rev. Mod. Phys. **63**, 239-311, (1991).

47. M. Sharma and I. Singh, "Electronic Specific Heat and Spin-Susceptibility of Heavy Fermion Systems: Role of s-f Interactions", Indian Journal of Pure & Applied Physics, **37**, 627, (1999).

48. D. N. Shi, and Z. Z. Li, "The Effect of Nonmagnetic Doping on the D-Wave Heavy Fermion Superconductivity", Communications in Theoretical Physics, **31**, 205, (1999).

49. D. N. Shi, Z. Z. Li, "The Effect of Hybridization Disorder on the D-Wave Heavy-Fermion Superconductivity", Physica C, **264**,120, (1996).

50. L. Soderholm and U. Staub, "The Role of Selected F Ions in the Supression of High-Tc Superconductivity", in Electron Correlations and Materials Properties, Ed. A.Gonis, N Kioussis and M. Ciftan, **115**, (1999).

51. F. Steglich, P. Gegenwart, C. Geibel, R. Helfrich, P. Hellmann, M. Lang, A. Link, R. Modler, G. Sparn, N. Buttgen and A. Loidl, "New Observations Concerning Magnetism and Supercomductivity in Heavy-Fermion Metals", Physica B, 224, 1, (1996).

52. F. Steglich, U. Ahlheim, C. D. Berdl, C. Geibel, M. Lang, A. Loidl, G. Sparn, "Superconductivity and Magnetism in Heavy-Fermion Compounds", Frontiers in Solid State Sciences, L. C. Cupta and M. S. Multani, eds. (World Scientific, Singapore), **1**, 527, (1992).

53. G. Swqtsky, "Experimental Studies of Electron Correlation Effects in Solids", in Electron Correlations and Materials Properties, Ed. A.Gonis, N Kioussis and M. Ciftan, **115**, (1999).

54. G. R. Stewart, "Heavy-Fermion Systems", Reviews of Modern Physics **56**, 755, (1984).

55. G. R. Stewart, "A Review Written in Near Record Time - A Citation Classic Commentary on Heavy-Fermion Systems" By G. R. Stewart, Current Contents/ Physical Chemical & Earth Sciences, 8, (1993).

56. L. Taillefer, "Current Issues In Heavy Fermion Superconductivity", Hyperfine Interactions, **85**, 379, (1994).

57. J. D. Thompson, J. M. Lawrence and Z. Fisk, "Scaling in Heavy-Fermion Systems", Journal of Low Temperature Physics, **95**, 59, (1994).

58. K. Ueda, "Recent Progress in Theoretical Understanding of Heavy Fermion Systems", Hyperfine Interactions, **104**, 141, (1997).

59. Y. J. Uemura, L. P. Le, G. M. Luke, B. J. Sternlieb and others, "Basic Similarities Among Cuprate, Bismuthate, Organic, Chevrel-Phase, and Heavy-Fermion Superconductors Shown by Penetration-Depth Measurements", Physical Review Letters, **66**, 2665, (1991).

60. H. Vonlohneysen, Recent Progress in the Study of Heavy-Fermion Systems". Hyperfine Interactions, **104**, 127, (1997).

61. U. Wyder, "Spin Fluctuations and Heavy Fermion Behavior", Physica B, **204**, 255, (1995).

62. B. Wolf, G. Bruls, D. Finsterbusch, I. Kouroudis and B. Luthi, "Superconductivity and High Field Phases in Heavy Fermion Systems", Physica B, **211**, 233, (1995).